T0142886

Lecture Notes in Applied and Computational Mechanics

Volume 90

Series Editors

This series aims to report new developments in applied and computational mechanics—quickly, informally and at a high level. This includes the fields of fluid, solid and structural mechanics, dynamics and control, and related disciplines. The applied methods can be of analytical, numerical and computational nature. The series scope includes monographs, professional books, selected contributions from specialized conferences or workshops, edited volumes, as well as outstanding advanced textbooks.

Indexed by EI-Compendex, SCOPUS, Zentralblatt Math, Ulrich's, Current Mathematical Publications, Mathematical Reviews and MetaPress.

More information about this series at http://www.springer.com/series/4623

Gernot Beer · Benjamin Marussig · Christian Duenser

The Isogeometric Boundary Element Method

Gernot Beer
Institute of Structural Analysis
Graz University of Technology
Graz, Austria

Benjamin Marussig
Graz Center of Computational Engineering
Graz University of Technology
Graz, Austria

Christian Duenser
Institute of Structural Analysis
Graz University of Technology
Graz, Austria

ISSN 1613-7736 ISSN 1860-0816 (electronic)
Lecture Notes in Applied and Computational Mechanics
ISBN 978-3-030-23341-9 ISBN 978-3-030-23339-6 (eBook)
https://doi.org/10.1007/978-3-030-23339-6

This Springer imprint is published by the registered company Springer Nature Switzerland AG
The registered company address is: Gewerbestrasse 11, 6330 Cham, Switzerland

Preface

The Boundary Element Method (BEM) has been the 'Cinderella' of numerical methods, dominated by the big sister, the Finite Element Method (FEM). Although the BEM offers significant advantages, especially for infinite domain problems and problems where surface stresses are important, the widespread use of it has been hampered because in its basic form, the method is restricted to linear, elastic and homogeneous domains. This has limited the practical application considerably, especially in geomechanics, where the consideration of geological features and material non-linear behaviour is important. However, in the last 25 years, some efforts have been made to overcome this limitation, details of which will be revealed in this book.

In 2009, the book 'Isogeometric Analysis: Toward Integration of CAD and FEA' [47], describing a new approach to simulation with a technology that is used by the Computer-Aided Design (CAD) community, was published. The ultimate aim was to avoid the effort of mesh generation, by taking geometry information directly from CAD programs and using the CAD functions for the simulation. It was found that this technology resulted in more accurate simulations with less numerical effort.

The potential of isogeometric analysis for the BEM was recognised by the authors a few years ago. Due to the fact that the BEM and CAD rely on a surface definition, they match each other well. Several papers have been published since then, exploiting this new technology with applications that varied from solid mechanics to fluid flow. It was discovered that the implementation of isogeometric concepts not only resulted in a substantial saving in mesh generation and analysis time but also provided the opportunity to tackle the problem of volume integration in the BEM. It was shown, for example, that geological features can be described in a similar way as in CAD programs by bounding surfaces.

The good results reported in several papers published in the journal *Computer Methods in Applied Mechanics and Engineering* and others have motivated us to write this book, mainly in order to widely disseminate the new approach and to show that the BEM can be even more beautiful than the 'big sister'. Here, we show how smooth geometries can be specified with few parameters, how accurate solutions can be obtained with a small number of unknowns and how non-linear problems can be solved without additional mesh generation. We hope that with this book, we are able to make a contribution towards the ultimate goal of a seamless integration of CAD and simulation, without the cumbersome requirement of mesh generation. The BEM, being based on a surface definition, is ideally suited for this. At the time of writing of this book, this ultimate goal has not been fully achieved. Research efforts are underway and this book is also meant to provide a contribution to these future developments.

Graz, Austria
May 2019

Gernot Beer
Benjamin Marussig
Christian Duenser

Acknowledgements

First and foremost, we would like to acknowledge the pioneering work on isogeometric methods of Prof. Tom Hughes, without which this book would not have been written. The first author became only aware of this exciting new way of doing analysis in 2012, after his keynote lecture at the coupled conference at the wonderful island of Kos and saw great potential for the Boundary Element Method (BEM).

Research work in applying this concept to the BEM started soon after. The first step was taken as part of the international research staff exchange programme *Numerical Simulation in Technical Sciences* and the second author spent 3 months at the Pontifical Catholic University of Rio de Janeiro, getting familiar with the method. The funding of the European Commission is gratefully acknowledged. In a 3-year Austrian Science Fund (FWF) project, the second author was then employed as a Ph.D. student, working on the seamless integration of design and the BEM together with postdoc Jürgen Zechner. The funding of the FWF, as well as the significant contribution by Jürgen, especially on the introduction of the fast solution methods that are presented in this book, is gratefully acknowledged. At the end of the project, the second author presented his Ph.D. thesis 'Seamless Integration of Design and Analysis through Boundary Integral Equations', which was passed with high distinction and was awarded the Dr.-Klaus-Körper Ph.D. Award by the Gesellschaft für Angewandte Mathematik und Mechanik (GAMM).

The first author became emeritus professor soon after the start of the FWF project and left Austria to take up a position as conjoint professor in Australia. The local supervision of the project was assisted by the incoming Prof. Thomas-Peter Fries. The authors are grateful for his engagement and enlightening discussions. Collaboration was started in 2016 with universities in Ferrara and Naples. We are thankful for the significant contributions of Vincenzo Mallardo and Eugenio Ruocco to the research on elasto-plastic analysis and viscous flow, the results of which appear in this book. The first author enjoyed a 2-month visit to the second University of Naples, partly funded by the university.

The second author was awarded an Erwin Schrödinger scholarship to spend 1 year with the group of Prof. Hughes at The University of Texas at Austin. His mentorship and the financial support of the FWF is gratefully acknowledged.

The software for most of the examples in this book was programmed in MATLAB using the NURBS toolkit developed by Mark Spink and Rafael Vazquez. The authors are grateful to them for making the toolkit freely available to developers.

Graz, Austria
May 2019

Gernot Beer
Benjamin Marussig
Christian Duenser

Contents

Chapter 1
Introduction

1.1 Motivation

The Boundary Element Method (BEM) started its life as the Boundary Integral Equation (BIE) method, based on a concept by Trefftz, who proposed an alternative to the Ritz method, on which the Finite Element Method (FEM) is based. The idea was to use fundamental solutions of the governing partial differential equation (PDE) inside the domain so that approximations only occur on the boundary. Results on the boundary are obtained by solving BIEs, which are obtained by transforming the PDE such that unknowns occur on the boundary only. The values inside the domain are obtained via post-processing. The boundary representation is in contrast to the FEM, where approximations of the solution occur throughout the domain and means that the order of the problem can be reduced by one dimension, as compared with the FEM (i.e. only surface integrals instead of volume integrals need to be evaluated). The method is very attractive for problems involving infinite domains since the radiation condition is implicitly fulfilled and as it turns out is also very suited for the interaction with Computer Aided Design (CAD).

As anyone involved in numerical simulation can attest, the generation of a mesh is the most cumbersome and error-prone operation in a simulation. In the FEM, the mesh is used not only to approximate the geometry of the actual domain but also for the approximation of the unknowns inside the domain. In large scale 3-D simulations, the verification of the quality of the computed results can become quite cumbersome.

Therefore an approach, where the mesh generation is restricted to the boundary, seems ideal for modelling. However, with a pure boundary discretisation, only problems with linear, elastic material behaviour and homogeneous domains can be solved. Also, in the BEM system matrices are fully populated, requiring more effort for the solution as compared with the FEM, where the sparsity can be exploited. Finally, the implementation of the method, especially the evaluation of singular integrals requires more skill and effort than is the case for the FEM.

G. Beer et al., *The Isogeometric Boundary Element Method*, Lecture Notes in Applied and Computational Mechanics 90,
https://doi.org/10.1007/978-3-030-23339-6_1

These are the reasons why the development of the two methods has been vastly different. Whereas the FEM is now widely used in simulation and the number of researchers working in this area is huge, the BEM is viewed more as an exotic method, the use of which is not widespread. However, as will be shown in this book, the main limitations of the BEM can be overcome. Fast solution methods can be applied that reduce the solution effort considerably and with the introduction of novel approaches to the evaluation of volume integrals, non-linear and non-homogeneous problems can be tackled.

In the last decade, the concept of Isogeometric Analysis (IGA) was introduced. The motivation behind it was that in most cases geometry is taken from data generated by Computer Aided Design (CAD) and then converted into a mesh suitable for the analysis. The question posed was: why can the geometrical data from CAD not be used directly for the simulation, without the requirement of mesh generation. This would not only save a considerable effort but is expected to result in more accurate simulations since the geometry is not approximated. The pathway to achieving this is through the use of the same basis functions that the CAD systems use, namely non-uniform rational B-splines also known as NURBS, instead of the Lagrange polynomials, currently used for simulation. If these NURBS are used in the geometrical description of the problem then it is possible to obtain geometrical data directly from the CAD program and avoid mesh generation altogether. However, seldom things are so easy and this is not straightforward, but more about this later. It turned out that using B-splines and NURBS for the approximation of the unknown also brings substantial benefits. As we will see later these are quite amazing functions.

The emergence of isogeometric analysis was a game changer for simulation and even more so for the BEM. Since both the BEM and CAD define domains by bounding surfaces, they are ideally suited to each other. First implementations of the isogeometric BEM showed very promising results. Early on the authors also discovered, that expanding the capabilities of the BEM to deal with non-linear, heterogeneous problems could be achieved without the requirement of coupling to FEM or additional mesh generation. It was decided that the time had come to disseminate these findings to a wider audience and therefore the motivation for writing this book. It seems appropriate to first have a look at the early developments of the BEM in order to show the basic principles behind the method.

1.2 A Short History of the Boundary Element Method

Early developments. The idea of using fundamental solutions of the governing differential equation for solving boundary value problems was proposed first by Trefftz [181] as early as 1926. To explain the method, assume that u is a function that satisfies the Laplace equation:

$$\nabla^2 u = 0 \tag{1.1}$$

throughout the domain Ω, subjected to known values on its boundary Γ of either u, i.e. Dirichlet boundary condition (BC) or its normal derivative $\frac{\partial u}{\partial n}$ (Neumann BC).

To obtain a solution of Eq. (1.1) for given BCs, we assume that u and $\frac{\partial u}{\partial n}$ can be approximated by

$$u(\hat{\boldsymbol{x}}) = \sum_{n=1}^{N} \mathsf{U}(\tilde{\boldsymbol{x}}_n, \hat{\boldsymbol{x}}) F_n \tag{1.2}$$

$$\frac{\partial u(\hat{\boldsymbol{x}})}{\partial n} = \sum_{n=1}^{N} \frac{\partial \mathsf{U}}{\partial n}(\tilde{\boldsymbol{x}}_n, \hat{\boldsymbol{x}}) F_n \tag{1.3}$$

where $\mathsf{U}(\tilde{\boldsymbol{x}}_n, \hat{\boldsymbol{x}})$ is a solution of the differential equation at a field point $\hat{\boldsymbol{x}}$ for an infinite domain due to a unit source at $\tilde{\boldsymbol{x}}_n$ (we will refer to this as a fundamental solution) and F_n are unknown fictitious[1] source intensities. We can now use (1.2) to specify Dirichlet boundary conditions or (1.3) to specify Neumann boundary conditions at a number of boundary points N and solve for F_n. For example, if $u(\hat{\boldsymbol{x}}_i)$ is specified at points $\hat{\boldsymbol{x}}_i$ on the boundary we have

$$\sum_{n=1}^{N} \mathsf{U}(\tilde{\boldsymbol{x}}_n, \hat{\boldsymbol{x}}_i) F_n = u(\hat{\boldsymbol{x}}_i) \quad i = 1, \ldots, N. \tag{1.4}$$

Once F_n is known, the values of u at any point $\boldsymbol{x}$ can be obtained by post-processing

$$u(\boldsymbol{x}) = \sum_{n=1}^{N} \mathsf{U}(\tilde{\boldsymbol{x}}_n, \boldsymbol{x}) F_n. \tag{1.5}$$

Although this seems to be an elegant method not requiring an integration, it has serious drawbacks when applied to practical problems. Firstly, there is no control over how the unknown varies between boundary points $\boldsymbol{x}_i$, which may lead to erroneous results. Secondly, the method is not very user-friendly since it requires the specification of source points in addition to boundary points.

A more user-friendly method, known as the indirect BIE method, was developed using a Galerkin procedure, where instead of satisfying the boundary conditions exactly at specific points the error in the satisfaction of BCs is minimised. However, the fact that fictitious source intensities have to be computed first still remains the main drawback of indirect methods, as they are called, because the unknown is not directly determined. This seems to be the main reason why the method has not been widely used in the engineering community.

The first mention of the direct BIE, where results are obtained directly, seems to be by Jaswon in 1963 [91]. In this paper, he proposed to use Green's reciprocal identity to transform a differential equation into an integral equation.

[1] The term fictitious is used here to indicate that they do not have physical meaning.

Applying Green's reciprocal identity (or Green's second identity) to sufficiently smooth functions ϕ, ψ we obtain

$$\int_{\Omega} (\phi \nabla^2 \psi - \psi \nabla^2 \phi) \, d\Omega = \int_{\Gamma} \left(\phi \frac{\partial \psi}{\partial n} - \psi \frac{\partial \phi}{\partial n} \right) d\Gamma. \tag{1.6}$$

Jaswon substituted $u(\hat{\boldsymbol{x}})$ for ψ and $\mathsf{U}(\tilde{\boldsymbol{x}}_n, \hat{\boldsymbol{x}})$ for ϕ to obtain an integral equation that now directly links the known and unknown values at the boundary

$$\int_{\Gamma} \left(\mathsf{U}(\tilde{\boldsymbol{x}}_n, \hat{\boldsymbol{x}}) \frac{\partial u(\hat{\boldsymbol{x}})}{\partial n} - u(\hat{\boldsymbol{x}}) \frac{\partial \mathsf{U}(\tilde{\boldsymbol{x}}_n, \hat{\boldsymbol{x}})}{\partial n} \right) d\Gamma = 0. \tag{1.7}$$

He proposed to write the equations (collocate them) at a finite number of points $\tilde{\boldsymbol{x}}_n$. He approximated the unknown by piecewise constant functions and performed the integration mainly analytically and "sometimes numerically".

Rizzo seems to have been the first to extend the idea of Jaswon to vector potential theory, i.e. the elasticity theory governed by the Navier–Cauchy differential equation [148]. Instead of Green's reciprocal identity, he used Betti's reciprocal theorem, which is its "engineering" equivalent. Rizzo emphasises that the original idea of boundary integral equations is based on calculus and had nothing to do necessarily with numerical computation. Indeed he explains in [23] that if, with the theorem of integral calculus, we integrate the slopes $\frac{\partial f}{\partial x}$ of a function $f(x)$ from a point x_A to another x_B we get the difference in the function values at those (boundary) points, i.e. $f(x_B) - f(x_A)$. Green's theorem is really only a generalisation of this idea. Rizzo admitted that he "regarded the computer at that time as a rather hostile beast [...] and I asked it to do only the most elementary and predictable things". So most integrations, especially the ones where singularities were involved, were performed analytically. He had sincere reservations regarding the suitability of numerical (Gaussian) integration.

At about the same time people at the Chamber of Mines in South Africa were struggling with the problem of simulating complex underground mine excavations. They discovered that the use of the FEM, due to the fact that the infinite rock mass needed to be discretised, would – with the computer hardware at that time – lead to impossible mesh generation and solution effort. Based on work by Salamon [155] in 1963 and Crouch [49] the displacement discontinuity method was developed, a special type of BIE method where the unknown is the displacement discontinuity between surfaces that are very close to each other (in fact they are assumed to be coincidental in the theory). This allowed the simulation of the extraction of gold or coal seams in underground mines. Some early examples can be seen in the paper by Deist et al. [53]. The research group at the Chamber of Mines also developed a three-dimensional BEM code with constant triangular facet elements and applied it to practical problems. The first author had the privilege to visit the group on behalf of Mount Isa mines in Australia, who were desperately looking for a simulation software that could tackle their complex configuration of underground excavations.

A good summary of the developments in the mining sector at that time can be found in [33].

Period of high activity. The use of constant elements to describe the boundary and the variation of unknowns put a rather severe limitation on the practical use of the method. Indeed, it should be pointed out that in contrast to the FEM where constant elements would not work because they violate the required compatibility condition, they are working in the BEM, as will be explained and exploited later in this book. However, simulations with constant elements were not very efficient and accurate and did not exploit the superior capability of the BEM to simulate stress concentrations on smooth boundaries. It was due to Lachat and Watson [103], who borrowed the concept of shape functions from the FEM, that this problem was tackled. The term Boundary Element Method (BEM) was born. It is interesting here to quote the reaction of Rizzo in [23] to the first publication of this idea: "I was shocked and chagrined on two counts when I saw this paper. First, I was still so unfamiliar in principle with the FEM in 1975 that I was surprised at the systematic role which shape functions played there, let alone the role they could play in BEM. Secondly, [...] I was erroneously under the impression that Gaussian quadrature, needed if shape functions were introduced, would not provide enough accuracy [...] to lead to acceptable solution accuracy. I couldn't have been more mistaken."

There is no doubt that integration, especially of the singular integrals, requires special consideration. The implementation of the advanced BEM, together with a detailed description of how the integrals need to be numerically evaluated was first described in detail by John Watson [2]. Indeed, this is where the first author learned about the BEM and joined with him to publish the first book that dealt with both the FEM and the BEM [21]. This development started a flurry of activities. Six books were published on the topic between 1980 and 1983 [3, 34, 35, 50, 106, 121] and a great number of journal publications appeared. It is interesting to note here that the authors of these books were engineers. Mathematicians made significant contributions (for a summary see for example [192]) especially in the area of error analysis and fast methods, discussed later. It was found that for the collocation method used by engineers at that time "the mathematical (error) analysis lagged seriously behind the practical experiences" which was not the case for the Galerkin method. However, it was shown in [21] that the additional numerical overhead due to the additional integration involved in the Galerkin method did not result in a significant improvement of the solution quality.

The main research activities at the time concentrated on the extension of the method to heterogeneous domains and non-linear material behaviour. Since in the original BEM only surface integrals are considered, such effects that occur inside the volume, cannot be considered, unless a volume integral is added. Several methods were proposed for the evaluation of this volume integral: domain integration via cells [178, 185], the dual reciprocity method [127] and the particular solution method [72]. Whereas the first method requires a volume mesh, the last two claim to be "meshless". However, even though no actual volume mesh is required there is an approximation introduced via radial basis functions and the anchor points of these functions have to be defined. The accuracy of the results depends on the choice of their locations. In addition, the methods are not applicable to infinite domain problems. A comparison of the three methods was done in [90] and came to the conclusion that in terms of accuracy and effort, domain integration via cells was superior. It appears that the cell integration method is currently the most popular method. The method is described in more detail in [65].

Regarding the treatment of heterogeneous domains also some progress was made. Early approaches concentrated on the solution of piecewise heterogeneous domains by connecting boundary element regions with different elastic properties satisfying equilibrium and compatibility conditions at the interface (similar to the coupling with the FEM discussed next). The practical application of this was not entirely satisfactory since it involved additional discretisation effort. A more elegant approach, where inclusions could be considered within the iterative elasto-plastic algorithm was published in [146]. This makes sense since at least in geomechanics the material behaviour of geological inclusions is rarely elastic. This topic will be discussed in more detail later.

Another possibility exploited in [71] was the coupling with the FEM where non-homogeneous and non-linear conditions were restricted to the FEM domain, with the BEM simulating the infinite domain. The first author has published on this topic (see for example [20, 89]). The method was applied to some complex simulations of underground excavations in civil engineering and mining, reported in [14]. However, this approach was not found to be entirely satisfactory since the mesh generation effort was still large and the error introduced at the coupling boundary was largely unknown.

The second main issue which was addressed was a reduction in the computational effort. Since in the BEM all system matrices are fully populated, matrix assembly and solution effort can become very large especially for 3-D problems. Fortunately one can exploit the fact that fundamental solutions decay rapidly with increasing distance between the source point and field point. To reduce the effort one can either take an engineering approach such as the *lumping* introduced by J. Watson in [21] or a more mathematical approach such as hierarchical matrices [78] or fast multipole [108]. All these approaches exploit the decay of the fundamental solutions by approximating the integrals in the *far field* considering a cluster of collocation points instead of individual points (more details will be provided later).

In [14], it was shown on a practical implementation in FORTRAN that the BEM is very suitable for parallel processing. Using an element by element concept, borrowed from the FEM community, and an iterative solver it was shown that a large 3-D problem could be solved during a "coffee break". These developments certainly helped to reduce the criticism of the method with regards to the effort required to solve a problem.

Stagnation and new resurgence. After the euphoria there seems to have been a stagnation, with a decline in publications. Despite being well-suited for infinite domain problems such as underground excavations the application of the BEM has been limited. The main reason is that an efficient treatment of heterogeneous domains and non-linear material behaviour has not been rigorously pursued. With the emergence of isogeometric analysis there is hope of a resurgence of the method since the BEM complements this new simulation paradigm perfectly, as we will outline in this book. The increase in papers on isogeometric analysis is exponential, similar to when the FEM emerged. It is hoped that the BEM can play an important role here.

1.3 Notation

In this book we use both tensor and matrix notation. Tensor notation is used because it results in short equations and is very useful for presenting fundamental solutions. However, matrix notation is required when discussing assembly and the solution of equations.

1.3.1 Cartesian Coordinates and Vector Components

We define the Cartesian coordinates with $\boldsymbol{x} = \{x_1, x_2, x_3\}$ and the components of a vector $\mathbf{u} = \{u_1, u_2, u_3\}$. Sometimes we use x, y, z to denote Cartesian axes.

1.3.2 Indices and Summation Convention

A comma indicates derivation, i.e.

$$\frac{\partial u}{\partial x_i} = u_{,i} \qquad \frac{\partial^2 u}{\partial x_i \partial x_j} = u_{,ij} \qquad \frac{\partial^2 u}{\partial x_i^2} = u_{,ii}. \tag{1.8}$$

The Einstein summation convention is used, for example

$$u_{,ii} = 0 \tag{1.9}$$

means an addition of terms $i = 1, 2, 3$:

$$u_{,11} + u_{,22} + u_{,33} = 0 \tag{1.10}$$

Rarely we use the Nabla operator:

$$\nabla^2 u = u_{,ii} \tag{1.11}$$

1.3.3 Tensor Algebra

The number of subscripts defines the order of a tensor. A vector is a tensor of order 1. Examples of a tensor of order 2 are the stress and strain tensor

$$\boldsymbol{\sigma} = \sigma_{ij} = \begin{pmatrix} \sigma_{11} & \sigma_{12} & \sigma_{13} \\ \sigma_{21} & \sigma_{22} & \sigma_{23} \\ \sigma_{31} & \sigma_{32} & \sigma_{33} \end{pmatrix} \qquad \boldsymbol{\epsilon} = \epsilon_{ij} = \begin{pmatrix} \epsilon_{11} & \epsilon_{12} & \epsilon_{13} \\ \epsilon_{21} & \epsilon_{22} & \epsilon_{23} \\ \epsilon_{31} & \epsilon_{32} & \epsilon_{33} \end{pmatrix}. \tag{1.12}$$

In this book we also use the *Voigt* notation, whereby the stresses are assembled in a vector

$$\{\boldsymbol{\sigma}\} = \begin{Bmatrix} \sigma_x \\ \sigma_y \\ \sigma_z \\ \tau_{xy} \\ \tau_{yz} \\ \tau_{xz} \end{Bmatrix} \tag{1.13}$$

where $\sigma_x = \sigma_{11}$, $\sigma_y = \sigma_{22}$, $\sigma_z = \sigma_{33}$, $\tau_{xy} = \sigma_{12} = \sigma_{21}$, $\tau_{yz} = \sigma_{23} = \sigma_{32}$, $\tau_{xz} = \sigma_{13} = \sigma_{31}$. The strains in Voigt notation are

$$\{\boldsymbol{\epsilon}\} = \begin{Bmatrix} \epsilon_x \\ \epsilon_y \\ \epsilon_z \\ \gamma_{xy} \\ \gamma_{yz} \\ \gamma_{xz} \end{Bmatrix} \tag{1.14}$$

where $\epsilon_x = \epsilon_{11}$, $\epsilon_y = \epsilon_{22}$, $\epsilon_z = \epsilon_{33}$, $\gamma_{xy} = \epsilon_{12} + \epsilon_{21}$, $\gamma_{yz} = \epsilon_{23} + \epsilon_{32}$, $\gamma_{xz} = \epsilon_{13} + \epsilon_{31}$.

An example of a fourth order tensor is for example the elasticity tensor

$$\mathbb{C} = C_{ijkl}. \tag{1.15}$$

An example of a multiplication of a fourth order tensor with a second order tensor for $i, j, k, l = 1, 2$ is given by

$$\boldsymbol{\sigma} = \mathbb{C} : \boldsymbol{\epsilon} = C_{ijkl}\,\epsilon_{kl} = C_{ij11}\,\epsilon_{11} + C_{ij21}\,\epsilon_{21} + C_{ij22}\,\epsilon_{22} + C_{ij12}\,\epsilon_{12}. \tag{1.16}$$

In Voigt notation this is

$$\{\boldsymbol{\sigma}\} = \mathbf{D}\,\{\boldsymbol{\epsilon}\}. \tag{1.17}$$

where $\mathbf{D}$ is the elasticity matrix.

The dot or scalar product of two vectors is defined as

$$a = \mathbf{u} \cdot \mathbf{v} = u_i\, v_i = u_1\, v_1 + u_2\, v_2 + u_3\, v_3. \tag{1.18}$$

The cross product of two vectors is given by

$$\mathbf{u} \times \mathbf{v} = \|\mathbf{u}\| \|\mathbf{v}\| \sin\theta\, \mathbf{n}. \tag{1.19}$$

where θ is the angle between vectors $\mathbf{u}$ and $\mathbf{v}$ in the plane defined by them and $\mathbf{n}$ is a unit vector normal to it.

1.3.4 Notes on Naming Convention

In general, we use bold upper case letters for matrices and bold lower case letters for vectors. Curly brackets are used to specify Voigt notation for stresses and strains. We use lower case Greek letters for local axes and upper case Greek letters to specify boundary and domain.

1.4 Glossary of Terms

B-splines and NURBS

Anchor	A point within the parameter space that is associated to a basis function.
B-splines	Piecewise polynomial basis functions.
Convex combination	A linear combination where all coefficients are non-negative and sum to 1.
Convex hull	The set formed by all convex combinations of a point set.
Control point	Spatial coefficients associated to a spline basis function.
Control polygon	Linear connection of the control points of a spline curve.
Control mesh	Linear connection of the control points of a spline surface.
Element	A non-zero knot span, which in contrast to a zero knot span, defines a polynomial segment of a spline.
Greville abscissae	Most commonly used anchors.
Hierarchical B-splines	A sequence of nested spline spaces allowing for local refinement.
Index-set	A set of indices, e.g., labels of basis functions.
Knot	Coefficient defining the parametric domain of a B-spline.
Knot multiplicity	Number indicating how often a specific knot value occurs in a given knot vector.
Knot span	Interval between two successive knots.

Knot vector	Non-decreasing sequence of knots which (together with the polynomial order) define a B-spline basis.
Local support	Emphasis that the support of a basis function is not global but restricted to a certain area of the parameter space.
Mesh	A collection of elements. We will usually use this term to refer to conventional faceted meshes rather than spline representations.
Nested spaces	If two spaces are nested, one space is fully contained within the other.
NURBS	Non-uniform rational B-splines.
Patch	A B-spline or NURBS object in model space.
Polynomial order	Highest exponent of a polynomial.[2]
Spline	A smooth piecewise polynomial function.
Support	Interval in which a basis function B_i is non-zero, i.e., $\operatorname{supp}\{B_i\} := \{\xi \in [a, b] : B_i(\xi) \neq 0\}$.
Spline space	Linear space of all possible spline functions of a given spline basis, i.e., $\mathbb{S}_{\Xi,p} := \operatorname{span}\{B_i\}$ for a given knot vector Ξ and order p.
Span of B-splines	The set of all finite linear combination of the sequence of B-spline basis functions with corresponding real coefficients, i.e., $\operatorname{span}\{B_i\} := \{\sum_i^I B_i c_i : c_i \in \mathbb{R}\}$.
Two-scale relation	The connection of a B-spline to a set B-splines of a finer nested space.
Weights	Coefficients associated to B-splines (or their control points) to obtain piecewise rational objects, i.e., NURBS patches.

Boundary Element Method

Cauchy principal value	The Cauchy principal value of a finite integral of a function f is given by $\int_a^b f(x)dx := \lim_{\varepsilon \to c}(\int_a^{c-\varepsilon} f(x)dx + \int_{c+\varepsilon}^b f(x)dx)$, where c denotes the location of a singularity of the integrand.
Field point	Spatial coordinate at which the value of a fundamental solution is measured.
Fundamental solution	A linear operator $\mathcal{G}$ of a differential operator $\mathcal{L}$ that satisfies $\mathcal{L}\mathcal{G}u = u = \mathcal{G}\mathcal{L}u$. If $\mathcal{G}$ is an integral operator with kernel $\mathsf{G}(\tilde{\boldsymbol{x}}, \hat{\boldsymbol{x}})$ then G is denoted as fundamental solution as well.
Kernel	The null space of a linear operator. The term is also used to refer to fundamental solutions.
Regular integral	Integrals with non-singular integrands.
Singular integral	Integrals where the integrand tends to infinity.
Source point	Spatial coordinate from which a fundamental solution emanates.

[2]It is noted that this is also referred to as the degree of a polynomial in the CAD literature.

1.5 Organisation of the Text

The book starts with a general introduction to the BIE method. Here we show how various differential equations can be converted into integral equations. The differential equations considered range from the Laplace equation to the Stokes equation. Together with the integral equations we also present the fundamental solutions, which are the basic building blocks for the BIE.

The aim of the following chapters is to introduce efficient methods for the numerical solution of the integral equations. A detailed description is given of the tools, required to achieve this. This starts in Chap. 3 with an introduction to basis functions in particular B-splines. The basis functions have a dual use: they are used to describe the geometry of the problem and to approximate the unknowns. Chapter 4 then introduces NURBS as a generalisation of B-splines and explains how they are used in CAD to describe complex geometries. The next chapter explains how data suitable for numerical simulation can be extracted from CAD data.

Chapter 6 deals with methods of solving the integral equations numerically. This involves basically two discretisation steps: the definition of the geometry by patches and the approximation of the unknown. It is a feature of the BEM that the methods presented here are used for all the applications discussed in this book, only the fundamental solutions change. In this chapter, we also discuss the solution of the resulting simultaneous equations and show how storage and computer time can be reduced.

Integration plays an important role in BEM and if this is not done properly bad results can be obtained. Therefore Chap. 7 deals with numerical integration techniques. Special care has to be taken and special methods need to be implemented because the integrands exhibit various degrees of singular behaviour.

Chapter 8 introduces the first type of applications that are discussed in the book, potential flow. Here we concentrate mainly on fluid flow and the examples presented range from flow around obstacles to the free surface flow through a dam. Here the advantage of the isogeometric approach can be demonstrated very well with accurate and sometimes even exact solution being obtained with very few unknowns. The next application is in elasticity. Up to this point in the book the simulation has not used trimmed models as they often occur in CAD. Therefore Chap. 10 introduces the reader into methods that can be used to make trimmed models fit for simulation.

Using boundary integrals alone only linear problems with homogeneous domains can be solved. As this would restrict the practical application, we discuss methods of overcoming this restriction. This means that we have to consider effects that occur inside the domain, the so-called body forces and that we have to deal with the evaluation of volume integrals. Chapter 11 is therefore concerned with methods that can be used for this, involving a mapping of subdomains.

The next chapter uses these methods to analyse problems where part of the domain has different material properties. Another problem that can be tackled with the consideration of body forces is non-linear material behaviour. Chapter 13 therefore shows how one type of non-linear behaviour, namely plasticity can be tackled. Chapter 14 deals with a special application in geomechanics, in particular underground excavation problems, where the methods of dealing with heterogeneous domains and non-linear material behaviour are applied.

Chapter 15 deals with problems in viscous flow, which is – due to the highly non-linear behaviour – a rather challenging problem. In the penultimate chapter we show how to deal with time dependent problems but this is limited to transient potential problems and acoustics.

A Summary and Outlook concludes this book.

Chapter 2
The Boundary Integral Equations

The possibility of transforming differential equations to integral equations has been explored very early. In 1860 for example Du Bois-Reymond wrote: "Since 1852, I have come across integral equations in the theory of partial differential equations so often that I am convinced that advances in this theory are linked to the treatment of integral equations, about which virtually nothing is known". The basic idea is indeed a very attractive one. If the equations are written on the boundary only, where the known conditions are specified, then essentially this means that the problem dimension is reduced by one order, i.e. a 3-D problem is reduced to a 2-D one. The result is a substantial reduction in the number of unknowns, especially for problems that involve a large or infinite ratio of volume to surface as they occur for example in geomechanics. The other advantage is that the functions used (fundamental solutions) exactly satisfy the differential equations. Therefore approximations only occur on the boundary. The only drawbacks of this strategy are that a fundamental solution of the governing equations is required and that effects that occur inside the domain are not considered. We will discuss in detail later how this last obstacle can be overcome.

In the following, we will show methods, which will be used to transform various differential equations into integral equations. In particular we will use Green's reciprocal theorem, the residual method and Betti's theorem.

Green's reciprocal identity allows us to transform volume integrals (over Ω) to surface integrals (over Γ). Applying it to sufficiently smooth functions ϕ, ψ we obtain

$$\int\limits_{\Omega} \left(\phi \nabla^2 \psi - \psi \nabla^2 \phi\right) d\Omega = \int\limits_{\Gamma} \left(\phi \frac{\partial \psi}{\partial n} - \psi \frac{\partial \phi}{\partial n}\right) d\Gamma. \tag{2.1}$$

The residual method states that the integral of a differential equation multiplied with a weight function W must be zero, i.e. if

$$\mathcal{L} u = 0 \tag{2.2}$$

where $\mathcal{L}$ is a differential operator then it holds that

G. Beer et al., *The Isogeometric Boundary Element Method*, Lecture Notes in Applied and Computational Mechanics 90,
https://doi.org/10.1007/978-3-030-23339-6_2

$$\int_{\Omega} W\,\mathcal{L}\,u\,d\Omega = 0. \tag{2.3}$$

The theorem of Betti is based on the reciprocal theorem by Maxwell in elasticity. If we have two solutions (load cases) that satisfy the differential equation then it can be proven that the work done by the loads of case I along the displacements of case II is the same as the work done by the loads of case II along the displacements of case I. We apply this to distributed loads (tractions) acting on a boundary Γ and state that the work done by the tractions of case I ($\mathbf{t}^{(I)}$) times the displacements of case II ($\mathbf{u}^{(II)}$) should be the same as the work done by the tractions of case II ($\mathbf{t}^{(II)}$) with the displacements of case I ($\mathbf{u}^{(I)}$), resulting in

$$\int_{\Gamma} \mathbf{u}^{(I)}\,\mathbf{t}^{(II)}\,d\Gamma = \int_{\Gamma} \mathbf{u}^{(II)}\,\mathbf{t}^{(I)}\,d\Gamma \tag{2.4}$$

Another important ingredient for the BEM are the fundamental solutions. The primary fundamental solution $\mathsf{U}(\tilde{\boldsymbol{x}}, \hat{\boldsymbol{x}})$ (also referred to as *Kernel*) of a differential operator $\mathcal{L}$ is a solution in an infinite domain due to point sources and relates to the effect at $\hat{\boldsymbol{x}}$ of a unit point source at $\tilde{\boldsymbol{x}}$.

The following notation is adopted:

- $\tilde{\boldsymbol{x}}$ coordinates of a source point
- $\hat{\boldsymbol{x}}$ field point
- $\boldsymbol{x}$ any point

Furthermore, we define $\mathbf{r}$ as the distance vector between the source point and field point, r as its length and

$$r_i = \hat{x}_i - \tilde{x}_i, \quad i = 1, \ldots, d \tag{2.5}$$

$$r_{,i} = \frac{1}{r}(\hat{x}_i - \tilde{x}_i) \tag{2.6}$$

$$\frac{\partial r}{\partial n} = r_{,i}\,n_i = \frac{1}{r}\mathbf{r}\cdot\mathbf{n} \tag{2.7}$$

with $\mathbf{n} = \{n_i\}_{i=1}^{d}$ defined as a unit vector normal to the boundary and d is the dimension.

The *Kronecker Delta* is

$$\delta_{ij} = \begin{cases} 1 & \text{when } i = j, \\ 0 & \text{when } i \neq j. \end{cases} \tag{2.8}$$

The *Dirac Delta* function is defined as

$$\delta(|\tilde{x}_i - \hat{x}_i|) = 0 \quad \text{when} \quad |\tilde{x}_i - \hat{x}_i| \neq 0 \tag{2.9}$$

$$\int_{\Omega} \delta(|\tilde{x}_i - \hat{x}_i|) d\Omega = 1 \tag{2.10}$$

The function can also be expressed in polar coordinates in which case it is a function of r only. In order to fulfil the requirement that this integral has to be 1 the relationship between $\delta(|\tilde{x}_i - \hat{x}_i|)$ and $\delta(r)$ must be as follows (see Fig. 2.1)

$$\delta(|\tilde{x}_i - \hat{x}_i|) = \delta(r)\frac{1}{4\pi r^2} \tag{2.11}$$

in 3-D and

$$\delta(|\tilde{x}_i - \hat{x}_i|) = \delta(r)\frac{1}{2\pi r}. \tag{2.12}$$

in 2-D.

We will also apply the Dirac Delta function in time as

$$\delta(t - \tau) = 0 \quad \text{when} \quad t \neq \tau \tag{2.13}$$

$$\int_{-\infty}^{+\infty} \delta(t - \tau) d\tau = 1. \tag{2.14}$$

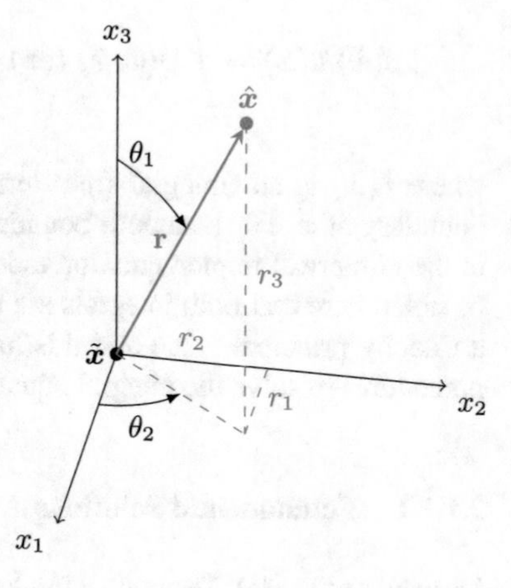

Fig. 2.1 Cartesian and polar coordinate systems

2.1 Potential Problems

Potential problems is a generic term that is used to describe scalar field problems, i.e. where the primary variable is a scalar.

2.1.1 *Steady State Flow*

The Laplace equation describes steady state heat conduction, potential or electrical flow. For an isotropic material we have

$$k\, u_{,ii}(\hat{\boldsymbol{x}}) = 0 \tag{2.15}$$

where the primary variable u may be the temperature, potential or voltage and k is the conductivity or permeability. Applying Green's reciprocal identity we obtain

$$\int\limits_{\Gamma} \mathsf{U}(\tilde{\boldsymbol{x}}, \hat{\boldsymbol{x}})\, t(\hat{\boldsymbol{x}})\, d\Gamma(\hat{\boldsymbol{x}}) = \int\limits_{\Gamma} \mathsf{T}(\tilde{\boldsymbol{x}}, \hat{\boldsymbol{x}})\, u(\hat{\boldsymbol{x}})\, d\Gamma(\hat{\boldsymbol{x}}) \tag{2.16}$$

where $\mathsf{U}(\tilde{\boldsymbol{x}}, \hat{\boldsymbol{x}})$ is the primary fundamental solution of the differential equation at point $\hat{\boldsymbol{x}}$ due to a source at $\tilde{\boldsymbol{x}}$, $\mathsf{T}(\tilde{\boldsymbol{x}}, \hat{\boldsymbol{x}}) = -k\frac{\partial \mathsf{U}(\tilde{\boldsymbol{x}}, \hat{\boldsymbol{x}})}{\partial \mathbf{n}(\hat{\boldsymbol{x}})}$ is the secondary fundamental solution and $t(\hat{\boldsymbol{x}}) = -k\frac{\partial u}{\partial \mathbf{n}}(\hat{\boldsymbol{x}})$ is the flow normal to the boundary. Because the fundamental solutions tend to infinity as $\hat{\boldsymbol{x}}$ approaches $\tilde{\boldsymbol{x}}$ a limiting process has to be applied to obtain the integral equation

$$c(\tilde{\boldsymbol{x}})\, u(\tilde{\boldsymbol{x}}) = \int\limits_{\Gamma} \mathsf{U}(\tilde{\boldsymbol{x}}, \hat{\boldsymbol{x}})\, t(\hat{\boldsymbol{x}})\, d\Gamma(\hat{\boldsymbol{x}}) - \int\limits_{\Gamma} \mathsf{T}(\tilde{\boldsymbol{x}}, \hat{\boldsymbol{x}})\, u(\hat{\boldsymbol{x}})\, d\Gamma(\hat{\boldsymbol{x}}) \tag{2.17}$$

where $c(\tilde{\boldsymbol{x}})$ is an (integral free) term whose value depends on the shape of the boundary at $\tilde{\boldsymbol{x}}$. For a smooth boundary its value is $\frac{1}{2}$. However, we will find later in the numerical implementation that this term needs not to be evaluated. It should be noted here that both integrals are singular and the second integral only exists as a Cauchy principal value and this has to be considered when applying numerical procedures to solve the integral equation. But more about this later.

2.1.1.1 Fundamental Solutions

Isotropic material. To obtain a fundamental solution we first express the differential operator in polar coordinates[1]

[1] We change u to upper case U to indicate that the result will be a fundamental solution.

$$\mathsf{U}_{,ii} = \frac{1}{r}(r \cdot \mathsf{U}_{,r})_{,r} + \frac{1}{r^2}\,\mathsf{U}_{,\theta\theta} \tag{2.18}$$

where subscripts $(\cdot)_{,r}$ and $(\cdot)_{,\theta}$ denote partial differentiation with respect to r and θ, respectively. The solution due to a point source must be radially symmetric and therefore the second term on the right side of Eq. (2.18) is zero. To obtain the fundamental solution, which is a solution due to a Dirac Delta function on the right-hand side, we need to solve the following differential equation in 2-D

$$k\,\frac{1}{r}(r \cdot \mathsf{U}_{,r})_{,r} = -\delta(r)\frac{1}{2\pi r}. \tag{2.19}$$

By integrating twice we obtain

$$k\,\mathsf{U}(r) = \frac{1}{2\pi}\ln\left(\frac{1}{r}\right) + C_1\,\ln(r) + C_2. \tag{2.20}$$

C_2 corresponds to a constant potential which is set to zero. C_1 can be determined using the impulse condition

$$\int_{\Omega} \mathsf{U}_{,ii}\,d\Omega = -\int_{\Omega} \delta(x)\,d\Omega = -1. \tag{2.21}$$

By applying the Gauss reciprocal identity the volume integral can be transformed into a surface integral

$$\int_{\Omega} \mathsf{U}_{,ii}\,d\Omega = \int_{\Gamma} \mathsf{U}_{,i}\,n_i\,d\Gamma = \int_{\Gamma} \mathsf{U}_{,r}\,d\Gamma. \tag{2.22}$$

The impulse condition can now be written as:

$$\int_{\Gamma} \mathsf{U}_{,r}\,d\Gamma = -1 \tag{2.23}$$

We can easily integrate this to obtain:

$$2\pi r\left(C_1 - \frac{1}{2\pi}\right)\frac{1}{r} = -1 \tag{2.24}$$

which results in $C_1 = 0$. For an isotropic material the solutions are therefore given by

$$\mathsf{U}(\tilde{\boldsymbol{x}}, \hat{\boldsymbol{x}}) = \frac{1}{2\pi k}\ln\left(\frac{1}{r}\right) \tag{2.25}$$

$$\mathsf{T}(\tilde{\boldsymbol{x}}, \hat{\boldsymbol{x}}) = -k\frac{\partial \mathsf{U}(\tilde{\boldsymbol{x}}, \hat{\boldsymbol{x}})}{\partial n} = -\frac{1}{2\pi}\frac{\partial r}{\partial n}\left(\frac{1}{r}\right). \tag{2.26}$$

In a similar way the 3-D fundamental solution can be obtained as

$$\mathsf{U}(\tilde{\boldsymbol{x}}, \hat{\boldsymbol{x}}) = \frac{1}{4\pi k}\left(\frac{1}{r}\right) \tag{2.27}$$

$$\mathsf{T}(\tilde{\boldsymbol{x}}, \hat{\boldsymbol{x}}) = -\frac{1}{4\pi}\frac{\partial r}{\partial n}\left(\frac{1}{r^2}\right). \tag{2.28}$$

The results for the potential at an interior point $\boldsymbol{x}$ can be obtained by

$$u(\boldsymbol{x}) = \int_\Gamma \mathsf{U}(\boldsymbol{x}, \hat{\boldsymbol{x}})\, t(\hat{\boldsymbol{x}})\, d\Gamma(\hat{\boldsymbol{x}}) - \int_\Gamma \mathsf{T}(\boldsymbol{x}, \hat{\boldsymbol{x}})\, u(\hat{\boldsymbol{x}})\, d\Gamma(\hat{\boldsymbol{x}}) \tag{2.29}$$

and for the flow vector we have

$$q_i(\boldsymbol{x}) = -k\frac{\partial u}{\partial x_i} = \int_\Gamma \mathsf{S}_i(\boldsymbol{x}, \hat{\boldsymbol{x}})\, t(\hat{\boldsymbol{x}})\, d\Gamma(\hat{\boldsymbol{x}}) - \int_\Gamma \mathsf{R}_i(\boldsymbol{x}, \hat{\boldsymbol{x}})\, u(\hat{\boldsymbol{x}})\, d\Gamma(\hat{\boldsymbol{x}}). \tag{2.30}$$

The derived fundamental solutions are:

$$\mathsf{S}_i(\boldsymbol{x}, \hat{\boldsymbol{x}}) = -k\frac{\partial \mathsf{U}(\boldsymbol{x}, \hat{\boldsymbol{x}})}{\partial x_i} = -\frac{r_{,i}}{2\pi r} \tag{2.31}$$

$$\mathsf{R}_i(\boldsymbol{x}, \hat{\boldsymbol{x}}) = -k\frac{\partial \mathsf{T}(\boldsymbol{x}, \hat{\boldsymbol{x}})}{\partial x_i} = \frac{k}{2\pi r^2}(2r_{,i}(r_{,j}n_j) - n_i) \tag{2.32}$$

in two dimensions and

$$\mathsf{S}_i(\boldsymbol{x}, \hat{\boldsymbol{x}}) = -k\frac{\partial \mathsf{U}(\boldsymbol{x}, \hat{\boldsymbol{x}})}{\partial x_i} = -\frac{r_{,i}}{4\pi r^2} \tag{2.33}$$

$$\mathsf{R}_i(\boldsymbol{x}, \hat{\boldsymbol{x}}) = -k\frac{\partial \mathsf{T}(\boldsymbol{x}, \hat{\boldsymbol{x}})}{\partial x_i} = -\frac{k}{4\pi r^3}(3r_{,i}(r_{,j}n_j) - n_i) \tag{2.34}$$

in three dimensions.

Anisotropic material. In the case where the conductivity is direction dependent, we have to solve the following differential equation

$$K_{ij}\, u_{,ij} = \delta(|\tilde{x}_i - \hat{x}_i|) \tag{2.35}$$

where K_{ij} are the coefficients of matrix **K** containing conductivities or permeabilities in the global coordinate directions. If the conductivities are not specified in the global directions then the following transformation from the local $\mathbf{K}'$ to the global $\mathbf{K}$ has to be applied

$$\mathbf{K} = \mathbf{T}^{\mathrm{T}}\mathbf{K}'\mathbf{T} \tag{2.36}$$

where $\mathbf{T}$ is the transformation matrix (see Appendix A.5.1). As has been shown in [27] fundamental solutions can be obtained by replacing r with $\bar{r}$ and k with $\bar{k}$ into the equations for the isotropic material where

$$\bar{k} = \sqrt{\det(\mathbf{K})}\ , \quad \bar{r} = \mathbf{r}^{\mathrm{T}}\mathbf{K}^{-1}\mathbf{r}. \tag{2.37}$$

The fundamental solutions and the derived solutions for anisotropic material are listed in the Appendix A.

2.1.2 Transient Flow

The time-dependent flow is governed by the diffusion differential equation

$$ku_{,ii} - \rho c_h \dot{u} = 0 \tag{2.38}$$

where ρ is the density and c_h the specific heat in case of thermal problems and dynamic viscosity in the case of fluid flow. $\dot{u}$ is the rate of change in temperature or potential and the overdot means differentiation to time. For simplicity, we do not consider any flow generated inside the volume.

Note that the primary variable u is now a function of space and time $\hat{\tau}$, i.e.

$$u = u(\boldsymbol{x}, \hat{\tau}). \tag{2.39}$$

Here we employ the residual method which states that the integral over time (from zero to a specified time $\tilde{T}$) and space of the differential equation multiplied with a *weight function* g must be zero

$$\int_0^{\tilde{T}}\int_\Omega \mathrm{g}(k\, u_{,ii} - \rho c_h \dot{u}) d\Omega\, d\tilde{\tau} = 0. \tag{2.40}$$

Applying integration by parts twice we obtain

$$\begin{aligned} 0 = &\int_0^{\tilde{T}}\int_\Gamma (\mathrm{g}\, ku_{,i}n_i + k\, \mathrm{g}_{,i}n_i\, u)\, d\Gamma d\tilde{\tau} \\ &- \int_\Omega |\mathrm{g}\, \rho c_h \tilde{\tau}|_0^{\tilde{T}} d\Omega + \int_0^{\tilde{T}}\int_\Omega (k\mathrm{g}_{,ii} + \rho c_h \dot{\mathrm{g}})\, u\, d\Omega\, d\tilde{\tau}. \end{aligned} \tag{2.41}$$

If we substitute g $= \mathsf{U}(\tilde{\boldsymbol{x}}, \tilde{\tau}, \hat{\boldsymbol{x}}, \hat{\tau})$, i.e., the primary fundamental solution of the differential equation for the effect at location $\hat{\boldsymbol{x}}$ at time $\hat{\tau}$ of a pulse source applied at $\tilde{\boldsymbol{x}}$ at time $\tilde{\tau}$, then the last volume integral collapses to a point, i.e.

$$\int_0^{\tilde{T}}\int_\Omega \left(k\mathsf{U}_{,ii} + \rho c_h \dot{\mathsf{U}}\right) u \, d\Omega \, d\tilde{\tau} = u(\tilde{\boldsymbol{x}}, \tilde{\tau}). \tag{2.42}$$

Note that the *causality* condition implies that

$$\mathsf{U}(\tilde{\boldsymbol{x}}, \tilde{\tau}, \hat{\boldsymbol{x}}, \hat{\tau}) = 0 \quad \text{for} \quad \hat{\tau} > \tilde{\tau}. \tag{2.43}$$

Then for $\tilde{\tau} < \tilde{T}$, the integral equation is obtained as

$$\begin{aligned} u(\tilde{\boldsymbol{x}}, \tilde{\tau}) = &\int_0^{\tilde{\tau}}\int_\Gamma [-\mathsf{U}(\tilde{\boldsymbol{x}}, \tilde{\tau}, \hat{\boldsymbol{x}}, \hat{\tau})\, t(\hat{\boldsymbol{x}}, \hat{\tau}) + \mathsf{T}(\tilde{\boldsymbol{x}}, \tilde{\tau}, \hat{\boldsymbol{x}}, \hat{\tau})\, u(\hat{\boldsymbol{x}}, \hat{\tau})]\, d\Gamma\, d\hat{\tau} \\ &+ \int_\Omega \mathsf{U}(\tilde{\boldsymbol{x}}, \tilde{\tau}, \hat{\boldsymbol{x}}, 0)\, \rho\, c_h u(\hat{\boldsymbol{x}}, 0)\, d\Omega \end{aligned} \tag{2.44}$$

where $\mathsf{T}(\tilde{\boldsymbol{x}}, \tilde{\tau}, \hat{\boldsymbol{x}}, \hat{\tau}) = -k\frac{\partial \mathsf{U}(\tilde{\boldsymbol{x}}, \tilde{\tau}, \hat{\boldsymbol{x}}, \hat{\tau})}{\partial x_i} n_i(\hat{\boldsymbol{x}})$.

We can visualise the meaning of the integral equation by displaying the process in space and time in Fig. 2.2. In the figure it is shown how a source pulse is applied at $\tilde{\boldsymbol{x}}$ at time $\tilde{\tau}$ and the effect it has on u at $\hat{\boldsymbol{x}}$ and time $\hat{\tau}$. The integral equation implies

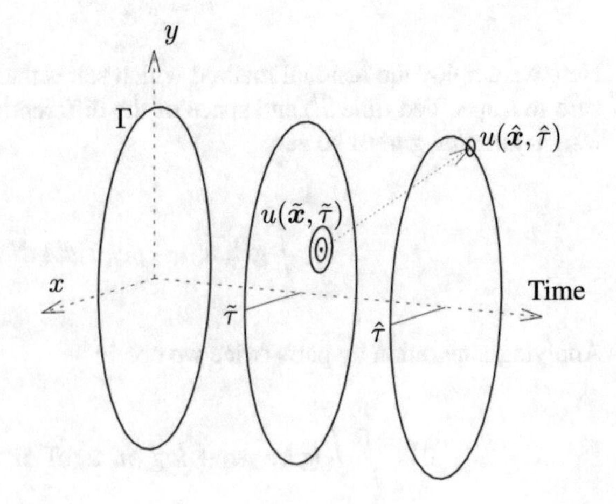

Fig. 2.2 Visualisation of the space-time domain showing the source point and the field point

that we sum (integrate) all source effects up to $\tilde{\tau}$. If $u(\hat{\boldsymbol{x}},0)=0$ then the domain integral disappears.

Moving the source point to the boundary with a limiting process and introducing the symbol $*$ for the Riemann convolution integral

$$\mathsf{U} * t = \int_0^{\tilde{\tau}} \mathsf{U}(\tilde{\boldsymbol{x}},\tilde{\tau},\hat{\boldsymbol{x}},\hat{\tau})\, t(\hat{\boldsymbol{x}},\hat{\tau}) d\hat{\tau} \tag{2.45}$$

$$\mathsf{T} * u = \int_0^{\tilde{\tau}} \mathsf{T}(\tilde{\boldsymbol{x}},\tilde{\tau},\hat{\boldsymbol{x}},\hat{\tau})\, u(\hat{\boldsymbol{x}},\hat{\tau}) d\hat{\tau} \tag{2.46}$$

we obtain the following integral equations

$$c(\tilde{\boldsymbol{x}}_n)\, u(\tilde{\boldsymbol{x}}_n,\tilde{\tau}) = \int_\Gamma [\mathsf{U} * t](\tilde{\boldsymbol{x}}_n,\tilde{\tau},\hat{\boldsymbol{x}}) d\Gamma(\hat{\boldsymbol{x}}) - \int_\Gamma [\mathsf{T} * u](\tilde{\boldsymbol{x}}_n,\tilde{\tau},\hat{\boldsymbol{x}}) d\Gamma(\hat{\boldsymbol{x}}) \tag{2.47}$$

2.1.2.1 Fundamental Solutions

The fundamental solution U is obtained by solving the following differential equation

$$\kappa \mathsf{U}_{,ii} - \dot{\mathsf{U}} = -\delta(|\tilde{x}_i - \hat{x}_i|)\delta(\tilde{\tau}-\hat{\tau}) \tag{2.48}$$

where the second Dirac Delta function has been introduced to represent a unit pulse source applied at $\tilde{\boldsymbol{x}}$ at time $\tilde{\tau}$. The fundamental solutions at $\hat{\boldsymbol{x}}$ at time $\hat{\tau}$ are

$$\mathsf{U}(\tilde{x},\tilde{\tau},\hat{x},\hat{\tau}) = \frac{1}{[4\pi\kappa(\hat{\tau}-\tilde{\tau})]^{d/2}} \exp\left(\frac{-r^2}{4\kappa(\hat{\tau}-\tilde{\tau})}\right) \tag{2.49}$$

$$\mathsf{T}(\tilde{x},\tilde{\tau},\hat{x},\hat{\tau}) = \frac{k\, r}{2^{(d+1)}\pi^{d/2}[\kappa(\hat{\tau}-\tilde{\tau})]^{(d+2)/2}} \frac{\partial r}{\partial n} \exp\left(\frac{-r^2}{4\kappa(\hat{\tau}-\tilde{\tau})}\right) \tag{2.50}$$

where $\kappa = k/(\rho c_h)$ and d is the dimensionality of the problem.

2.1.3 *Scalar Wave Problems, Acoustics*

Another problem with a scalar primary variable is described by the Helmholtz equation

$$u_{,ii} - \frac{1}{c^2}\ddot{u} = 0. \tag{2.51}$$

One application is acoustics. The primary variable u is the acoustic potential, the double overdot signifies double differentiation in time and c is the speed of sound. The integral equation can be obtained using the theorem of Betti

$$c(\tilde{\boldsymbol{x}})u(\tilde{\boldsymbol{x}}) = \int_{\Gamma} \mathsf{U}(\tilde{\boldsymbol{x}}, \hat{\boldsymbol{x}})t(\hat{\boldsymbol{x}})d\Gamma(\hat{\boldsymbol{x}}) - \int_{\Gamma} \mathsf{T}(\tilde{\boldsymbol{x}}, \hat{\boldsymbol{x}})u(\hat{\boldsymbol{x}})d\Gamma(\hat{\boldsymbol{x}}) \tag{2.52}$$

where $t = \frac{\partial u}{\partial n}$ is the acoustic velocity. We also define the acoustic pressure as

$$p = i\omega\rho u \tag{2.53}$$

where ω is the angular frequency and ρ is the mass density.

2.1.3.1 Fundamental Solutions

The fundamental solutions for 3-D are given by

$$\mathsf{U} = \frac{e^{i\alpha r}}{4\pi r} \qquad \text{and} \qquad \mathsf{T} = \frac{e^{i\alpha r}}{4\pi r^2}\left(i\alpha r - \frac{\partial r}{\partial n}\right) \tag{2.54}$$

where α denotes the wave number.

2.2 Elasto-Statics – Navier–Cauchy Equations

2.2.1 Isotropic Material

Problems in elasto-statics are governed by the Navier–Cauchy differential equation, which can be viewed as a kind of vector Laplace equation

$$\frac{\mu}{1-2\nu} u_{j,ij} + \mu\, u_{i,jj} = 0. \tag{2.55}$$

$\mu = \frac{E}{2(1+\nu)}$ is the shear modulus also referred to as G, E is the modulus of elasticity and ν the Poisson's ratio. u_j are components of the primary variable, namely the elastic displacement vector.

The following relationship exist between the displacements and strains

$$\boldsymbol{\epsilon} = \epsilon_{ij} = \frac{1}{2}(u_{i,j} + u_{j,i}) \tag{2.56}$$

between strains and stresses

$$\boldsymbol{\sigma} = \sigma_{ij} = \lambda \delta_{ij} \epsilon_{kk} + 2\mu \epsilon_{ij} \tag{2.57}$$

where $\lambda = \frac{E\,\nu}{(1+\nu)(1-2\nu)}$ and between boundary tractions and stresses

$$t_i = \sigma_{ij}\, n_j. \tag{2.58}$$

To convert this differential equation into an integral equation we use the theorem of Betti as explained at the beginning of this chapter. We now substitute the fundamental solutions U, T for $\mathbf{u}^{(I)}, \mathbf{t}^{(I)}$ and the actual solutions $\mathbf{u}, \mathbf{t}$ for $\mathbf{u}^{(II)}, \mathbf{t}^{(II)}$ into Eq. (2.4) and obtain

$$\int_\Gamma \mathsf{U}(\tilde{\boldsymbol{x}}, \hat{\boldsymbol{x}})\, \mathbf{t}(\hat{\boldsymbol{x}})\, d\Gamma(\hat{\boldsymbol{x}}) = \int_\Gamma \mathsf{T}(\tilde{\boldsymbol{x}}, \hat{\boldsymbol{x}})\, \mathbf{u}(\hat{\boldsymbol{x}})\, d\Gamma(\hat{\boldsymbol{x}}). \tag{2.59}$$

In the absence of body forces the only other work done is by the point forces at source point $\tilde{\boldsymbol{x}}$. Placing all source points on the boundary and applying a limiting process at the singularity we obtain

$$\mathbf{c}(\tilde{\boldsymbol{x}})\, \mathbf{u}(\tilde{\boldsymbol{x}}) = \int_\Gamma \mathsf{U}(\tilde{\boldsymbol{x}}, \hat{\boldsymbol{x}})\, \mathbf{t}(\hat{\boldsymbol{x}})\, d\Gamma(\hat{\boldsymbol{x}}) - \int_\Gamma \mathsf{T}(\tilde{\boldsymbol{x}}, \hat{\boldsymbol{x}})\, \mathbf{u}(\hat{\boldsymbol{x}})\, d\Gamma(\hat{\boldsymbol{x}}) \tag{2.60}$$

where $\mathbf{c}(\tilde{\boldsymbol{x}})$ is an (integral free) term that depends on shape of the boundary. For a smooth boundary $\mathbf{c}(\tilde{\boldsymbol{x}}) = \frac{1}{2}\mathbf{I}$, where $\mathbf{I}$ is the identity matrix, but as we will see later this term needs not be evaluated. Again it should be noted that the second integral only exists as a Cauchy principal value.

2.2.1.1 Fundamental Solutions

For the case of isotropic material the fundamental solutions are obtained by solving the differential equation for unit point forces in an infinite domain, i.e.

$$\frac{1}{1-2\nu} \mathsf{U}_{ik,kj} + \mathsf{U}_{ij,kk} = -\frac{1}{\mu} \delta(|\tilde{x}_i - \hat{x}_i|)\delta_{ij}. \tag{2.61}$$

For the solution we introduce a tensor potential $G_{ij} = G\delta_{ij}$ such that

$$\mathsf{U}_{ij} = G_{ij,mm} - \frac{1}{2(1-\nu)} G_{im,jm} \tag{2.62}$$

so that Eq. (2.61) is replaced by

$$G_{,mmkk} = -\frac{1}{\mu} \delta(|\tilde{x}_i - \hat{x}_i|). \tag{2.63}$$

The solution is then obtained by consecutively solving 2 Laplace equations

$$F_{,kk} = -\frac{1}{\mu}\delta(|\tilde{x}_i - \hat{x}_i|) \tag{2.64}$$

and

$$G_{,mm} = F. \tag{2.65}$$

The solution for the Laplace equation was derived previously and it is for 2-D problems equal to

$$F = \frac{1}{2\pi\mu}\ln\left(\frac{1}{r}\right). \tag{2.66}$$

By substituting this result into Eq. (2.65) and integrating twice we obtain

$$G_{ij} = \frac{r^2}{8\pi\mu}\ln\left(\frac{1}{r}\right) + \frac{r^2}{8\pi\mu}. \tag{2.67}$$

The last term is equivalent to a rigid body motion and can thus be disregarded leading to

$$G_{ij} = \frac{r^2}{8\pi\mu}\ln\left(\frac{1}{r}\right)\delta_{ij}. \tag{2.68}$$

By differentiating G twice and by substituting the result into Eq. (2.62) we obtain the fundamental solution as

$$\mathsf{U}_{ij} = \frac{1}{8\pi\mu(1-\nu)}\left((3-4\nu)\ln\left(\frac{1}{r}\right)\delta_{ij} + r_{,i}r_{,j}\right). \tag{2.69}$$

This solution is for plane strain. The general 2-D solution is given by

$$\mathsf{U}(\tilde{\boldsymbol{x}}, \hat{\boldsymbol{x}}) = \mathsf{U}_{ij}(\tilde{\boldsymbol{x}}, \hat{\boldsymbol{x}}) = C\left(C_1 \ln\left(\frac{1}{r}\right)\delta_{ij} + r_{,i}r_{,j}\right). \tag{2.70}$$

The solution for 3-D follows a similar process and results in

$$\mathsf{U}(\tilde{\boldsymbol{x}}, \hat{\boldsymbol{x}}) = \mathsf{U}_{ij}(\tilde{\boldsymbol{x}}, \hat{\boldsymbol{x}}) = C\frac{1}{r}(C_1\delta_{ij} + r_{,i}r_{,j}). \tag{2.71}$$

The solution for the traction is obtained by inserting Eq. (2.70) or (2.71) into Eq. (2.56) and applying Eqs. (2.57) and (2.58)

$$\begin{aligned}\mathsf{T}(\tilde{\boldsymbol{x}}, \hat{\boldsymbol{x}}) = \mathsf{T}_{ij}(\tilde{\boldsymbol{x}}, \hat{\boldsymbol{x}}) = -\frac{C_2}{r^d}\Big[&(C_3\delta_{ij} + (d+1)r_{,i}r_{,j})\frac{\partial r}{\partial n} \\ &- C_3(n_j r_{,i} - n_i r_{,j})\Big].\end{aligned} \tag{2.72}$$

Table 2.1 Constants for fundamental solutions. Note that the more familiar notation for the shear modulus G has been used instead of μ

Constant	Plane strain	Plane stress	3-D
d	1	1	2
C	$\frac{1}{8\pi G(1-\nu)}$	$\frac{(1+\nu)}{8\pi G}$	$\frac{1}{16\pi G(1-\nu)}$
C_1	$3-4\nu$	$\frac{3-\nu}{1+\nu}$	$3-4\nu$
C_2	$\frac{1}{4\pi(1-\nu)}$	$\frac{1+\nu}{4\pi}$	$\frac{1}{8\pi(1-\nu)}$
C_3	$1-2\nu$	$\frac{1-\nu}{1+\nu}$	$1-2\nu$
C_4	2	2	3
C_5	4	4	5

The constants are given in Table 2.1 for plane strain, plane stress and 3-D cases.

The displacement vector at interior points $\boldsymbol{x}$ is obtained by

$$\mathbf{u}(\boldsymbol{x}) = \int_\Gamma \mathsf{U}(\boldsymbol{x}, \hat{\boldsymbol{x}})\, \mathbf{t}(\hat{\boldsymbol{x}})\, d\Gamma(\hat{\boldsymbol{x}}) - \int_\Gamma \mathsf{T}(\boldsymbol{x}, \hat{\boldsymbol{x}})\, \mathbf{u}(\hat{\boldsymbol{x}})\, d\Gamma(\hat{\boldsymbol{x}}). \tag{2.73}$$

The strain can be computed by

$$\begin{aligned}\epsilon_{ij}(\boldsymbol{x}) = 0.5(u_{i,j} + u_{j,i}) = &\int_\Gamma \hat{\mathsf{S}}_{ijk}(\boldsymbol{x}, \hat{\boldsymbol{x}})\, t_k(\hat{\boldsymbol{x}})\, d\Gamma(\hat{\boldsymbol{x}}) \\ &- \int_\Gamma \hat{\mathsf{R}}_{ijk}(\boldsymbol{x}, \hat{\boldsymbol{x}})\, u_k(\hat{\boldsymbol{x}})\, d\Gamma(\hat{\boldsymbol{x}})\end{aligned} \tag{2.74}$$

where

$$\hat{\mathsf{S}}_{ijk}(\boldsymbol{x}, \hat{\boldsymbol{x}}) = \frac{C}{r^d}\left[C_3(r_{,j}\delta_{ik} + r_{,i}\delta_{jk}) - r_{,k}\delta_{ij} + C_4 r_{,i} r_{,j} r_{,k}\right] \tag{2.75}$$

and

$$\begin{aligned}\hat{\mathsf{R}}_{ijk}(\boldsymbol{x}, \hat{\boldsymbol{x}}) = \frac{C_2}{r^{d+1}}\Big[&C_4 \frac{\partial r}{\partial n}(\nu(r_{,j}\delta_{ik} + r_{,i}\delta_{jk}) + r_{,k}\delta_{ij} - C_5 r_{,i} r_{,j} r_{,k}) \\ &+ C_3(n_j\delta_{ik} - n_k\delta_{ij} + n_i\delta_{jk} + C_4 r_{,i} r_{,j} n_k) \\ &+ C_4\nu(n_j r_{,i} r_{,k} + n_i r_{,j} r_{,k})\Big].\end{aligned} \tag{2.76}$$

Using Hooke's law the boundary integral equation for the stress at an interior point $\boldsymbol{x}$ is obtained by

$$\sigma_{ij}(\boldsymbol{x}) = \int_\Gamma \mathsf{S}_{ijk}(\boldsymbol{x}, \hat{\boldsymbol{x}})\, t_k(\hat{\boldsymbol{x}})\, d\Gamma(\hat{\boldsymbol{x}}) - \int_\Gamma \mathsf{R}_{ijk}(\boldsymbol{x}, \hat{\boldsymbol{x}})\, u_k(\hat{\boldsymbol{x}})\, d\Gamma(\hat{\boldsymbol{x}}) \tag{2.77}$$

where

$$\mathsf{S}_{ijk}(\boldsymbol{x},\hat{\boldsymbol{x}}) = \frac{C_2}{r^d}\left[C_3(r_{,j}\delta_{ik} + r_{,i}\delta_{jk} - r_{,k}\delta_{ij}) + C_4 r_{,i}r_{,j}r_{,k}\right] \tag{2.78}$$

and

$$\begin{aligned}\mathsf{R}_{ijk}(\boldsymbol{x},\hat{\boldsymbol{x}}) = \frac{2GC_2}{r^{d+1}}\Big[C_4\frac{\partial r}{\partial n}\Big(C_3 r_{,k}\delta_{ij} + \nu(r_{,j}\delta_{ik} + r_{,i}\delta_{jk}) - C_5 r_{,i}r_{,j}r_{,k}\Big)\\ + C_4\nu(n_i r_{,j}r_{,k} + n_j r_{,i}r_{,k}) - (1-4\nu)n_k\delta_{ij}\\ + C_3(C_4 n_k r_{,i}r_{,j} + n_j\delta_{ik} + n_i\delta_{jk})\Big].\end{aligned} \tag{2.79}$$

2.2.2 Anisotropic Material

The fundamental solutions for anisotropic material have to obey the following differential equation

$$C_{ijkl}\,\mathsf{U}_{mk,lj} = -\delta(|\tilde{x}_i - \hat{x}_i|)\delta_{im}. \tag{2.80}$$

For a general anisotropic material, the constitutive tensor C_{ijkl} contains 21 independent elastic constants in 3-D. However, for the special case of a *transversely isotropic* or *orthotropic* material 5 or 9 constants need to be specified. Closed form solutions for special cases of transversely isotropic materials can be found in [54, 125] and for special cases of orthotropic materials for example in [119, 190].

The fundamental solutions for general anisotropy are impossible to obtain in closed form. However, an integral representation of the fundamental solution is available and the solutions can be obtained using a *Radon transform*, which is described in detail in [1, 67]. The primary fundamental solution in 3-D can be expressed in polar coordinates as follows

$$\mathsf{U}_{mk}(r,\theta_1,\theta_2) = \frac{1}{r}\,G^u_{mk}(\theta_1,\theta_2) \tag{2.81}$$

and the derivative is:

$$\mathsf{U}_{mk,l}(r,\theta_1,\theta_2) = \frac{1}{r^2}\,G^u_{mkl}(\theta_1,\theta_2). \tag{2.82}$$

By multiplying $\mathsf{U}_{mk,l}(r,\theta_1,\theta_2)$ with the constitutive tensor and the outward normal, the traction fundamental solution is obtained

$$\mathsf{T}_{mi}(r,\theta_1,\theta_2) = \frac{1}{r^2}\,G^t_{mij}(\theta_1,\theta_2)n_j \tag{2.83}$$

with

$$G^t_{mij}(\theta_1, \theta_2) = C_{ijkl}\, G^u_{mkl}(\theta_1, \theta_2). \tag{2.84}$$

Because they are separated from the singular terms $\frac{1}{r}$ and $\frac{1}{r^2}$, the Kernels G^u_{mk}, G^u_{mkl} and G^t_{mij} are non-singular and depend on the polar angles (defined in Fig. 2.1) only. The computation of these Kernels for the anisotropic case is possible numerically but very expensive. Since, as we will see later, Kernels have to be evaluated at a large number of points it is suggested in [66] to precompute values of G^u_{mk} and G^t_{mij} prior to the BEM analysis at a grid of points depending on the polar angles θ_1 and θ_2. During the BEM analysis the fundamental solution for specific values of θ_1, θ_2 is then interpolated between precomputed values at grid points.

Calculation of values at grid points. G^u_{mk} and its derivative G^u_{mkl} [5] can be calculated by using the *inverse Radon transform* which is given by

$$G^u_{mk}(\theta_1, \theta_2) = \frac{1}{8\pi^2} \oint_0^{2\pi} (M^{xx}_{mk})^{-1}\, d\phi \tag{2.85}$$

$$G^u_{mkl}(\theta_1, \theta_2) = \frac{1}{8\pi^2} \oint_0^{2\pi} \left(-r_{,l} (M^{xx}_{mk})^{-1} + x_l F_{mk}\right) d\phi \tag{2.86}$$

where

$$F_{mk} = (M^{xx}_{mn})^{-1} \left(M^{rx}_{nq} + M^{xr}_{nq}\right) \left(M^{xx}_{qk}\right)^{-1}. \tag{2.87}$$

The tensors M^{xx}_{mn}, M^{rx}_{nq} and M^{xx}_{qk} are defined as

$$M^{xx}_{ik} := C_{ijkl}\, x_j\, x_l, \quad M^{xr}_{ik} := C_{ijkl}\, x_j\, r_{,l} \quad \text{and} \quad M^{rx}_{ik} := C_{ijkl}\, r_{,j}\, x_l. \tag{2.88}$$

The functions G_{mk} and G_{mkl} in Eqs. (2.85) and (2.86) are integrated along a circle that is created by intersecting a unit sphere with a plane perpendicular to vector $\frac{1}{r}\boldsymbol{r}$, shown in Fig. 2.3.

The tensors in Eqs. (2.85) and (2.86) are transformed into matrices by replacing Cartesian coordinates by spherical coordinates.

$$\boldsymbol{r} = r \begin{bmatrix} \sin\theta_1 \cos\theta_2 \\ \sin\theta_1 \sin\theta_2 \\ \cos\theta_1 \end{bmatrix} \quad \rightsquigarrow \quad \frac{1}{r}\boldsymbol{r} = \begin{bmatrix} \sin\theta_1 \cos\theta_2 \\ \sin\theta_1 \sin\theta_2 \\ \cos\theta_1 \end{bmatrix} \tag{2.89}$$

As vector $\frac{1}{r}\boldsymbol{r}$ is only a function of the polar angles θ_1 and θ_2, vector $\boldsymbol{x}$ has to be expressed by these angles also. Thus, two arbitrary orthogonal vectors $\boldsymbol{a}$ and $\boldsymbol{b}$ are chosen which lie in the plane defined by vectors $\boldsymbol{x}$ and $\frac{1}{r}\boldsymbol{r}$.

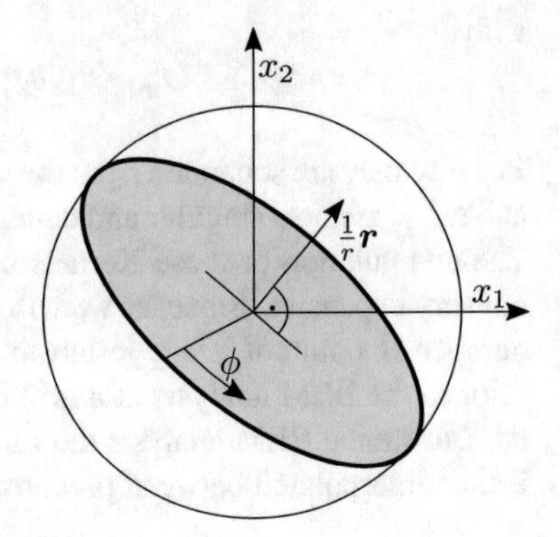

Fig. 2.3 Unit circle resulting from the intersection of a plane perpendicular to the unit vector $\frac{1}{r}\boldsymbol{r}$ (between $\tilde{\boldsymbol{x}}$ and $\hat{\boldsymbol{x}}$) with the unit sphere

$$\boldsymbol{a} = \begin{bmatrix} \sin\theta_2 \\ -\cos\theta_2 \\ 0 \end{bmatrix} \qquad \boldsymbol{b} = \begin{bmatrix} \cos\theta_1 \cos\theta_2 \\ \cos\theta_1 \sin\theta_2 \\ -\sin\theta_1 \end{bmatrix} \tag{2.90}$$

The vector $\boldsymbol{x}$ expressed by vectors $\boldsymbol{a}$, $\boldsymbol{b}$ and angle ϕ is

$$\boldsymbol{x}(\phi, \theta_1, \theta_2) = \boldsymbol{a}(\theta_1, \theta_2)\, \cos\phi + \boldsymbol{b}(\theta_1, \theta_2)\, \sin\phi. \tag{2.91}$$

In order to solve the integrals of Eqs. (2.85) and (2.86) the tensors of (2.88) are transformed to matrices and as a function of the integration variable ϕ. To do this, matrix $\mathbf{B}$ is introduced, which depends on the vectors $\boldsymbol{x}$ and $\frac{1}{r}\boldsymbol{r}$

$$\mathbf{B}(x_i(\phi)) = \begin{bmatrix} x_1 & 0 & 0 \\ 0 & x_2 & 0 \\ 0 & 0 & x_3 \\ 0 & x_3 & x_2 \\ x_3 & 0 & x_1 \\ x_2 & x_1 & 0 \end{bmatrix} \qquad \mathbf{B}(r_{,i}(\phi)) = \begin{bmatrix} r_{,1} & 0 & 0 \\ 0 & r_{,2} & 0 \\ 0 & 0 & r_{,3} \\ 0 & r_{,3} & r_{,2} \\ r_{,3} & 0 & r_{,1} \\ r_{,2} & r_{,1} & 0 \end{bmatrix}. \tag{2.92}$$

The tensors in Eq. (2.88) are now transformed to matrices as follows

$$\begin{aligned} \mathbf{M}^{xx}(\phi) &= \mathbf{B}^{\mathrm{T}}(x_i(\phi))\, \mathbf{D}\, \mathbf{B}(x_i(\phi)) \\ \mathbf{M}^{r,ix}(\phi) &= \mathbf{B}^{\mathrm{T}}(x_i(\phi))\, \mathbf{D}\, \mathbf{B}(r_{,i}(\phi)) \\ \mathbf{M}^{xr,i}(\phi) &= \mathbf{B}^{\mathrm{T}}(r_{,i}(\phi))\, \mathbf{D}\, \mathbf{B}(x_i(\phi)) \end{aligned} \tag{2.93}$$

where $\mathbf{D}$ is the elasticity matrix. Inserting this into Eqs. (2.85) and (2.86) the kernel functions in matrix notation are given by:

$$\mathbf{G}^{u}(\theta_1, \theta_2) = \frac{1}{8\pi^2} \oint_0^{2\pi} \left(\mathbf{M}^{xx}(\boldsymbol{x}(\phi, \theta_1, \theta_2))\right)^{-1} d\phi \tag{2.94}$$

$$\mathbf{G}_l^u(\theta_1,\theta_2) = \frac{1}{8\pi^2} \oint_0^{2\pi} \left(-r_{,l}(\mathbf{M}^{xx})^{-1} + x_l\mathbf{F}\right)^{-1} d\phi \tag{2.95}$$

where

$$\mathbf{G}_l^u = \begin{bmatrix} G_{11l}^u & G_{12l}^u & G_{13l}^u \\ G_{21l}^u & G_{22l}^u & G_{23l}^u \\ G_{31l}^u & G_{32l}^u & G_{33l}^u \end{bmatrix} \tag{2.96}$$

and the matrix $\mathbf{F}$ is:

$$\mathbf{F} = (\mathbf{M}^{xx})^{-1} \left(\mathbf{M}^{r,ix} + (\mathbf{M}^{r,ix})^{\mathrm{T}}\right) (\mathbf{M}^{xx})^{-1}. \tag{2.97}$$

The anisotropic fundamental solution is calculated prior to the BE analysis on a grid of points and a suitable interpolation scheme is used to get specific values for the fundamental solution as required.

As shown in Fig. 2.4, θ_1 ranges from 0 to π and θ_2 from 0 to 2π. This is subdivided into a grid of $N \times M$ grid points and stored in memory. During the integration the fundamental solution $\mathbf{G}(\theta_1, \theta_2)$ is interpolated using the stored grid point values at neighbouring grid-points as shown in Fig. 2.4. The interpolation scheme was first introduced by Wilson and Cruse [193] using a cubic Lagrangian interpolation and employing a grid of 33×65 points. Our experience is, that linear interpolation with a larger number of grid points significantly speeds up the execution of the integration, but at the cost of higher storage requirements. Numerical studies have shown that for a linear interpolation a grid size of 100×200 is sufficient for most engineering applications.

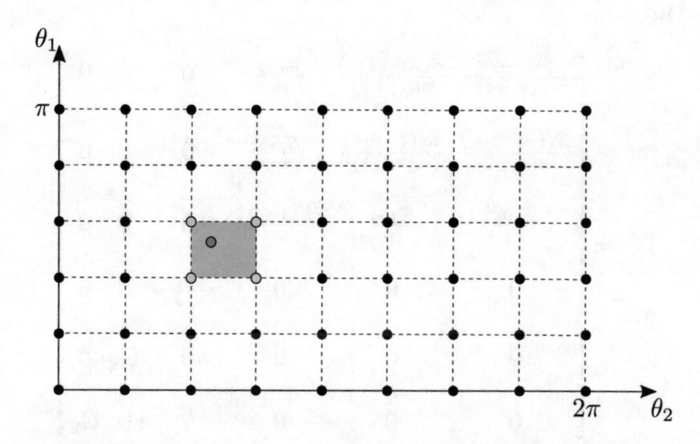

Fig. 2.4 Figure showing an example of a computational grid for a linear interpolation scheme. Interpolation points used for a given point (θ_1, θ_2, depicted by a red circle) are depicted by cyan circles

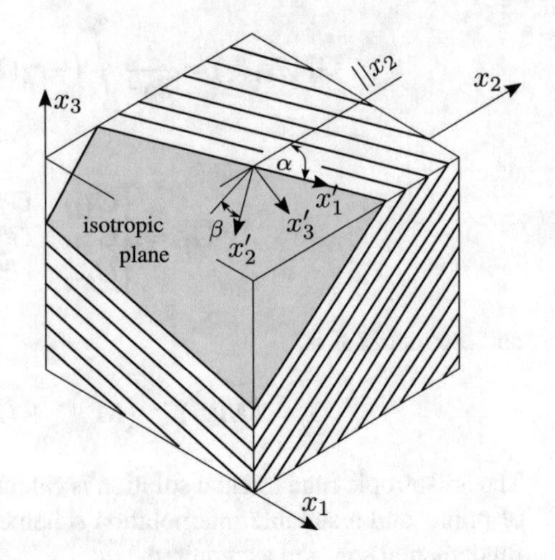

Fig. 2.5 Stratified rockmass – transversal isotropic material. The specified plane is described by the local coordinate system x_i' linked to the global coordinate system via the angles α and β

Example. As an example in geomechanics a stratified rock mass is modelled as a transversely isotropic material. In Fig. 2.5 the stratification of the rock is shown, where the isotropic plane is described by the local coordinate system x_i'. The transformation between the local and the global coordinate system is specified using the angles α and β. The material constants for the stratified rock are given in Table 2.2 for a transversely isotropic material with the Young's modulus E_1 and Poisson's ratio ν_1 parallel to the specified plane and E_2 and Poisson's ratio ν_2 normal to the plane. G_2 is the shear modulus related to shear stress in the specified plane, which is independent from the other parameters.

With these parameters, the matrix $\mathbf{D}'$ related to the local coordinate system can be defined as:

$$\mathbf{D}' = \begin{bmatrix} \frac{E_1(1-n\nu_2^2)}{m(\nu_1+1)} & \frac{E_1(n\nu_2^2+\nu_1)}{m(\nu_1+1)} & \frac{E_1\nu_2}{m} & 0 & 0 & 0 \\ \frac{E_1(n\nu_2^2+\nu_1)}{m(\nu_1+1)} & \frac{E_1(1-n\nu_2^2)}{(m(\nu_1+1))} & \frac{E_1\nu_2}{m} & 0 & 0 & 0 \\ \frac{E_1\nu_2}{m} & \frac{E_1\nu_2}{m} & \frac{E_2(1-\nu_1)}{m} & 0 & 0 & 0 \\ 0 & 0 & 0 & \frac{E_1}{2(\nu_1+1)} & 0 & 0 \\ 0 & 0 & 0 & 0 & G_2 & 0 \\ 0 & 0 & 0 & 0 & 0 & G_2 \end{bmatrix} \tag{2.98}$$

where $n = E_1/E_2$ and $m = 1 - \nu_1 - 2n\nu_2^2$.

However, for the computation of the fundamental solutions, the constitutive matrix has to be expressed in the global coordinate system and therefore the following

Table 2.2 Material parameters for transversely isotropic rock material

$E_1 = 40\,\mathrm{MN/m^2}$	$\nu_1 = 0.15$
$E_2 = 10\,\mathrm{MN/m^2}$	$\nu_2 = 0.08$
$G_2 = 5\,\mathrm{MN/m^2}$	
$\alpha = 0°$	$\beta = 30°$

coordinate transformation is necessary:

$$\mathbf{D} = \mathbf{T}^{\mathrm{T}}\,\mathbf{D}'\,\mathbf{T} \tag{2.99}$$

The transformation matrix $\mathbf{T}$ for this case is shown in Appendix A.5.2.

After the transformation we obtain

$$\mathbf{D} = \begin{bmatrix} 29.78 & 6.74 & 8.70 & 0 & 0 & -9.59 \\ 6.74 & 42.43 & 4.92 & 0 & 0 & -1.58 \\ 8.70 & 4.92 & 13.89 & 0 & 0 & -4.17 \\ 0 & 0 & 0 & 14.29 & -5.37 & 0 \\ 0 & 0 & 0 & -5.37 & 8.10 & 0 \\ -9.59 & -1.58 & -4.17 & 0 & 0 & 9.70 \end{bmatrix}. \tag{2.100}$$

Using this matrix, the fundamental solutions are calculated at grid points. In Fig. 2.6, we show values of G^u_{22} and G^t_{23} as function of θ_1 and θ_2.

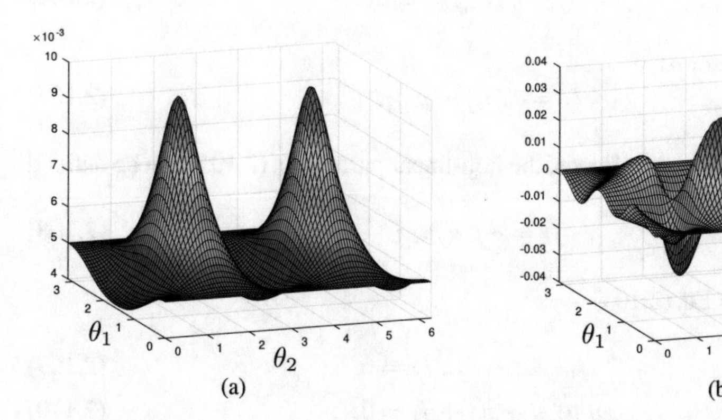

Fig. 2.6 Fundamental solutions for transversely isotropic rock: (a) G^u_{22} and (b) G^t_{23}

2.3 Stokes Flow

The dimensionless form of the Stokes equations for a Newtonian viscous fluid of constant density and viscosity can be written as

$$u_{i,i} = 0 \tag{2.101}$$

$$\frac{\partial u_i}{\partial t} + u_i u_{i,j} + p_{,i} - \frac{1}{Re} u_{i,ij} = 0 \tag{2.102}$$

where the primary variable u_i is the fluid velocity vector, p is the pressure and *Re* is the *Reynolds number* defined as

$$Re = \frac{\rho\, uH}{\mu} \tag{2.103}$$

where ρ is the density of the fluid, u is the characteristic velocity, H is a characteristic length and μ is the viscosity. In this book, we restrict ourselves to steady incompressible flow, which reduces the equations to

$$u_{i,i} = 0 \tag{2.104}$$

$$\mu\, u_{i,jj} + p_{,i} - \rho\, u_j u_{i,j} = 0. \tag{2.105}$$

We define the fluid stresses as:

$$\sigma_{ij} = \mu(u_{i,j} + u_{j,i}) \tag{2.106}$$

and the traction vector as

$$t_i = \sigma_{ij} n_j - p\, n_i. \tag{2.107}$$

In the absence of other body forces, the non-linear part of Eq. (2.105) can be defined as a body force

$$f_i = -\rho\, u_j u_{i,j} \tag{2.108}$$

and substituted into Eq. (2.105)

$$u_{i,i} = 0 \tag{2.109}$$

$$\mu\, u_{i,jj} + p_{,i} + f_i = 0. \tag{2.110}$$

For flow problems, it is convenient to express the total velocity u_i as a perturbation velocity $\dot{u}_i$ and a free stream velocity u^0

$$u_i = \dot{u}_i + u^0. \tag{2.111}$$

Substitution into Eq. (2.108) gives

$$f_i = -\rho(\dot{u}_j + u_j^0)\dot{u}_{i,j}. \tag{2.112}$$

To get the integral equations, we apply the same reciprocal theorem, that has been used to derive the integral equations for elasticity

$$\int_\Gamma \left[t_i^{(I)}u_i^{(II)} - t_i^{(II)}u_i^{(I)}\right] d\Gamma + \int_\Omega \left[f_i^{(I)}u_i^{(II)} - f_i^{(II)}u_i^{(I)}\right] d\Omega = 0 \tag{2.113}$$

where the superscripts (I) and (II) denote any two unrelated states that satisfy Eqs. (2.109) and (2.110). We now select state (I) to be the fundamental solution of the equations and state (II) the boundary value problem of interest. This leads to the following integral equation

$$\begin{aligned} c_{ij}(\tilde{\boldsymbol{x}})\, u_i(\tilde{\boldsymbol{x}}) = &\int_\Gamma [\mathsf{U}_{ij}(\tilde{\boldsymbol{x}},\hat{\boldsymbol{x}})t_i(\hat{\boldsymbol{x}}) - \mathsf{T}_{ij}(\tilde{\boldsymbol{x}},\hat{\boldsymbol{x}})u_i(\hat{\boldsymbol{x}})]\, d\Gamma \\ &- \int_\Omega \mathsf{U}_{ij}\, \rho(\dot{u}_k + u_k^0)\dot{u}_{i,k} d\Omega. \end{aligned} \tag{2.114}$$

In the volume integral derivatives of the velocity appear. These can be obtained by post-processing, using derived fundamental solutions of high singularity. Alternatively, we can use the divergence theorem on the volume integral resulting in

$$\begin{aligned} \int_\Omega \mathsf{U}_{ij}\, \rho(\dot{u}_k + u_k^0)\dot{u}_{i,k} d\Omega = &\int_\Gamma \mathsf{U}_{ij}\, \rho(\dot{u}_k + u_k^0)\dot{u}_i n_k d\Gamma \\ &- \int_\Omega \mathsf{U}_{ij,k}\, \rho(\dot{u}_k + u_k^0)\dot{u}_i d\Omega. \end{aligned} \tag{2.115}$$

Substitution into Eq. (2.114) gives

$$\begin{aligned} c_{ij}(\tilde{\boldsymbol{x}})\, \dot{u}_i(\tilde{\boldsymbol{x}}) = &\int_\Gamma [\mathsf{U}_{ij}(\tilde{\boldsymbol{x}},\hat{\boldsymbol{x}})t_i(\hat{\boldsymbol{x}}) - \mathsf{T}_{ij}(\tilde{\boldsymbol{x}},\hat{\boldsymbol{x}})\dot{u}_i(\hat{\boldsymbol{x}})]\, d\Gamma \\ &- \int_\Gamma \mathsf{U}_{ij}(\tilde{\boldsymbol{x}},\hat{\boldsymbol{x}})\, \rho(\dot{u}_k + u_k^0)\dot{u}_i n_k d\Gamma \\ &+ \int_\Omega \mathsf{U}_{ij,k}(\tilde{\boldsymbol{x}},\hat{\boldsymbol{x}})\, \rho(\dot{u}_k + u_k^0)\dot{u}_i d\Omega. \end{aligned} \tag{2.116}$$

2.3.1 Fundamental Solutions

To obtain the fundamental solution, we must solve the following differential equations

$$p_{,i} - \mu\, u_{i,jj} = \delta_{ij}\delta(|\tilde{x}_i - \hat{x}_i|), \quad u_{i,i} = 0. \tag{2.117}$$

We proceed in a similar way as for the elasticity problem and introduce a tensor potential G that reduces Eq. (2.117) to

$$G_{,mmkk} = -\frac{1}{\mu}\delta(|\tilde{x}_i - \hat{x}_i|). \tag{2.118}$$

In a similar way as for the elasticity problem, we obtain the solutions as

$$\mathsf{U}_{ij}(\tilde{\boldsymbol{x}}, \hat{\boldsymbol{x}}) = \frac{1}{4\pi\mu}\left(r_{,i}\, r_{,j} + \delta_{ij}\, \ln\left(\frac{1}{r}\right)\right) \tag{2.119}$$

$$\mathsf{T}_{ij}(\tilde{\boldsymbol{x}}, \hat{\boldsymbol{x}}) = \frac{1}{\pi r}\, r_{,i}\, r_{,j}\, r_{,k}\, n_k \tag{2.120}$$

for plane problems and

$$\mathsf{U}_{ij}(\tilde{\boldsymbol{x}}, \hat{\boldsymbol{x}}) = \frac{1}{8\pi\mu r}(r_{,i}\, r_{,j} + \delta_{ij}) \tag{2.121}$$

$$\mathsf{T}_{ij}(\tilde{\boldsymbol{x}}, \hat{\boldsymbol{x}}) = \frac{1}{4\pi r^2} r_{,i}\, r_{,j}\, r_{,k}\, n_k. \tag{2.122}$$

for 3-D problems.

The derived fundamental solution is given by

$$\mathsf{U}_{ij,k}(\tilde{\boldsymbol{x}}, \hat{\boldsymbol{x}}) = \frac{1}{4\pi\mu r}\left(\delta_{jk}\, r_i + \delta_{ik}\, r_j - \delta_{ij}\, r_k - 2\, r_i\, r_j\, r_k\right) \tag{2.123}$$

for 2-D and

$$\mathsf{U}_{ij,k}(\tilde{\boldsymbol{x}}, \hat{\boldsymbol{x}}) = \frac{1}{8\pi\mu r^2}\left(\delta_{jk}\, r_i + \delta_{ik}\, r_j - \delta_{ij}\, r_k - 3\, r_i\, r_j\, r_k\right). \tag{2.124}$$

for 3-D.

2.4 Summary

In this chapter, we have shown how differential equations can be converted into equivalent boundary integral equations. For the transformation, we presented three methods that can be applied, depending on the nature of the differential equation. Only the differential equations for the problems, that are discussed in this book, were considered which range from potential problems to elasticity and Stokes flow.

The fundamental solutions of the differential equation were also presented. The two items form the essential building blocks for building the isogeometric BEM framework, presented in the rest of the book.

Chapter 3
Basis Functions, B-splines

3.1 Introduction

Basis functions have two applications in numerical simulation: they are used to define the domain of interest and to approximate physical quantities of the problem considered, i.e., its boundary conditions and the solutions sought. In general, a function $f(\xi)$ of the coordinate ξ can be approximated by a polynomial $f_p(\xi)$ of order p

$$f(\xi) \approx f_p(\xi) = \sum_{i=0}^{I-1} N_i(\xi) c_i \tag{3.1}$$

where N_i are linearly independent basis functions, I is the number of parameters and c_i are the coefficients related to each N_i. ξ is the local coordinate in the *parameter space* which is often defined from 0 to 1 or -1 to 1. We will define the *order* of the polynomial as the highest exponent of its terms. It is, however, noted that this is also referred to as the *degree* of a polynomial in some publications.

For a given target function $f(\xi)$ and a set of N_i, the unknown coefficients c_i can be determined by different approximation methods such as *interpolation* or L_2-*projection*. The former forces $f_p(\xi)$ to coincide with $f(\xi)$ at distinct interpolation sites $\tilde{\xi}_i$ with $\tilde{\xi}_i \neq \tilde{\xi}_j$ for $i \neq j$. This yields the following system of equations

$$\mathbf{A}\,\mathbf{c} = \mathbf{f} \quad \text{with} \quad \mathbf{A}[i,j] = N_j(\tilde{\xi}_i) \quad \text{and} \quad \mathbf{f}[i] = f(\tilde{\xi}_i) \tag{3.2}$$

for $i, j = 0, \ldots, I-1$, where $\mathbf{c}$ denotes the vector containing all unknown coefficients c_i. In case of projection, the parameters c_i are determined by minimising the error along the function

$$\int_a^b (f_p(\xi) - f(\xi))^2 \, d\xi = 0 \iff \int_a^b \left(\sum_{i=0}^{I-1} N_i(\xi) c_i - f(\xi) \right)^2 d\xi = 0 \tag{3.3}$$

G. Beer et al., *The Isogeometric Boundary Element Method*, Lecture Notes in Applied and Computational Mechanics 90,
https://doi.org/10.1007/978-3-030-23339-6_3

where a and b are the limits of the parameter range, i.e., $\xi \in [a, b]$. The corresponding system of equations reads

$$\mathbf{M}\,\mathbf{c} = \mathbf{f}_h \tag{3.4}$$

with

$$\mathbf{M}[i,j] = \int_a^b N_i(\xi) N_j(\xi) d\xi \quad \text{and} \quad \mathbf{f}_h[i] = \int_a^b N_i(\xi) f(\xi) d\xi. \tag{3.5}$$

The system matrix $\mathbf{M}$ is usually referred to as *mass matrix*. The name comes from the usage in dynamics.

It is not surprising that the choice of the basis functions N_i has a great impact in both approaches. They significantly affect the properties of the corresponding system of equations and the quality and stability of the approximation scheme. In the following we focus on two specific basis functions types, namely Lagrange polynomials and B-splines.

3.2 Lagrange Polynomials

Lagrange functions are the most common basis functions used in numerical simulations. They are defined as

$$L_{i,p}(\xi) = \prod_{j=0, j\neq i}^{I-1} \frac{\xi - \tilde{\xi}_j}{\tilde{\xi}_i - \tilde{\xi}_j}, \qquad j = 0, \ldots, I-1. \tag{3.6}$$

with $I = p + 1$. Since the denominator is constant, $L_{i,p}$ is a polynomial of order p. These basis functions form a *partition of unity*

$$\sum_{i=0}^{I-1} L_{i,p}(\xi) = 1. \tag{3.7}$$

Furthermore, they fulfil the Kronecker delta property

$$L_{i,p}\left(\tilde{\xi}_j\right) = \delta_{ij}. \tag{3.8}$$

In other words, $L_{i,p}$ is 1 at its interpolation site $\tilde{\xi}_i$ and is zero at all the others. Such basis functions are *interpolatory*, since the parameters c_i associated to $L_{i,p}$ are the actual function values at the specific locations $\tilde{\xi}_i$. Thus, an interpolating polynomial to a function $f(\xi)$ is defined as

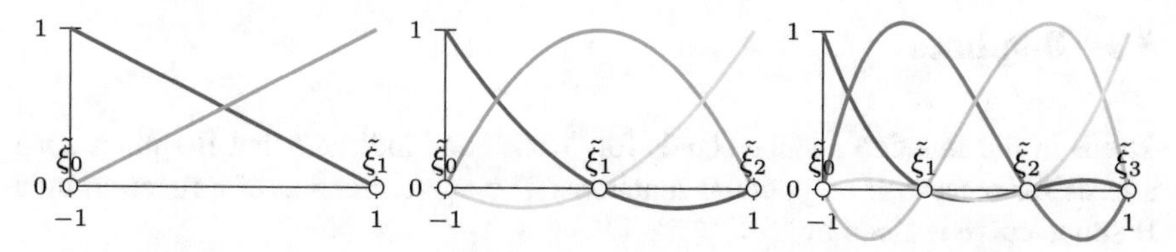

Fig. 3.1 Examples of Lagrange basis functions of order 1, 2 and 3 (from left to right)

$$f(\xi) \approx f_p(\xi) = \sum_{i=0}^{I-1} L_{i,p}(\xi) f\left(\tilde{\xi}_i\right) . \tag{3.9}$$

In fact, the Lagrange representation is one of the few forms in which the data points $f\left(\tilde{\xi}_i\right)$ appear explicitly. This may also be a key reason why these basis functions are so popular in numerical simulation, because the prescription of function values (e.g., for boundary conditions) and the interpretation of the coefficients (e.g., the results of a simulation) is straightforward.

Another interesting aspect is that the system matrix $\mathbf{A}$ of an interpolation turns into the identity matrix, when $L_{i,p}$ are employed as basis functions. Hence, solving Eq. (3.2) is trivial for arbitrary right-hand sides, and, since the linear system (3.2) is square, this implies that the solution (3.9) is unique [28], that means, it represents the *only* interpolant of order p to $f(\xi)$ at the interpolation sites $\tilde{\xi}_i$.

For numerical simulations, the Lagrange basis functions (3.6) are usually defined in the reference interval $\xi \in [-1, 1]$ with equally-spaced sites $\tilde{\xi}_i$. Figure 3.1 illustrates the resulting $L_{i,p}$ for $p = 1, 2, 3$. Note that most $L_{i,p}$ have positive as well as negative values; changing the sign at the sites $\tilde{\xi}_j$ that are not associated to them.

Lagrange polynomials are quite elegant, but they also have shortcomings, e.g.:

- Increasing the number of functions I of a single Lagrange basis $\{L_{i,p}\}_{i=0}^{I-1}$ implies an elevation of the order, since $I = p + 1$.
- High-order Lagrange polynomials can lead to instabilities, in particular for equally-spaced interpolation sites. This is discussed in more detail in Sect. 3.4.4.
- Alternatively, several Lagrange bases can be combined to form a piecewise-polynomial curve, but in this case no control over the smoothness between the distinct polynomial segments of the curve is provided.
- Since Lagrange functions form polynomials, they cannot represent conic sections such as circular arcs (which are often used in design, as outlined in Chap. 4).
- Defining a function by data points $f(\tilde{\xi}_i)$ may lead to oscillations and therefore, it is considered to be not shape-preserving.

These shortcomings motivate the use of B-splines and their generalisation which is called Non-Uniform Rational B-splines (NURBS).

3.3 B-splines

The B in the function's name stands for "basis" and indicates that B-splines span a basis for piecewise polynomial functions. The approximation of a function by a B-spline curve is given by

$$f(\xi) \approx f_{p,\Xi}(\xi) = \sum_{i=0}^{I-1} B_{i,p}(\xi) c_i. \tag{3.10}$$

To derive the equations for B-splines $B_{i,p}$, we start with a *knot vector*, Ξ. This is a vector containing a series of non-decreasing values of parametric coordinates, i.e.,

$$\Xi = (\xi_0, \xi_1, \dots, \xi_{I_\Xi}) \qquad \text{with} \qquad \xi_i \leqslant \xi_{i+1}. \tag{3.11}$$

The entries in Ξ are denoted as *knots* and the interval between the knots as *knot spans*. In particular, the ith knot span refers to the interval $[\xi_i, \xi_{i+1})$. In case the size of the knot span is non-zero, it may be referred to as an *element*. With the knot vector and the desired order p of the functions a recursive formula is applied for computing the functions.

We explain the generation of B-splines in Fig. 3.2 for a knot vector $\Xi = (1, 2, 3, 4, 5)$. First we compute the functions for order $p = 0$ (constant) by

$$B_{i,0}(\xi) = \begin{cases} 1 & \text{if} \quad \xi_i \leqslant \xi < \xi_{i+1}, \\ 0 & \text{otherwise.} \end{cases} \tag{3.12}$$

Higher order basis functions are defined by referencing lower order functions:

$$B_{i,p}(\xi) = \frac{\xi - \xi_i}{\xi_{i+p} - \xi_i} B_{i,p-1}(\xi) + \frac{\xi_{i+p+1} - \xi}{\xi_{i+p+1} - \xi_{i+1}} B_{i+1,p-1}(\xi) \tag{3.13}$$

It can be seen that the higher order functions are a combination of the lower order functions. As illustrated in Fig. 3.2, this leads to non-negative basis functions. Furthermore, the functions are non-zero only within certain portions of the overall parameter space $[a, b]$. To be precise, each $B_{i,p}$ has a local support, supp $\{B_{i,p}(\xi)\}$, given by $[\xi_i, \xi_{i+p+1})$. The coordinates in the knot vector may be non-uniformly spaced and we refer to the corresponding functions as *non-uniform B-splines*.

First derivatives are computed by

$$\frac{d}{d\xi} B_{i,p}(\xi) = \frac{p}{\xi_{i+p} - \xi_i} B_{i,p-1}(\xi) - \frac{p}{\xi_{i+p+1} - \xi_{i+1}} B_{i+1,p-1}(\xi). \tag{3.14}$$

In general, kth-derivatives are given by

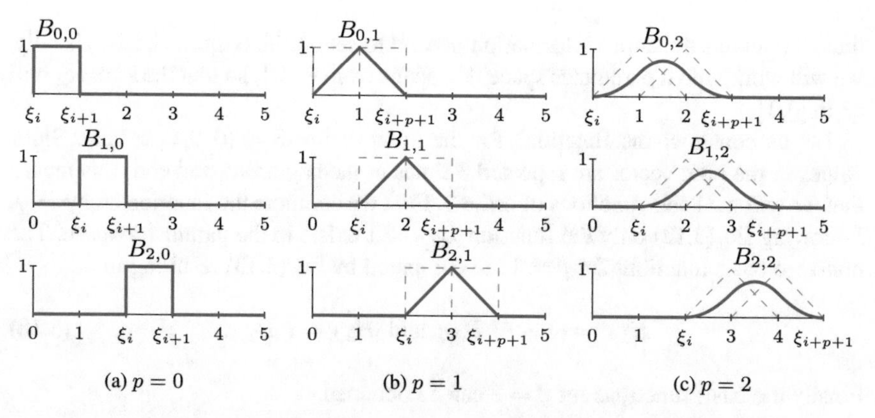

Fig. 3.2 B-spline functions of different orders p defined by (a) Eq. (3.12) or (b, c) the recursion (3.13). Dashed lines indicate the lower order functions used in the recursion

$$\begin{aligned}\frac{d^k}{d\xi^k} B_{i,p}(\xi) &= \frac{p}{\xi_{i+p} - \xi_i} \cdot \frac{d^{k-1}}{d\xi^{k-1}} B_{i,p-1}(\xi) \\ &\quad - \frac{p}{\xi_{i+p+1} - \xi_{i+1}} \cdot \frac{d^{k-1}}{d\xi^{k-1}} B_{i+1,p-1}(\xi).\end{aligned} \tag{3.15}$$

The number of values in the knot vector I_Ξ is linked to the order p and the number of parameters I, i.e., $I_\Xi = I + p + 1$. For details on efficient algorithms for evaluating B-splines and their derivatives, the interested reader is referred to the seminal book by Piegl and Tiller [132].

The knot vector defines not only the parameter space, but provides control over the continuity of the basis. To be precise, the continuity at a knot is C^{p-m} with m denoting the multiplicity of the knot value. The linear span of the set of linearly independent B-splines $\{B_{i,p}\}_{i=0}^{I-1}$ – the space of all possible spline functions on $[a, b]$ that are C^{p-m}-continuous at the knots – is completely defined upon the choice of the polynomial order p and the knot vector Ξ. In particular, the splines space on $[a, b]$ is defined by

$$\mathbb{S}_{\Xi,p}([a,b]) := \left\{ \sum_{i=0}^{I-1} B_{i,p}(\xi) c_i \;\middle|\; \xi \in [a,b],\ c_i \in \mathbb{R},\ i = 0, \ldots, I-1 \right\}. \tag{3.16}$$

A B-spline basis forms a positive partition of unity, that is,

$$B_{i,p}(\xi) \geq 0 \qquad \text{and} \qquad \sum_{i=0}^{I-1} B_{i,p}(\xi) = 1, \qquad \forall \xi \in [a,b]. \tag{3.17}$$

In the following, we will consider only *open knot vectors* where the values are repeated $p + 1$ times at the beginning and the end. Therefore, the knot multiplicity at the beginning and the end of a knot vector indicates the order of the basis functions and thus, it contains the entire information needed to set up the B-spline basis. Usually, we will work with a parameter space that spans from 0 to 1, so that the knots ξ_i will be in $[0, 1]$.

Let us construct the functions for the knot vector $\Xi = (0, 0, 0, 1, 1, 1)$. Since values in the knot vector are repeated 3 times at the beginning and end, this means that we will get basis functions of order 2. First we compute the functions for $p = 0$. Following Eq. (3.12) only the function $B_{2,0} = 1$ exists in the parameter space. The non-zero basis functions for $p = 1$ are computed by Eq. (3.13) resulting in

$$B_{1,1} = (1 - \xi)\, B_{2,0} \text{ and } B_{2,1} = \xi\, B_{2,0}. \tag{3.18}$$

Finally the basis functions for $p = 2$ can be obtained

$$B_{0,2} = (1 - \xi)\, B_{1,1},\ B_{1,2} = \xi\, B_{1,1} + (1 - \xi)\, B_{2,1} \text{ and } B_{2,2} = \xi\, B_{2,1}. \tag{3.19}$$

These 3 basis functions are C^∞-continuous between $\xi = 0$ and $\xi = 1$ (see Fig. 3.3). In this particular setting of an open knot vector with a single element, B-splines prescribe classical Bernstein polynomials.

We note that in case of Lagrange polynomials, each basis function can be directly associated with a specific point in the parameter, due to the Kronecker delta property (3.8). Since B-splines do not have this property *anchors* are introduced. The location of the ith anchor in the parameter space can be computed using a formula by Greville [75]:

$$\tilde{\xi}_i = \frac{\xi_{i+1} + \xi_{i+2} + \cdots + \xi_{i+p}}{p} \qquad i = 0, 1, \ldots, I - 1. \tag{3.20}$$

These anchors are commonly used as interpolation sites. Note that Eq. (3.20) results in an equally-spaced distribution of anchors for the special case of an open knot vector with a single element (cf., Figs. 3.3 and 3.4).

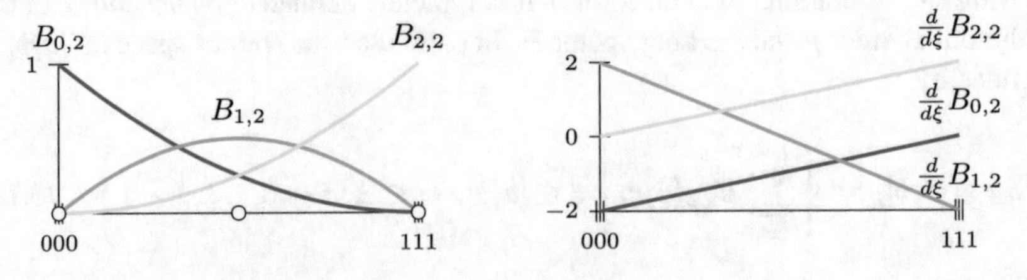

Fig. 3.3 Bernstein/B-spline functions of order 2 (left) and their first derivatives (right). The circles in the left figure indicate the corresponding anchors

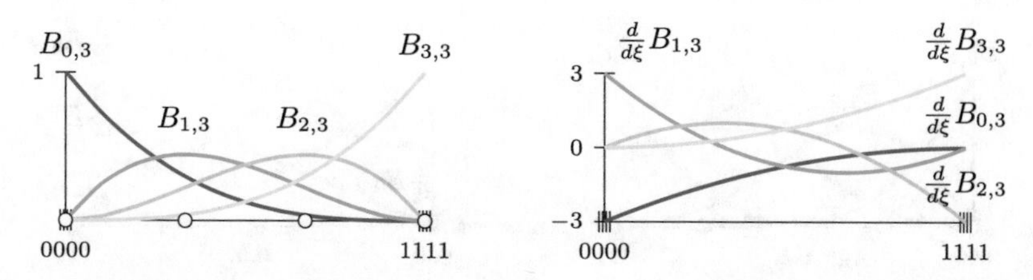

Fig. 3.4 Bernstein/B-spline functions of order 3 (left) and their first derivatives (right). The circles in the left figure indicate the corresponding anchors

3.3.1 Increasing the Order

To increase the order of the basis functions we insert knots at the beginning and end of the knot vector, i.e., to increase the order by one from 2 to 3 the new knot vector reads $\Xi = (0, 0, 0, 0, 1, 1, 1, 1)$. The resulting basis functions are shown in Fig. 3.4. Of course, changing Ξ affects the number of basis functions and the location of the corresponding anchors $\tilde{\xi}_i$.

We can continue to increase the order of the functions by inserting more zeros and ones. Note that the introduction of high-order B-splines requires only a change in the input parameters. The evaluation algorithms do not need to be altered, which makes the implementation of high-order basis functions trivial.

3.3.2 Control over the Continuity

In contrast to Lagrange polynomials we can now change the continuity inside the parameter space by insertion of knots. Here we show how this can be done using the B-splines of order 2 shown in Fig. 3.3 as starting point.

In a first step, the introduction of a C^1-continuity is considered. To reduce the continuity at a particular location we insert one knot (for example at $\xi = 0.5$). We can easily see from Eq. (3.12) that the only non-zero functions are $B_{2,0} = 1$ and $B_{3,0} = 1$. The following non-zero basis functions for $p = 1$ can be computed using Eq. (3.13):

$$\begin{aligned} B_{1,1} &= \quad 0 \quad + \frac{0.5-\xi}{0.5-0}\,B_{2,0} = (1-2\xi)\,B_{2,0} \\ B_{2,1} &= \frac{\xi-0}{0.5-0}\,B_{2,0} + \frac{1-\xi}{1-0.5}\,B_{3,0} = 2\xi B_{2,0} + 2(1-\xi)B_{3,0} \\ B_{3,1} &= \frac{\xi-0.5}{1-0.5}\,B_{3,0} + \quad 0 \quad = (2\xi-1)\,B_{3,0} \end{aligned} \tag{3.21}$$

The final quadratic basis functions are given by

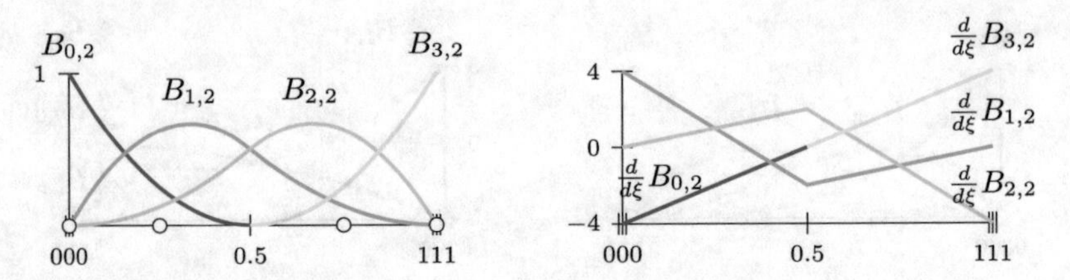

Fig. 3.5 B-spline functions and derivatives for $\Xi = (0, 0, 0, 0.5, 1, 1, 1)$

$$\begin{aligned}
B_{0,2} &= \quad 0 \quad + \frac{0.5-\xi}{0.5-0} B_{1,1} = (1-2\xi)\, B_{1,1}, \\
B_{1,2} &= \frac{\xi-0}{0.5-0} B_{1,1} + \frac{1-\xi}{1-0} B_{2,1} = 2\xi B_{1,1} + (1-\xi) B_{2,1}, \\
B_{2,2} &= \frac{\xi-0}{1-0} B_{2,1} + \frac{1-\xi}{1-0.5} B_{3,1} = \xi B_{2,1} + 2(1-\xi) B_{3,1}, \\
B_{3,2} &= \frac{\xi-0.5}{1-0.5} B_{3,1} + \quad 0 \quad = (2\xi-1)\, B_{3,1}.
\end{aligned} \tag{3.22}$$

These basis functions and their derivatives are shown in Fig. 3.5. We can see that functions $B_{0,2}$ and $B_{3,2}$ only span half of the parameter space and that the derivatives of $B_{1,2}$ and $B_{2,2}$ have a kink at 0.5 indicating the C^1-continuity introduced.

To further reduce the continuity to C^0 at that location, we simply repeat the knot value 0.5, thereby increasing its multiplicity from 1 to 2, and evaluate the basis again: now $B_{2,0}$ and $B_{4,0}$ are non-zero leading to

$$\begin{aligned}
B_{1,1} &= \quad 0 \quad + \frac{0.5-\xi}{0.5-0} B_{2,0} = (1-2\xi)\, B_{2,0}, \\
B_{2,1} &= \frac{\xi-0}{0.5-0} B_{2,0} + \quad 0 \quad = 2\xi B_{2,0}, \\
B_{3,1} &= \quad 0 \quad + \frac{1-\xi}{1-0.5} B_{4,0} = 2(1-\xi) B_{4,0}, \\
B_{4,1} &= \frac{\xi-0.5}{1-0.5} B_{4,0} + \quad 0 \quad = (2\xi-1) B_{4,0}
\end{aligned} \tag{3.23}$$

and

$$\begin{aligned}
B_{0,2} &= \quad 0 \quad + \frac{0.5-\xi}{0.5-0} B_{1,1} = (1-2\xi)\, B_{1,1}, \\
B_{1,2} &= \frac{\xi-0}{0.5-0} B_{1,1} + \frac{0.5-\xi}{0.5-0} B_{2,1} = 2\xi B_{1,1} + (1-2\xi)\, B_{2,1}, \\
B_{2,2} &= \frac{\xi-0}{0.5-0} B_{2,1} + \frac{1-\xi}{1-0.5} B_{3,1} = 2\xi B_{2,1} + 2(1-\xi)\, B_{3,1}, \\
B_{3,2} &= \frac{\xi-0.5}{1-0.5} B_{3,1} + \frac{1-\xi}{1-0.5} B_{4,1} = (2\xi-1)\, B_{3,1} + 2(1-\xi) B_{4,1}, \\
B_{4,2} &= \frac{\xi-0.5}{1-0.5} B_{4,1} + \quad 0 \quad = (2\xi-1) B_{4,1}.
\end{aligned} \tag{3.24}$$

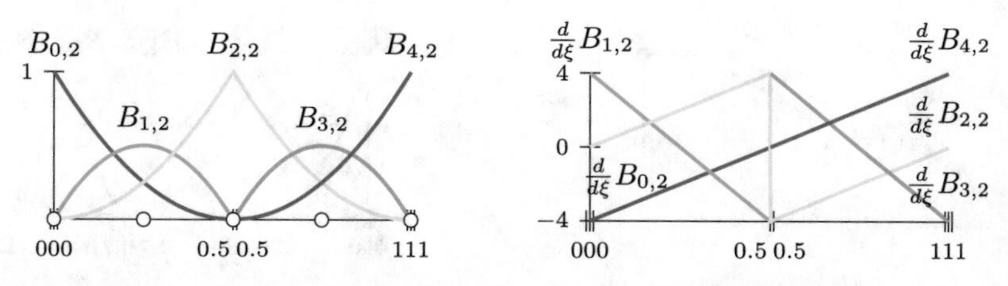

Fig. 3.6 B-spline functions and derivatives for $\Xi = (0, 0, 0, 0.5, 0.5, 1, 1, 1)$

The results are shown Fig. 3.6. Note that the third basis function, i.e., $B_{2,2}$, has a C^0-continuity at $\xi = 0.5$ and all other basis functions only span half of the parameter space.

In conclusion, the continuity of a B-spline basis can be controlled by adding internal knots into Ξ and changing their multiplicity. Hence, B-splines provide a very flexible and handy tool that allows specifying (i) the order of basis functions, (ii) the total number functions, (iii) the distribution of the knot spans, and (iv) the continuity between the resulting piecewise polynomial segments. The recursive evaluation (3.13) takes care of the rest. At this point it may be worth to take a closer look at the systematic of this formula. It is apparent that a basis function $B_{i,p}$ $(p > 0)$ is determined by a linear combination of B-splines of the previous order $B_{j,p-1}$ with $j = \{i, i+1\}$. The coefficients multiplied with these $B_{j,p-1}$ define linear functions that go from 0 to 1 within the support of $B_{i,p-1}$ and from 1 to 0 within the support of $B_{i+1,p-1}$. Note that increasing the multiplicity of knots results in a smaller overlap of these supports, which leads to the reduced continuity. Consider $B_{2,2}$ of Eqs. (3.22) and (3.24) for instance. In the former case the basis functions consists of B-splines which are non-zero in $[0, 1)$ and $[0.5, 1]$, respectively. On the other hand, these supports are $[0, 0.5)$ and $[0.5, 1]$ in the latter case and hence, they do not overlap.

3.3.3 *Representation of Functions*

Here we show that B-splines offer a great flexibility for representing functions. In general, a B-spline object inherits the properties of its basis. This is illustrated in Fig. 3.7 for a curve with B-splines specified by $\Xi = (0, 0, 0, 0.3, 0.3, 0.7, 0.7, 0.7, 1, 1, 1)$ and $p = 2$. Note that the interpolatory and discontinuous points ($\mathbf{c}_2$, $\mathbf{c}_4$ and $\mathbf{c}_5$) correspond directly to the C^0- and C^{-1}-continuities in the parameter space. The linear interpolation between the *control points* $\mathbf{c}_i$ forms the so-called *control polygon*.

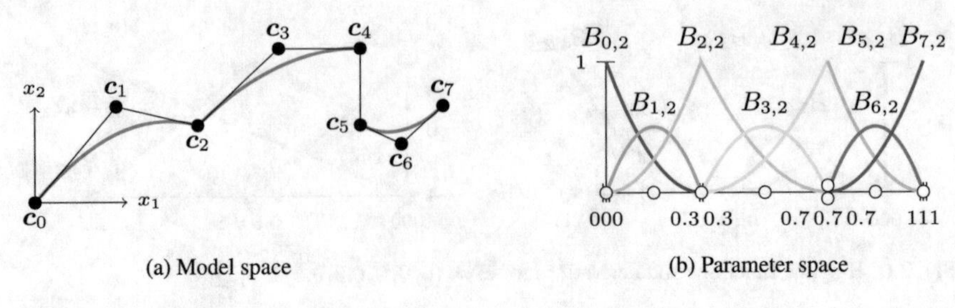

(a) Model space (b) Parameter space

Fig. 3.7 A piecewise polynomial curve being defined by one B-spline basis with control points c_i. The corresponding basis functions are shown on the right and their Greville abscissae are depicted by white circles. Note the coinciding anchors at the discontinuous knot at 0.7

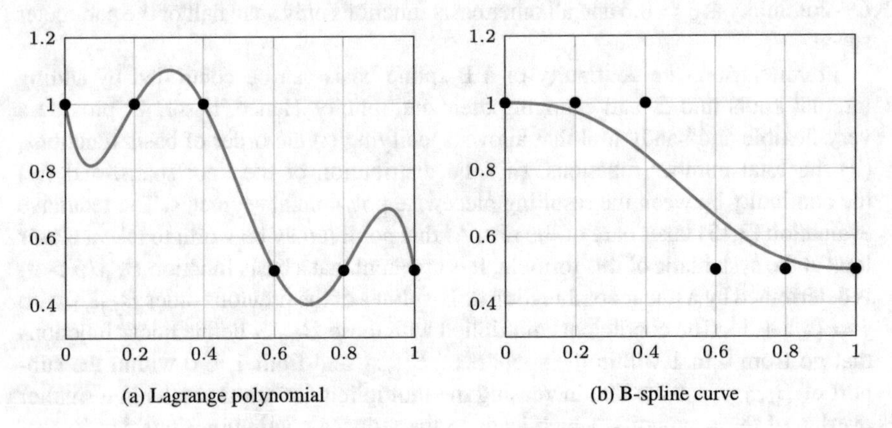

(a) Lagrange polynomial (b) B-spline curve

Fig. 3.8 Approximation of a discontinuous function by (a) nodal points and (b) control parameters

Let us compare the concept of control variables versus interpolatory nodal points used in case of Lagrange polynomials. As mentioned in Sect. 3.2, using nodal points is considered as not shape-preserving and here we illustrate what this statement means. Consider a discontinuous function defined by 3 successive points with the value 1.0 followed by another 3 points with the value 0.5. The interpolation of these points with Lagrange polynomials of order 5 is shown in Fig. 3.8(a), whereas Fig. 3.8(b) displays a B-spline curve of the same order using these points as control variables.

It is apparent that the former has an oscillatory behaviour. The latter captures the actual step function much better, since B-spline curves have the property of *variation diminution*. Put simply, this means that the output of a curve or surface scheme "wiggles less" than the data from which it is constructed [59]. Hence, control variables provide a closer relation to the corresponding object than nodal points. This can be exploited for the controlling and modelling shapes, as will be discussed in more detail later on in Chap. 4.

It should have become clear now that B-splines are quite special functions that differ from Lagrange polynomials by the following:

- They work with control parameters instead of nodal values, i.e., parameters are usually not located on the curve.
- Therefore, setting up the matrix $\mathbf{A}$ for an interpolation problem (3.2) does not lead to the identity matrix.
- Rather than specifying nodal sites $\tilde{\xi}_i$, all properties of a B-spline basis are determined by a knot vector Ξ and the polynomial order p (which may also be encoded in Ξ in case of open knot vectors).
- This provides more control over the parameter space, e.g., it allows controlling the continuity of a set of basis functions.

3.4 Refinement Options

B-splines provide more flexibility in setting up a basis, a property that is also reflected in the refinement options which include order elevation, knot insertion and k-refinement. Order elevation and knot insertion are similar to p- and h-refinements of Lagrange basis functions used in conventional simulations. However, the subtle difference is that the B-spline refinement operations involve changing the knot vector, whereas the traditional approaches involve a (re-)arrangement of nodal points. The best way to understand the refinement with B-splines is by an example where a non-trivial function shall be approximated. Therefore, we employ the function

$$f(\xi) = \sin(5\pi\xi) + \frac{1}{1.05 - \xi} \qquad \xi \in [0, 1]. \tag{3.25}$$

In the following, the required parameters c_i of the approximation to $f(\xi)$ will be computed by L_2-projection (3.5), if not stated otherwise.

Remark In this section, the focus lies on the comparison of the options for the manipulation of B-spline and Lagrange parameter spaces. A key feature of the B-spline related refinement schemes is that they lead to nested spline spaces. Dealing with basis functions only, this property is simply established by guaranteeing that a refined knot vector $\hat{\Xi}$ forms a super-set of the initial one Ξ, that is, $\Xi \subset \hat{\Xi}$. As soon as B-spline curves and surfaces are considered, the refinement procedures are a bit more involved, because a proper update of the control coefficients is required as well. This will be discussed in detail later on in Sect. 4.1.3.

3.4.1 Order Elevation

Order elevation involves increasing the order of the basis functions. This is performed by increasing the multiplicity of all existing knots of the knot vector. In Fig. 3.9 the approximation of the function by Lagrange polynomials and by B-splines (when

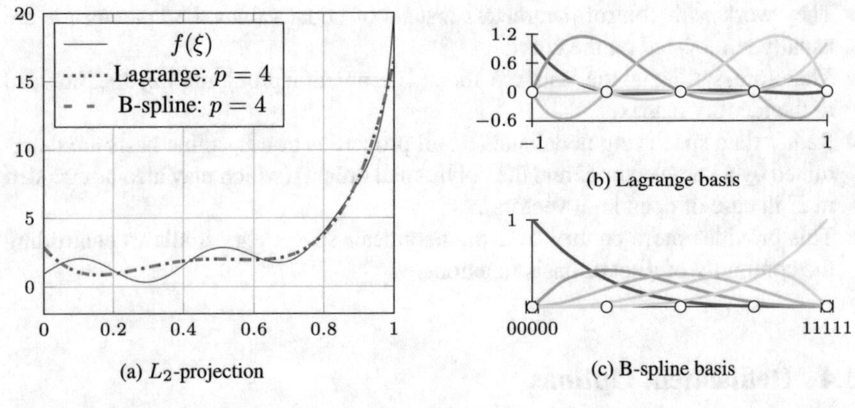

Fig. 3.9 Approximation of function (3.25) with B-splines and Lagrange functions of order 4 (left) and the corresponding basis functions (right)

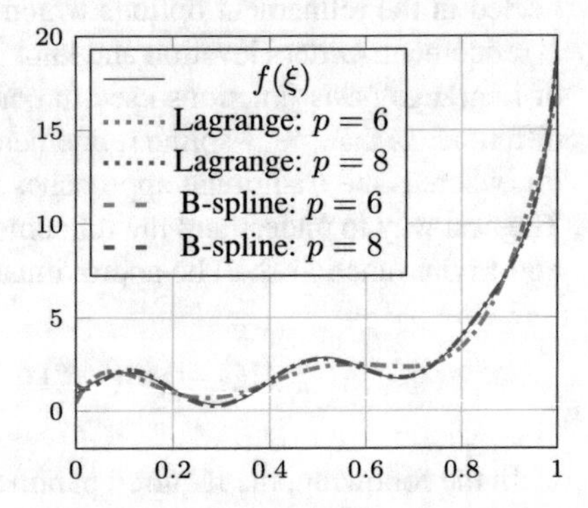

Fig. 3.10 Approximation of function (3.25) with Lagrange polynomials and B-splines with increased order, i.e., $p = \{6, 8\}$

the order is elevated from 3 to 4) is shown. Note that the number and location of evaluation sites in Fig. 3.9(c, b) coincide.

In addition, the improvement of the approximation by increasing the order is shown in Fig. 3.10. Not much difference can be seen between both basis function types. In fact, the quality of the approximation is identical. The reason for this is that Lagrange polynomials and B-splines span different bases, but for the same polynomial space. However, this holds solely true in this particular setting and changes as soon as the B-spline basis contains interior knots which are not C^0-continuous. Finally, we remark that order elevation of B-splines has particular interesting properties when interior knots are present. Usually, this is the case and we will come back to these properties in Sect. 3.4.3.

3.4.2 Knot Insertion

Knot insertion is a refinement scheme for B-splines, but there is an equivalent approach for Lagrange polynomials known as h-refinement. The common feature of these refinements is that they increase the number of basis functions, while their polynomial order remains unaltered. There is, however, a substantial difference: for B-splines, this refinement adds new values to an existing parameter space, whereas h-refinement accumulates several parameter spaces (finite elements) to a joined set. Figure 3.11 shows approximations with 6 Lagrange as well as B-spline basis function of order 2, together with the related basis.

Note that the h-refinement concept yields C^0-continuity between the two elements, which leads to a kink in the approximation (see dotted blue graph shown in Fig. 3.11(a). It is emphasised that this effect is inherent to this refinement approach and therefore independent of the order. With knot insertion, on the other hand, the smoothness between elements can be controlled. Recall that the continuity of a B-spline basis at a knot is C^{p-m} where m denotes the multiplicity of the knot value. Hence, the basis in Fig. 3.11(c) (as well as the corresponding approximation) is C^1-continuous.

Figure 3.12 illustrates the improvement of the approximations due to h-refinement and knot insertion. It is noted that the number of basis function is no longer the same for the Lagrange and the B-spline examples; the finest Lagrange basis ($n_{el} = 8$) consists of 24 functions, while the finest B-spline basis ($n_{knots} = 8$) has merely 11 functions. Furthermore, it is emphasised the kinks occur even in case of the finest Lagrange approximation of the smooth target function $f(\xi)$.

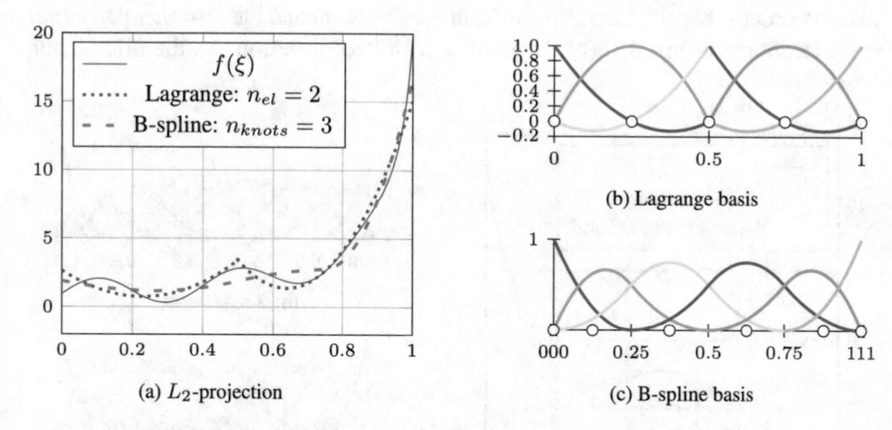

Fig. 3.11 Approximation of function (3.25) with B-splines and Lagrange functions of order 2 (left) and the corresponding basis functions (right). The Lagrange basis consists of two elements, $n_{el} = 2$, and the B-spline basis has three interior knots with multiplicity 1, $n_{knots} = 3$, which yields 4 elements. Both bases possess 6 basis functions

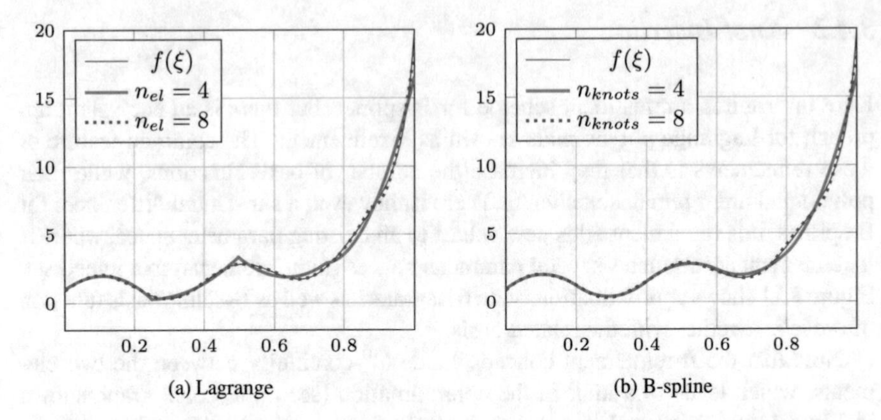

Fig. 3.12 L_2-projection using Lagrange polynomials with increasing number of elements, n_{el}, (left) and B-splines with increasing number of interior knots with multiplicity 1, n_{knots}, (right). In all cases the order is 2

3.4.3 *k*-Refinement

k-refinement is a scheme that is unique to B-splines. In principle, it is a combination of knot insertion and order elevation, and affects the smoothness between elements, i.e., non-zero knot spans. The key point to consider is that it is *not* arbitrary in which order these two refinement techniques are applied. Consider Fig. 3.13: starting from a quadratic B-spline basis given by $\Xi = (0, 0, 0, 1, 1, 1)$, the basis functions are refined by one order elevation and the insertion of four knots, i.e., 0.2, 0.4, 0.6, and 0.8. In case the order is elevated first, we perform k-refinement and the resulting parameter space is shown in Fig. 3.13(b). Beginning with knot insertion, on the other hand,

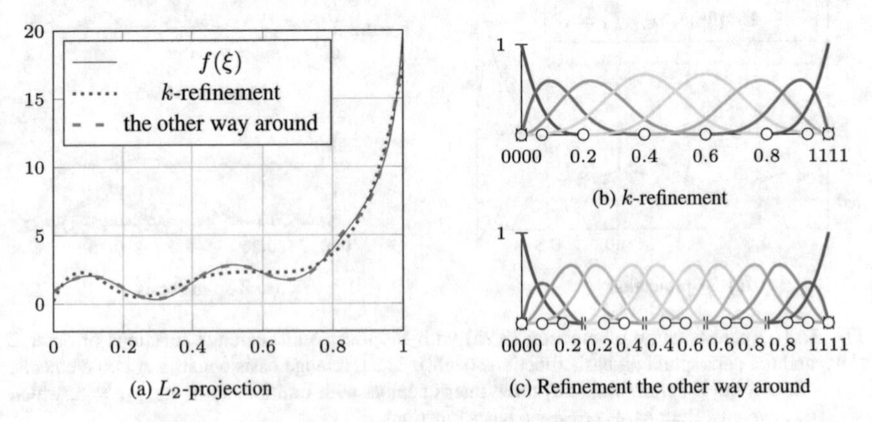

Fig. 3.13 The combination of one order elevation and knot insertion of four knots applied in different order. In (b), the order is elevated first leading to k-refinement; (c) shows the basis function when the refinement schemes are applied the other way around

yields another set of basis functions illustrated in Fig. 3.13(c). Note that the number of elements is the same, but for k-refinement (i) the total number of basis is less and (ii) the continuity is higher, i.e., C^2 rather than C^1. These different properties are also reflected in the difference of the approximations illustrated in Fig. 3.13(a). The distinguishing feature of k-refinement is that it facilitates an enrichment of a parameter space that guarantees *maximal smoothness* between elements, i.e., C^{p-1}-continuity.

The dependence on the sequence of refinement steps occurs because order elevation (as well as knot insertion) are designed so that they lead to nested spline spaces. Hence, a C^1-continuity introduced by a knot insertion must not be altered by a subsequent order elevation and in turn, the multiplicity of *every* existing knot has to be increased (cf., Fig. 3.13(c)). This may seem a bit restrictive at this point, since our focus lies on the basis functions. However, this property of the B-spline refinement options is actually a very powerful one and in Chap. 4, we will demonstrate that it is of paramount importance for the representation of geometric models and a big advantage of isogeometric analysis over conventional simulation methods.

3.4.4 *Comparison*

So far, we showed that the different refinement options of B-splines and Lagrange polynomials yield different approximations $\tilde{f}(\xi)$ of a target function $f(\xi)$. Here, the quality of these options is compared by computing the L_2-error of the approximation. The error is defined as

$$\epsilon = \sqrt{\frac{\int_0^1 \left(f(\xi) - \tilde{f}(\xi)\right)^2 d\xi}{\int_0^1 (f(\xi))^2}} \tag{3.26}$$

and it will be related to the degrees of freedom, i.e., the number of basis functions.

For all examples of this section, the initial parameter space is given by a single element ($n_{el} = 1$) defined by quadratic basis functions. Hence, order elevation results refer to B-splines defined over a single non-zero knot span. For k-refinement, on the other hand, the number of elements is set to $n_{el} = p + 1$. It is explicitly noted if the multiplicity m of the corresponding interior knots is not equal to 1.

Approximating a smooth function. In this example, the target function (3.25) is approximated by L_2-projection. We compare the performance of the different refinement options. To be precise, Fig. 3.14 shows the convergence behaviour of h-refinement, knot insertion, p-refinement, order elevation, and k-refinement with maximal smoothness (multiplicity $m = 1$), whereas Fig. 3.15 illustrates the effect on k-refinement when smoothness is reduced. Note that the case $m = p - 1$ refers to the situation when knot insertion and order elevation are applied in the reversed order. The various orders of the resulting basis functions are given in the legends.

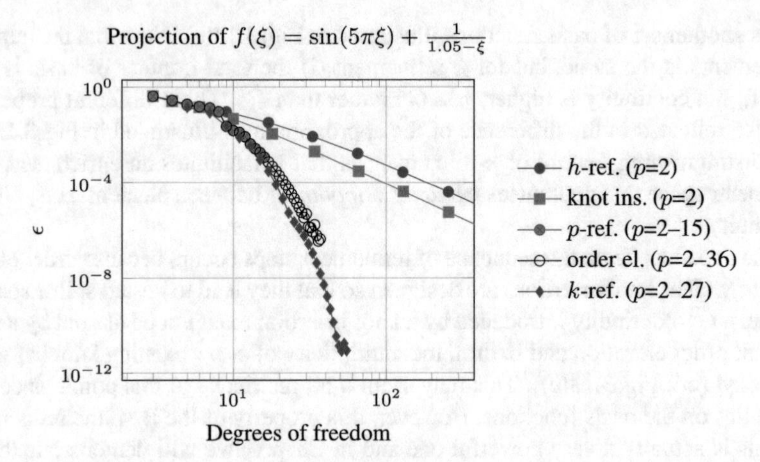

Fig. 3.14 Left: Approximation error ϵ of the different refinement options for Lagrange functions (h-ref. and p-ref.) and B-splines (order el., knot ins., and k-ref.)

Based on the graphs given in Fig. 3.14 and Fig. 3.15 several interesting aspects can be observed:

- The convergence rate of h-refinement and knot insertion is identical, but there is a significant offset in favour of the latter due to the higher smoothness of B-splines.
- The beneficial effect of higher smoothness can also be observed in Fig. 3.15.
- Order elevated B-splines, that span a single element, yield the same results as their Lagrangian counterparts obtained by p-refinement.
- k-refinement performs similar to p-refinement and both are powerful strategies to obtain high accuracy with a low number of degrees of freedom.

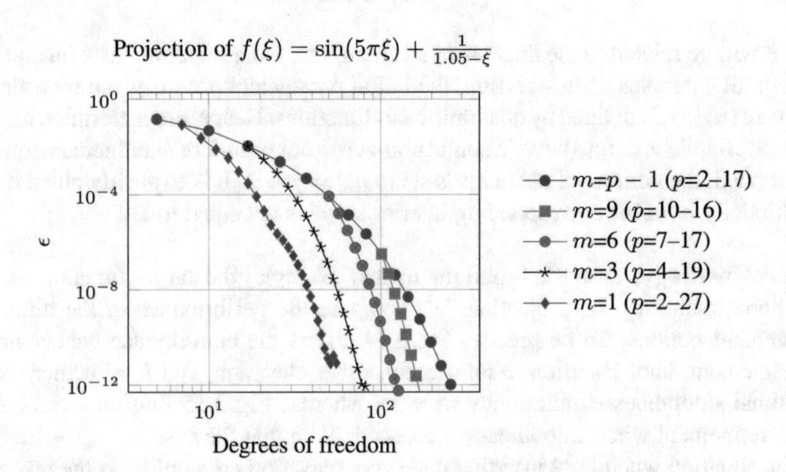

Fig. 3.15 Comparison of k-refinement with maximal smoothness ($m = 1$) and versions with reduced smoothness ($m > 1$). The graph '$m = p - 1$' refers to the case when knot insertion is performed before order elevation. For all approaches, the employed order(s) p are listed in the legend

- Due to numerical rounding errors, the computations become unstable beyond a certain order (which is about 25 in this example).

Another, more subtle, aspect is that p-refinement already stops at order 15, while order elevation continues until order 36. The reason for this is merely a pragmatic decision: employing high-order B-splines is trivial, because it only affects a variable for the definition (3.13) of the functions. In case of Lagrange polynomials, on the other hand, every order is usually implemented explicitly and we chose to opt out at $p = 15$. At this point it should be noted that in conventional simulation orders of Lagrange polynomials are seldom increased beyond order 4.

To sum up, higher orders as well as higher smoothness improves the convergence behaviour when smooth functions are approximated. There is, however, a computational threshold where pure order elevation is no longer beneficial, because numerical rounding errors can become dominant as indicated by the oscillations of the last points of the graph of Fig. 3.14 related to order elevation. A powerful feature of k-refinement is that it is not limited by this threshold, since it enables "order-elevation-like" convergence behaviour for arbitrarily fine B-spline spaces.

Runge phenomenon. The stability issues of high-order approximations discussed in the previous example occur mainly due to numerical rounding errors. Here, we focus on another problem related to high-order basis functions, namely the famous Runge phenomenon. In 1901, Runge [153] revealed that polynomial interpolation using *equally-spaced points* is extremely ill-conditioned. That is, it diverges exponentially as $p \to \infty$, even if the function is analytic, and even in exact arithmetic [180].

To illustrate this effect, the following problem is considered: the so-called Runge function

$$f(\xi) = \frac{1}{1 + (10\xi - 5)^2} \qquad \xi \in [0, 1] \tag{3.27}$$

is interpolated by solving Eq. (3.2). The interpolation sites $\tilde{\xi}_i$ are determined by the Greville abscissae (3.20), which leads to equally-spaced anchors in case of order elevation of a basis with a single element (cf., Fig. 3.4). In the following we refer to this refinement option as p-refinement in order to emphasise that the resulting basis consists of a single non-zero knot span only. In addition to the B-spline approximations, we apply so-called barycentric interpolation in Chebyshev points (for details see [180]). In general, Chebyshev interpolation sites $\tilde{\xi}_i^c$ are given by

$$\tilde{\xi}_i^c = \cos\left(\frac{i\pi}{I - 1}\right), \qquad i = 0, \ldots, I - 1, \qquad \xi \in [-1, 1] \tag{3.28}$$

and for this example, they are mapped to the target interval $\xi \in [0, 1]$. The barycentric approach is employed for two reasons: (i) it is a numerically stable process for interpolation and untroubled by rounding errors, and (ii) the use of Chebyshev points is a remedy to the divergence phenomenon of high-order polynomials. In short,

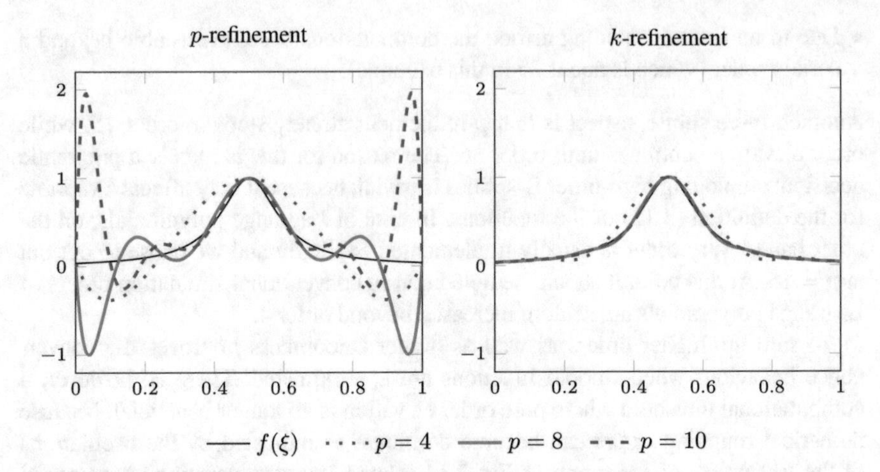

Fig. 3.16 Runge phenomenon: interpolants to the Runge function (3.27) obtained by p-refinement (left) and k-refinement (right)

barycentric interpolation in Chebyshev points is designed to deal with the issues that high-order polynomials induce and thus, it provides an ideal reference to assess the performance of the different refinement options.

Figure 3.16 displays some interpolants to the Runge function, and Fig. 3.17 illustrates the convergence of knot insertion, k-refinement, p-refinement, and the barycentric approach. In the latter figure, the graphs focus on even orders; odd orders yield slightly better results, but the overall behaviour is the same. The number of elements in case of k-refinement is set to $n_{el} = p + 1$ and the multiplicity of all interior knots is 1.

It is apparent that p-refinement with equally-spaced interpolation points diverges. This holds true for Lagrange polynomials as well as B-splines; or more precisely Bernstein polynomials. On the other hand, k-refinement converges almost as well as the barycentric scheme, allowing a stable interpolation using B-splines with orders greater than 70. To be clear, we do not recommend to use such high orders for computations, but this example emphasises that high orders can be very easily realised in case B-splines, and k-refinement provides a stable scheme to utilise them.

Gibbs phenomenon. Polynomials oscillate when they interpolate *discontinuous* data. This so-called Gibbs phenomenon has already been observed in Sect. 3.3.3, where we compared the behaviour of nodal point in contrast to control variables (see Fig. 3.8). Here, we demonstrate that this oscillatory behaviour is not restricted to Lagrange polynomials, but can arise for B-splines as well.

Consider the following step function

$$f(\xi) = \text{sign}(\xi - 0.5), \qquad \xi \in [0, 1] \tag{3.29}$$

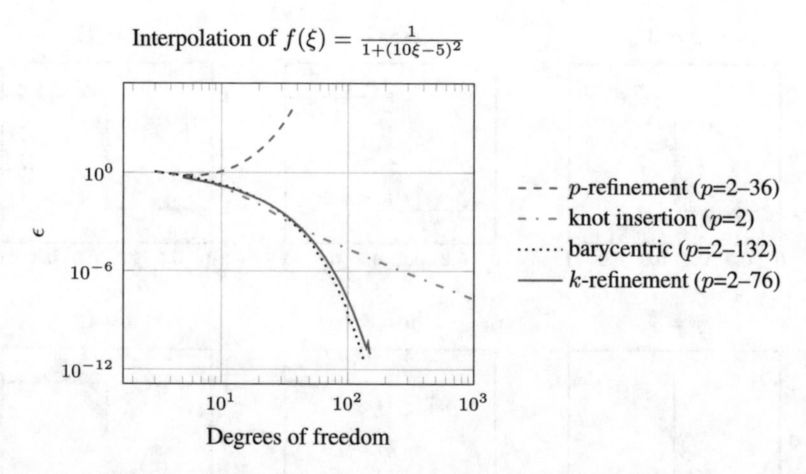

Fig. 3.17 Runge phenomenon: divergence in case of p-refinement, but convergence of the other B-spline refinement options as well as the barycentric interpolation in Chebyshev points. The employed orders p are listed in the legend

Table 3.1 Maximal height, $\max(|y|)$, of the functions shown in Fig. 3.18

p	p-refinement	k-refinement	Barycentric
3	1.5034	1.2084	1.1880
4	1.1877	1.0497	1.0455
7	2.5487	1.2491	1.2633
8	1.3438	1.0505	1.0617
11	11.6625	1.2565	1.2746
12	2.6388	1.0514	1.0640

which has a jump at $\xi = 0.5$. This function is interpolated by B-spline curves and the related control coefficients are obtained by interpolation (3.2). Note that this is not the same as assigning values of $f(\xi)$ directly to the control variables; it rather forces a B-spline curve to coincide with the data points of the interpolation problem. The initial discretisation is given by a single element and the number of basis functions is increased by p-refinement (i.e., order elevation for a basis with a single element) and k-refinement. Barycentric interpolation at Chebyshev points is again used as a reference.

Figure 3.18 illustrates various interpolants to the target function (3.29). In case of k-refinement, the number of elements is set to $n_{el} = p+1$ or $n_{el} = p+2$, so that the jump is always located in the middle of an element. In addition, the maximal height, $\max(|y|)$, of the computed interpolations are listed in Table 3.1.

Note that p-refinement diverges (as it did in the previous example), which is further evidence that interpolation at equally-spaced data points is very ill-conditioned. However, the results related to the other approaches demonstrate a stable behaviour.

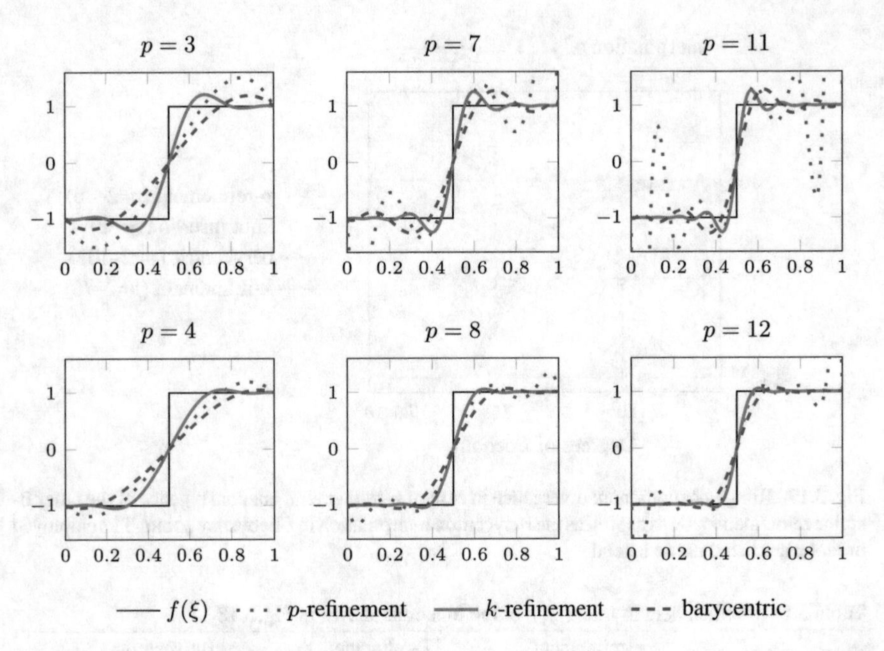

Fig. 3.18 Gibbs phenomenon: interpolation of a step functions $f(\xi)$ using various odd (above) and even (below) orders, and different refinement schemes

Albeit, there is a substantial overshoot near the jump, especially for odd orders. These overshoots gets narrower as p increases, but their height hardly changes. In fact, these oscillations will not go away even if $p \to \infty$, and for interpolation at Chebyshev points the height of the overshoot converges to ≈ 1.2823 and ≈ 1.0658 for odd and even orders, respectively [180]. Based on the numerical results provided in Table 3.1, k-refinement shows a similar behaviour with a slightly smaller $\max(|y|)$.

The effect of a higher number of elements n_{el} and the smoothness of the basis is demonstrated in Fig. 3.19 for various odd orders. In particular, the multiplicity m of inner knots is set to 1 or p, which leads to C^{p-1}-continuous and C^0-continuous parameter spaces, respectively. Recall that the C^0-continuous case yields the same results as a Lagrange discretisation. The corresponding convergence plots are shown in Fig. 3.20.

It is quite apparent that the higher smoothness is beneficial for the stability of the interpolation problem. When $m = 1$ all orders perform similarly with a smooth and rapid decay of the overshoots, and increasing the number of elements shifts these overshoots towards the discontinuity. In the non-smooth setting, on the other hand, high orders result in instabilities manifested as extreme oscillations at the ends of the element that contains the discontinuity. These oscillations can dominate the overall error of the interpolation problem. In Fig. 3.20, there are significant offsets between the dashed graphs related to the non-smooth basis functions which become worse with increasing order, whereas the convergence behaviour of the smooth discretisations does not change with the order. For the sake of completeness, we point out

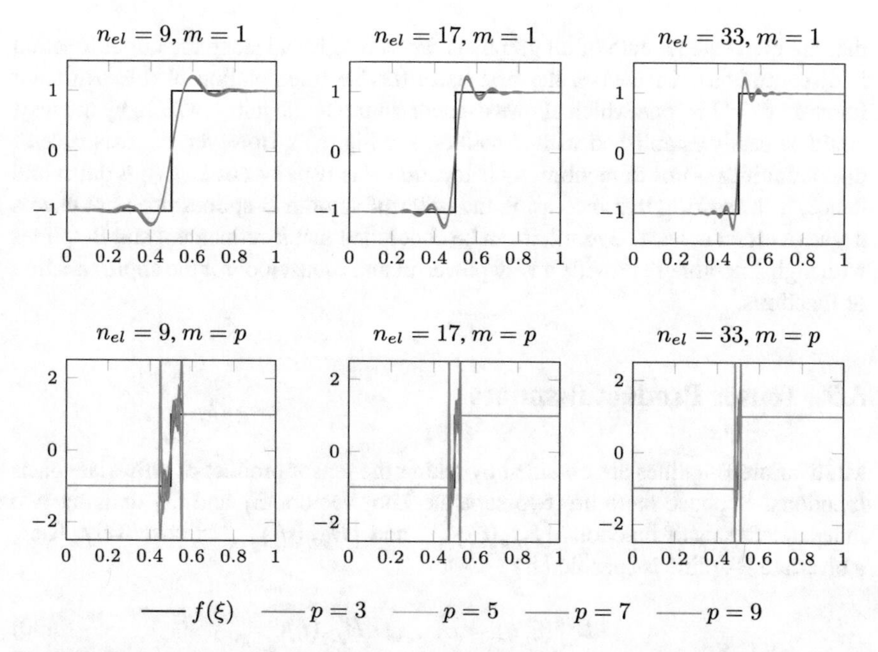

Fig. 3.19 Gibbs phenomenon: interpolation of a step functions $f(\xi)$ using different numbers of elements n_{el}, orders p, and smoothness due to the knot multiplicity m (i.e., C^{p-1}-continuous (above) and C^0-continuous (below))

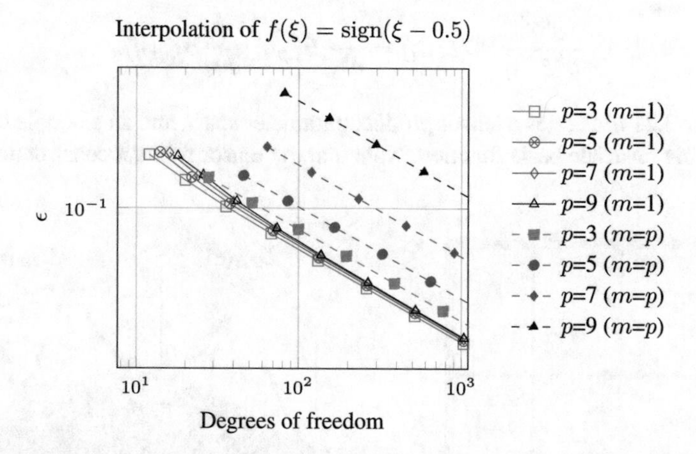

Fig. 3.20 Gibbs phenomenon: convergence of interpolations of various order p to a discontinuous step function. The refinement is carried out by uniform knot insertion with different multiplicity m of the inserted knots; solid lines refer to parameter space with maximal smoothness, while dashed lines present the results of C^0-continuous basis functions

that the convergence rate of all graphs is far from optimal since the target function is discontinuous. Of course, the best basis for the interpolation of this particular function would be one which allows a discontinuity at the jump (which, by the way, could be easily established with B-splines, see Fig. 3.7). However, the existence of discontinuities – not to mention their location – is usually not known a priori and hence, it is assuring that increasing the order of smooth B-splines does not have a negative effect at least. Overall, it can be concluded that k-refinement and B-splines with high smoothness provide a very powerful and robust tool for the approximation of functions.

3.5 Tensor Product B-splines

Multivariate B-splines are obtained by taking the tensor product of univariate basis functions. Suppose there are two separate knot vectors Ξ_1 and Ξ_2 defining two independent sets of functions $\{B_{i,p}(\xi)\}_{i=0}^{I-1}$ and $\{B_{j,q}(\eta)\}_{j=0}^{J-1}$, respectively. Then, a bivariate B-spline is specified by

$$B_{ij}^{pq}(\xi,\eta) = B_{i,p}(\xi)\cdot B_{j,q}(\eta). \tag{3.30}$$

with Ξ_1 and Ξ_2 spanning a two-dimensional parameter space of orders p and q in the parametric directions ξ and η, respectively. Derivatives of bivariate B-splines are analogously defined by

$$\frac{\partial^{k+l}}{\partial^k\xi\partial^l\eta}B_{ij}^{pq}(\xi,\eta) = \frac{d^k}{d^k\xi}B_{i,p}(\xi)\frac{d^l}{d^l\eta}B_{j,q}(\eta). \tag{3.31}$$

Figure 3.21 illustrates a tensor product parameter space and an example of a corresponding bivariate basis function. Note that the tensor product construction leads

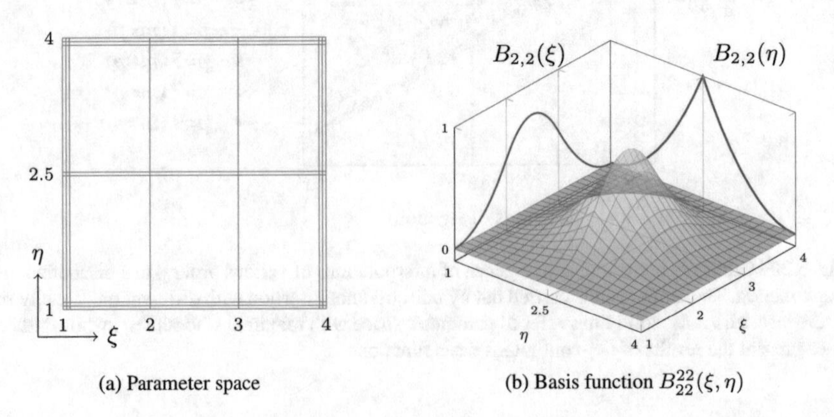

(a) Parameter space (b) Basis function $B_{22}^{22}(\xi,\eta)$

Fig. 3.21 A bivariate basis defined by the knot vectors $\Xi_1 = (1,1,1,2,3,4,4,4)$ related to ξ and $\Xi_2 = (1,1,1,2.5,2.5,4,4,4)$ for η: (a) shows the bivariate parameter space spanned by Ξ_1 and Ξ_2, whereas (b) depicts the construction of a corresponding tensor product B-spline

to a set of rectangular knot spans which determine the elements for the isogeometric simulation. Furthermore, it is highlighted that the properties of the univariate B-splines (e.g., the C^1-continuity of $B_{2,2}(\eta)$ shown in Fig. 3.21(b)) are passed on to their bivariate counterparts. The tensor product concept is very general and can be easily extended to higher dimensional multivariate basis functions. However, we will focus on the bivariate case in this book.

3.6 Refinement of Tensor Product B-splines

The different refinement options of univariate B-splines have been presented in Sect. 3.4. Here, we discuss their application and limitation in the context of multivariate tensor product B-splines. Without loss of generality, we focus on the two-dimensional case.

3.6.1 *General Refinement*

Knot insertion, order elevation and k-refinement can be applied to bivariate B-splines in the same way as in the one-dimensional setting – by adaptation of the basis functions' knot vectors Ξ_1 and Ξ_2. However, the tensor product construction (3.30) implies that a modification of Ξ_1 or Ξ_2 propagates through the entire parameter space as shown in Fig. 3.22. In particular for knot insertion, this is a limiting factor, since it prohibits *local* refinement. Especially for numerical simulations, where changes in the physical fields may occur rapidly, local refinement is very beneficial. Fortunately, a lot of research has been devoted to this issues so that various remedies are available.

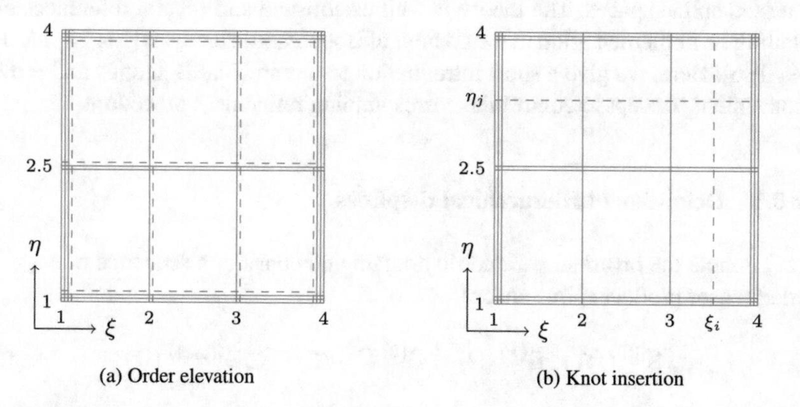

Fig. 3.22 Global refinement of a bivariate B-spline basis: (a) order elevation of p and q from 2 to 3 and (b) insertion of two knots, i.e., $\xi_i = 3.5$ and $\eta_j = 3.25$. Dashed lines indicate the additional knots introduced, while solid lines refer to the initial discretisation

3.6.2 Local Refinement Options

Local refinement of multivariate splines is an active area of research and several techniques have been developed, such as T-splines [8, 164], locally refined B-splines [56, 92], and hierarchical B-splines [31, 63, 98, 186]. An adaptation of the latter are truncated hierarchical B-splines [69, 117]; or THB-splines in short. Some of these concepts were first presented in the context of CAD (e.g., T-splines and hierarchical B-splines). However, their application in an analysis setting has provided a huge impetus to their further development. In fact, these concepts have become so technically mature that the question is no longer if local refinement of multivariate spline is feasible, but what technique do you prefer.

In general, refinement of splines is based on the concept of *function refinement*. Refinement of the underlying set of elements (i.e., non-zero knot spans) is implicit in its construction. We already discussed the corresponding fundamental technique, namely *knot insertion*. However, an additional aspect of knot insertion will be outlined in the following, namely the coefficients of the two-scale relation which links the initial set of B-splines to the refined ones. A local multivariate interpretation of knot insertion is used in the context of T-splines and locally refined B-splines. Other local refinement techniques, such as hierarchical B-splines, are based on *subdivision*, a technique that can be derived from knot insertion. In the following, we will focus on these hierarchical concepts, since, in the authors' opinion, they provide a very natural extension to conventional B-splines.

3.6.3 Hierarchical B-splines

Multilevel hierarchical B-splines [98] provide local refinement through a hierarchy of nested spline spaces. The theory is well established and several references exist detailing its implementation in the context of isogeometric analysis [31, 32, 83, 109, 159, 186]. Here, we give a short introduction to hierarchical B-splines and provide a convenient concept to control the corresponding refinement procedure.

3.6.3.1 Definition of Hierarchical B-splines

Let Ω denote the bivariate parametric domain and consider a sequence of $\ell_{\max} + 1$ nested tensor product spline spaces

$$\mathbb{S}^{(0)}(\Omega) \subset \mathbb{S}^{(1)}(\Omega) \subset \mathbb{S}^{(2)}(\Omega) \subset \cdots \subset \mathbb{S}^{(\ell_{\max})}(\Omega) \tag{3.32}$$

where $\mathbb{S}^{(\ell)} := \operatorname{span}\left\{ B_{ij}^{(\ell)} \text{ for } ij \in \beta^{(\ell)} \right\}$. The index-set $\beta^{(\ell)}$ contains all indices of the basis functions on level ℓ. For the sake of simplicity, we assume these spaces have the same polynomial order in each parametric direction and are defined by a sequence of nested knot vectors $\Xi_I^{(0)} \subset \Xi_I^{(1)} \subset \cdots \subset \Xi_I^{(\ell_{\max})}$, $I = 1, 2$.

A multilevel spline space is defined as follows,

$$\mathcal{MS}(\Omega) := \operatorname{span}\left\{B_{ij}^{(\ell)} \text{ for } ij \in \beta_A^{(\ell)},\ \ell = 0, 1, \ldots, \ell_{\max}\right\}. \tag{3.33}$$

Here $\beta_A^{(\ell)}$ is an index-set that refers to *active* basis functions on level ℓ. Its complement, $\beta_D^{(\ell)}$, such that $\beta_A^{(\ell)} \cap \beta_D^{(\ell)} = \emptyset$ and $\beta_A^{(\ell)} \cup \beta_D^{(\ell)} = \beta^{(\ell)}$, refers to all B-splines that are *inactive* on level ℓ. Starting from an initial spline space, $\mathbb{S}^{(0)}(\Omega)$, hierarchical refinement proceeds by deactivating a basis functions on level ℓ and replacing it by finer B-splines on level $\ell + 1$. Therefore, a relation between the two scales has to be established and the refinement process has to maintain global linear independence and nestedness of the spaces. These aspects pose constraints on the index-sets $\{\beta_A^{(0)}, \ldots, \beta_A^{(\ell_{\max})}\}$. The transition from one level to another, and a proper balancing of the refinement across the different levels are further elements to be considered. We shall discuss each of these topics below.

3.6.3.2 Two-Scale Relation

Consider two nested one-dimensional spline spaces that define the basis functions on the levels ℓ and $\ell + 1$, respectively. There is a special relation between the B-splines defined on these two scales (i.e., the two nested levels): a coarse-scale function $B_j^{(\ell)}$ can be exactly represented by a linear combination of fine-scale functions $B_i^{(\ell+1)}$ by

$$B_{j,p}^{(\ell)}(\xi) = \sum_{i \in \beta^{(\ell+1)}} \mathbf{S}_{ij}^{(\ell)}\, B_{i,p}^{(\ell+1)}(\xi) \tag{3.34}$$

with the *subdivision matrix* of the coarse level $\mathbf{S}^{(\ell)}$. In other words, $\mathbf{S}^{(\ell)}$ defines the relation between the two scales, ℓ and $\ell + 1$. The corresponding matrix-entries are computed as follows: first consider the situation that the knot vector related to the finer level has only a single knot more than the coarse one, that is, $\Xi^{(\ell+1)} = \Xi^{(\ell)} \cup \hat{\xi}$ with $\hat{\xi} \in [\xi_s, \xi_{s+1})$. This is equivalent to knot inserting $\hat{\xi}$ into the sth knot span of $\Xi^{(\ell)}$, and the non-zero entries of the corresponding matrix $\hat{\mathbf{S}}$ are determined by

$$\begin{cases} \hat{\mathbf{S}}(k, k-1) & = 1 - \alpha_k \\ \hat{\mathbf{S}}(k, k) & = \alpha_k \end{cases} \quad \text{with} \quad \alpha_k = \begin{cases} 1 & k \leqslant s - p \\ \frac{\hat{\xi} - \xi_k}{\xi_{k+p} - \xi_k} & s - p + 1 \leqslant k \leqslant s \\ 0 & k \geqslant s + 1 \end{cases} \tag{3.35}$$

Remark The coefficients α_k and $1 - \alpha_k$ are the same values that will occur in the context of knot insertion of B-spline curves and surfaces, where they define the proper linear combination of existing control points to generate new ones (see Sect. 4.1.3).

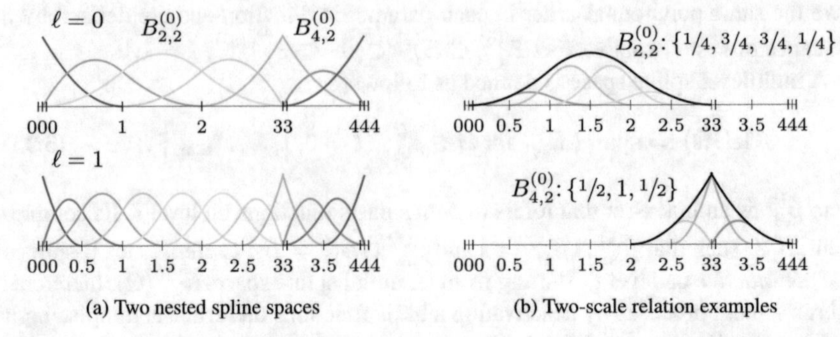

Fig. 3.23 A hierarchy of B-spline spaces with two scales (left) and examples of the related correlations between their basis functions (right). In (b), the listed subdivision coefficients correspond to the highlighted columns of (3.36), and define the linear combination of the child functions on the finer level 1 (shown in colour) to obtain the parent functions $B^{(0)}_{2,2}$ and $B^{(0)}_{4,2}$ (shown in black)

In case $\Xi^{(\ell+1)}$ consists of r additional knots, Eq. (3.35) can be subsequently applied. Each time the initial $\Xi^{(\ell)}$ is updated, until it is equal to $\Xi^{(\ell+1)}$ and the corresponding $\hat{\mathbf{S}}_i$, $i = 1, \dots, r$ are computed. The final subdivision matrix $\mathbf{S}^{(\ell)}$ is obtained by $\mathbf{S}^{(\ell)} = \hat{\mathbf{S}}_r \, \hat{\mathbf{S}}_{r-1} \dots \hat{\mathbf{S}}_1$.

We illustrate the two-scale relation and its connection to the subdivision matrix by the following example: starting from a parameter space given by $\Xi^{(0)} = (0,0,0,1,2,3,3,4,4,4)$, a uniformly refined level is introduced by knot insertion at $\hat{\xi}_j = \{0.5, 1.5, 2.5, 3.5\}$. The resulting hierarchy of the spline spaces is shown in Fig. 3.23(a). The resulting subdivision matrix is given by

$$\mathbf{S}^{(0)} = \begin{pmatrix} 1 & & & & & & \\ 1/2 & 1/2 & & & & & \\ & 3/4 & 1/4 & & & & \\ & 1/4 & 3/4 & & & & \\ & & 3/4 & 1/4 & & & \\ & & 1/4 & 3/4 & & & \\ & & & 1/2 & 1/2 & & \\ & & & & 1 & & \\ & & & & 1/2 & 1/2 & \\ & & & & & 1/2 & 1/2 \\ & & & & & & 1 \end{pmatrix}. \tag{3.36}$$

Note that the number of columns (7) and rows (11) correspond to the number of basis functions on level 0 and 1, respectively. Furthermore, all entries of $\mathbf{S}^{(0)}$ are positive and the rows sum to one; these properties are directly inherited from Eq. (3.35). The columns encode the relation between the *parent* function on the coarse scale and its *children* on the fine scale. For the highlighted columns, these two-scale relations are illustrated in Fig. 3.23(b).

In the multivariate setting, two-scale relations are obtained by tensor products of univariate ones. They are encoded in a subdivision matrix $\mathbf{S} := \mathbf{S}_1 \otimes \mathbf{S}_2$, where

$\mathbf{S}_i$, $i = 1, 2$, denote the univariate subdivision matrices in each of the parametric directions. With these preliminaries, the presented material is independent of the spatial dimension and hence, we can limit our discussion to the univariate setting without loss of generality.

3.6.3.3 Local Refinement Procedure

Based on the two-scale relation, we are now able to refine an individual basis function, rather than an entire set of B-splines, as it is the case for general knot insertion. However, we have to maintain nested spaces and linear independence of the B-splines within the hierarchical basis. Nested refinements imply polynomial reproducibility, which is important for the accuracy of the method. A necessary and sufficient condition to maintain nested spaces is to guarantee that

$$\text{span}\left\{B_i^{(\ell)},\ i \in \beta_D^{(\ell)}\right\} \subset \text{span}\left\{B_i^{(k)},\ \text{For } i \in \beta_A^{(k)},\ k = \ell+1, \ldots, \ell_{\max}\right\}. \tag{3.37}$$

In words: functions that are *deactivated* on level ℓ should be representable by a linear combination of active functions on levels $\ell+1, \ldots, \ell_{\max}$. Suppose $B_j^{(\ell)}(\xi)$ is deactivated. A sufficient condition to satisfy (3.37) is to activate all its children, i.e.,

$$\{B_i^{(\ell+1)}(\xi),\ \mathbf{S}_{ij}^{(\ell)} \neq 0\}. \tag{3.38}$$

The subdivision matrices give all the information required to successfully perform this operation. Due to their sparsity and the fact that they are univariate operators, one may choose to explicitly store these matrices in a sparse format. This allows efficient access to the parent-child relation at any time.

The transition from one active region of a level to another can be treated different ways as illustrated in Fig. 3.24: in case of (a) classical hierarchical B-splines, active basis functions at the region where the supports of B-splines of different levels overlap

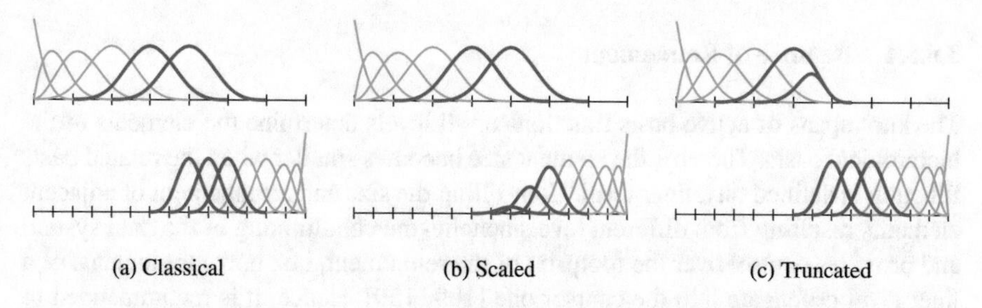

(a) Classical (b) Scaled (c) Truncated

Fig. 3.24 Different types of a hierarchical B-splines basis. Basis functions corresponding to the overlapping area of two levels (coarse-scale on the top; fine-scale at the bottom) are highlighted in red. Inactive basis functions are not displayed. The basis in (b) and (c) form a partition of unity. In (b), this is achieved by scaling on the finer level, whereas truncation on the coarser level is used in (c)

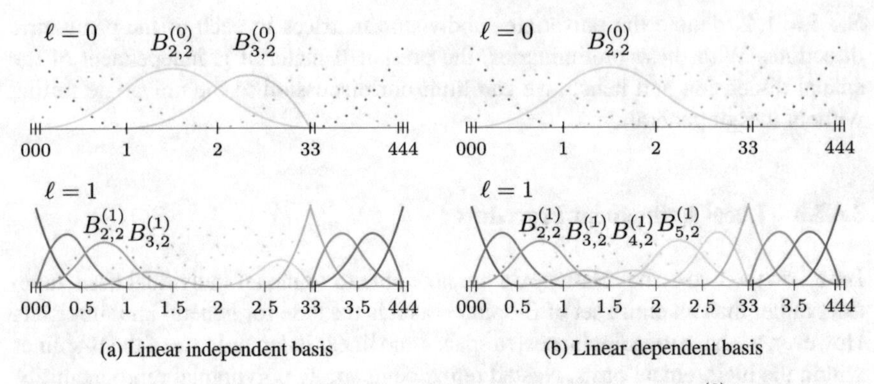

Fig. 3.25 The two-level spline basis in (a) is linear independent. Deactivation $B^{(0)}_{3,2}$ leads to a linear dependency shown (b), since the remaining $B^{(0)}_{2,2}$ can be represented in terms of $B^{(1)}_{i,2}, i = 2, 3, 4, 5$

are not altered, whereas an adaptation is applied in the (b) scaled and (c) truncated version in order to establish a partition of unity. These adaptations are directly linked to the entries of the subdivision matrix as we will discuss in the subsequent sections. In principle, every hierarchical B-spline type may be used in an analysis context, but we will often use the scaled version for illustrations since it has a particular interesting connection to the subdivision matrix. Further, the truncated version has computational benefits which will be outlined in Sect. 3.6.4.1.

Linear independence is lost when an active basis function on level ℓ can be represented as a linear combination of active functions on levels $\ell + 1, \ldots, \ell_{\max}$. Although local refinement is driven by deactivating a parent function and activating only its children, linear dependencies can still occur, as is illustrated in Fig. 3.25. To avoid this from happening the linear dependency needs to be explicitly removed. This is done by checking for each active parent function on level $\ell < \ell_{\max}$ if all its children are active as well. If this is the case, then the parent function is deactivated, thereby removing the linear dependency.

3.6.3.4 Balance of Refinement

The knot spans of active basis functions of all levels determine the elements of the hierarchical basis. Thereby, the element size becomes smaller when the related basis function is defined on a finer level. Controlling the size and arrangement of adjacent elements resulting from different levels benefits the conditioning of the final system and provides control over the footprint of the refinement, i.e., how far elements of a finer level propagate into the coarser one [109, 159]. Hence, it is recommended to appropriately *balance* the refinements across the separate levels. Here we present a simple concept introduced in [117].

In general, each level can be associated to a region corresponding to the supports of its active functions, that is,

$$\Omega^{(\ell)} := \bigcup_{i \in \beta_A^{(\ell)}} \operatorname{supp}\left\{B_i^{(\ell)}(\xi)\right\}. \tag{3.39}$$

Figure 3.26 depicts an example of a (scaled) three-level hierarchical spline space and illustrates its regions $\Omega^{(0)}, \Omega^{(1)}, \Omega^{(2)}$. We can assign a value to each basis function, yielding a vector of weights $\mathbf{w}^{(\ell)} = \left\{w_0^{(\ell)}, w_1^{(\ell)}, \ldots, w_{n^\ell}^{(\ell)}\right\}$ per level. We shall refer to the entries in $\mathbf{w}^{(\ell)}$ as the *subdivision weights* and they are determined by

$$w_i^{(0)} = \begin{cases} 1 & i \in \beta_A^{(0)} \\ 0 & i \in \beta_D^{(0)} \end{cases} \tag{3.40}$$

and when $\ell > 0$,

$$w_i^{(\ell)} = \begin{cases} \sum_{j \in \beta_D^{(\ell-1)}} \mathbf{S}_{ij}^{(\ell-1)} & i \in \beta_A^{(\ell)} \\ 0 & i \in \beta_D^{(\ell)}. \end{cases} \tag{3.41}$$

In words: subdivision weights of inactive functions are zero, while those related to active functions on level ℓ are in the range of (0, 1], and they are computed as the row-sum of the columns of $\mathbf{S}^{(\ell-1)}$ corresponding to deactivated functions on level $\ell - 1$. Consider the example given in Fig. 3.26. The coefficients of $\mathbf{w}^{(0)}$ are $\{1, 1, 1, 0, 0, 0, 0\}$ according to Eq. (3.40). Those of $\mathbf{w}^{(1)}$ are computed as follows,

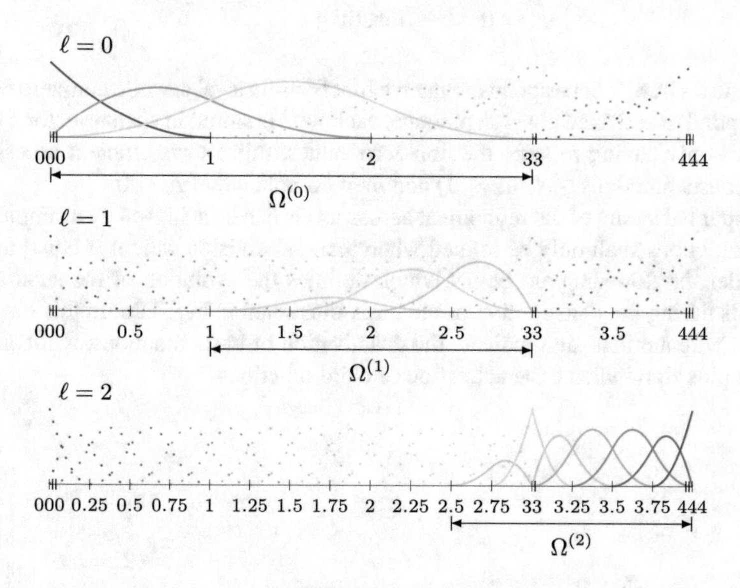

Fig. 3.26 Three-level hierarchical B-spline basis. The functions are scaled using the subdivision weights defined in (3.41) to form a partition of unity. Active basis functions are represented by solid lines

$$\mathbf{w}^{(1)} = \begin{pmatrix} 0 \\ 0 \\ 0 \\ 0 \\ 1/4 \\ 3/4 \\ 1 \\ 0 \\ 0 \\ 0 \\ 0 \end{pmatrix} \quad \text{corresponding to} \quad \mathbf{S}^{(0)} = \left(\begin{array}{ccc|cccc} 1 & & & & & & \\ 1/2 & 1/2 & & & & & \\ & 3/4 & 1/4 & & & & \\ & 1/4 & 3/4 & & & & \\ \hline & & 3/4 & 1/4 & & & \\ & & 1/4 & 3/4 & & & \\ & & & 1/2 & 1/2 & & \\ \hline & & & & 1 & & \\ & & & & 1/2 & 1/2 & \\ & & & & & 1/2 & 1/2 \\ & & & & & & 1 \end{array}\right). \tag{3.42}$$

Only entries in $\mathbf{w}^{(1)}$ corresponding to active functions, highlighted in red, need to be computed. Their value corresponds to the row-sum of columns 3–6 of $\mathbf{S}^{(0)}$. Note that the columns 0–2 are ignored since they refer to active basis functions on level 0. The remaining values in $\mathbf{w}^{(1)}$ are zero by definition (3.41), because these indices are in $\beta_D^{(1)}$. It is emphasised that a non-zero subdivision weight defines the scaling factor of the related basis function needed to obtain the partition of unity. Thus, the change of the height of the scaled B-splines shown in Fig. 3.26 corresponds directly to the values stored in $\mathbf{w}^{(\ell)}$.

The subdivision weights can also be used to control the local refinement process. First of all, we can divide B-splines of a hierarchical basis into three classes:

$$\begin{cases} w_i = 1 & \text{(active, refineable)} \\ 0 < w_i < 1 & \text{(active, not refineable)} \\ w_i = 0 & \text{(inactive)} \end{cases} . \tag{3.43}$$

Zero entries in $\mathbf{w}^{(\ell)}$ correspond to inactive functions on level ℓ, while non-zero entries correspond to activated ones. In practice, each $\mathbf{w}^{(\ell)}$ is stored in sparse vector format. Besides only having to store the non-zero values, this allows efficient access into active basis functions ($0 < w_i \leq 1$) and their complement ($w_i = 0$).

Proper balancing of the refinement across levels can be achieved by a simple rule: basis functions shall only be refined when their subdivision weight is equal to one. Consider the flow-diagram below which outlines the evolution of the subdivision weights during the construction of the basis illustrated in Fig. 3.26. In this diagram, arrows with a minus sign indicate the deactivation of basis functions, while arrows with a plus sign indicate the activation of child functions.

$$\mathbf{w}^{(0)} = \begin{pmatrix} 1 \\ 1 \\ 1 \\ 1 \\ 1 \\ 1 \\ 1 \end{pmatrix} \xrightarrow{-} \mathbf{w}^{(0)} = \begin{pmatrix} 1 \\ 1 \\ 1 \\ 0 \\ 0 \\ 0 \\ 0 \end{pmatrix} \xrightarrow{+} \mathbf{w}^{(1)} = \begin{pmatrix} 0 \\ 0 \\ 0 \\ 0 \\ 1/4 \\ 3/4 \\ 1 \\ 1 \\ 1 \\ 1 \\ 1 \end{pmatrix}$$

Following the diagram from left to right, the first step is that the basis functions 3–6 corresponding to level 0 are refined, that is, they are deactivated and their children are activated. This is encoded in the subdivision weights as follows: $\{w_3^{(0)}, w_4^{(0)}, w_5^{(0)}, w_6^{(0)}\}$ are set to zero and the entries of $\mathbf{w}^{(1)}$ are computed as given in (3.41). As a result, the entries 6–10 of $\mathbf{w}^{(1)}$ become equal to 1 and are allowed to be refined. Hence, in a following step, we can refine the functions 7–10 of level 1,

$$\mathbf{w}^{(1)} = \begin{pmatrix} 0 \\ 0 \\ 0 \\ 0 \\ 1/4 \\ 3/4 \\ 1 \\ 1 \\ 1 \\ 1 \\ 1 \end{pmatrix} \xrightarrow{-} \mathbf{w}^{(1)} = \begin{pmatrix} 0 \\ 0 \\ 0 \\ 0 \\ 1/4 \\ 3/4 \\ 1 \\ 0 \\ 0 \\ 0 \\ 0 \end{pmatrix} \xrightarrow{+} \mathbf{w}^{(2)} = \begin{pmatrix} 0 \\ 0 \\ 0 \\ 0 \\ \vdots \\ 1/2 \\ 1 \\ 1 \\ 1 \\ 1 \\ 1 \\ 1 \end{pmatrix}$$

leading to the three-level hierarchical spline space depicted in Fig. 3.26.

In conclusion, the sparse vectors $\mathbf{w}^{(\ell)}$ with $\ell = \{0, 1, \ldots, \ell_{\max}\}$ encode the sets of active and inactive basis functions, $\beta_A^{(\ell)}$ and $\beta_D^{(\ell)}$. Furthermore, the non-zero subdivision weights, $w_i^{(\ell)}$, can be used effectively to provide a very simple balancing criterion. This criterion assures a proper transition between refinement levels so that no hierarchical level is omitted during the local refinement. In other words, B-splines on level ℓ are merely surrounded by basis functions related to the levels $\ell - 1$, ℓ, or $\ell + 1$. This balancing minimises the detrimental effect on conditioning of the multilevel spline basis. Moreover, the subdivision weights provide the scaling factors to form a partition of unity in the hierarchical basis by

$$\sum_{\ell=0}^{\ell_{\max}} \sum_{i \in \beta_A^{(\ell)}} w_i^{(\ell)} \cdot B_i^{(\ell)} = 1. \tag{3.44}$$

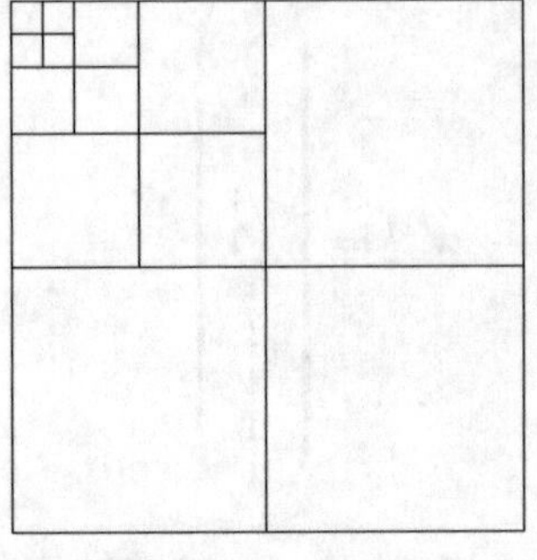

(a) Refinement of a corner

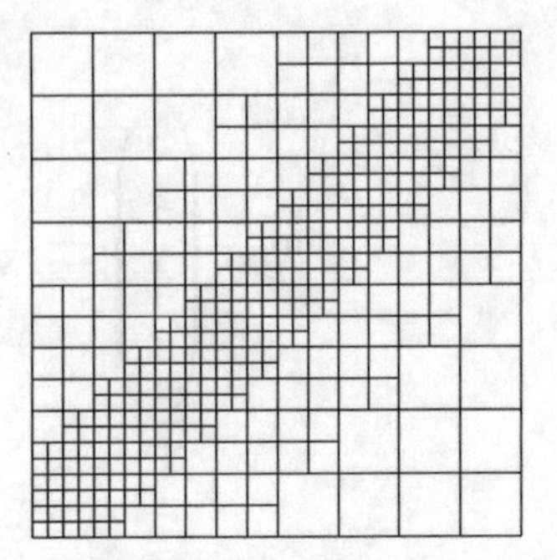

(b) Refinement along the diagonal

Fig. 3.27 Illustration of the refinement footprint of the balancing scheme based on Eq. (3.43). Both cases show the elements due to local refinement of a maximal continuity spline spaces defined by open knot vectors and degree $p = q = 2$

These weights are also essential for the definition of truncated hierarchical B-splines, as will be discussed in Sect. 3.6.4.1. Finally, a drawback of the presented balancing is that the region affected by the local refinement could be large, especially in the setting of maximal continuity spline spaces. Two examples of this refinement footprint obtained are displayed in Fig. 3.27.

3.6.3.5 Interpolation with Hierarchical B-splines

The Greville abscissae $\tilde{\xi}_i$ defined in Eq. (3.20) are in general good points for interpolation problems [28]. However, the application of the Greville abscissae to a hierarchical basis requires special considerations since anchors of different levels may coincide as illustrated in Fig. 3.28(a).

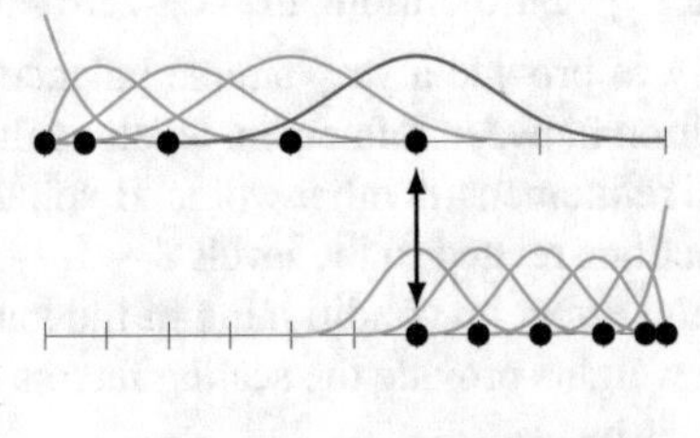

(a) Coinciding Greville abscissae

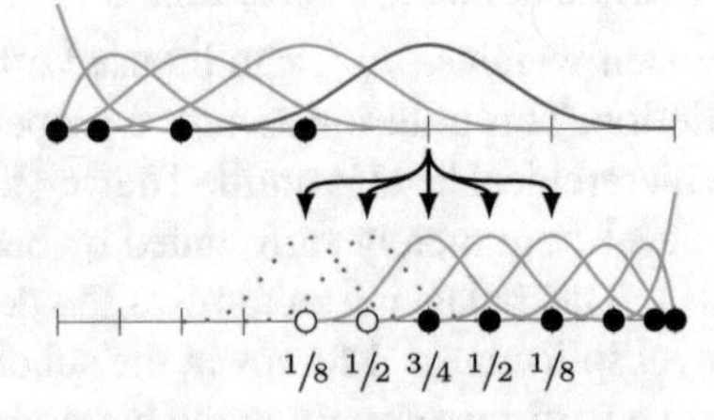

(b) Adaptive isogeometric collocation

Fig. 3.28 Greville abscissae of hierarchical B-splines: (a) potential problems at the transition from one level to another and (b) substitution of the troublesome point by a set of fine-scale abscissae $\tilde{\xi}_j^{(\ell+1)}$ weighted by corresponding coefficients of the related two-scale relation. In (b), the additionally introduced interpolation sites are marked by white circles and the related inactive fine-scale B-splines are illustrated by dotted lines. The accumulation of their values is indicted by arrows

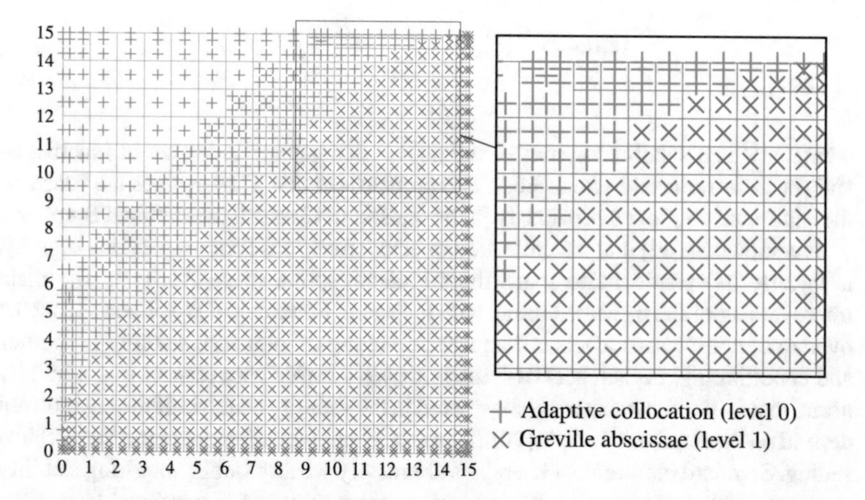

Fig. 3.29 Anchors of the bivariate hierarchical basis of order $p = 4$ due to adaptive isogeometric collocation

To overcome this problem, adaptive isogeometric collocation has been proposed in [157]. Its idea is the following: use the Greville abscissae $\tilde{\xi}_i^{(\ell)}$ of active basis functions if they lie outside (or on the boundary) of $\Omega^{(k)}$, $k > \ell$, that is, a support of an active basis function of a finer level; apply a weighted collocation scheme otherwise. In the latter case, the B-spline of interest, $B_i^{(\ell)}$, is associated with a set of interpolation sites, derived from the Greville abscissae $\tilde{\xi}_j^{(\ell+1)}$ of its children on the subsequent level. Figure 3.28(b) shows the set of anchors of a univariate cubic hierarchical B-spline basis due to this adaptive collocation scheme. It is pointed out that the fine-scale abscissae related to the overlapping $B_i^{(\ell)}$ and the listed numbers are derived from the two-scale relation. In other words, the accumulation of these weighted values yields the final contribution of $B_i^{(\ell)}$. Note that some abscissae may still coincide with anchors of the coarse level, but there will always be at least one independent point in the overall set of interpolation sites associated to $B_i^{(\ell)}$, which guarantees that the resulting system of equation is not singular. Figure 3.29 shows the distribution of anchors due to the adaptive scheme for a bivariate basis.

3.6.4 Truncated Hierarchical B-splines

3.6.4.1 Definition of Truncated Hierarchical B-splines

Truncated hierarchical B-splines (THB-spline) were introduced in [69] to improve the numerical properties of the multilevel spline space. Truncated basis functions are given by

$$\operatorname{trunc} B_j^{(\ell)}(\xi) = \sum_{i \in \beta_D^{(\ell+1)}} \mathbf{S}_{ij}^{(\ell)} B_i^{(\ell+1)}(\xi) \tag{3.45}$$

where $\mathbf{S}^{(\ell)}$ is a subdivision matrix and $\beta_D^{(\ell+1)}$ denotes the index-set of inactive B-splines, as introduced in Sect. 3.6.3. It is emphasised that THB-splines are based on the same data employed previously in the context of classical hierarchical B-splines.

It is shown in [69] that the collection of active functions forms a partition of unity, is linearly independent and spans the multilevel spline space $\mathcal{MS}(\Omega)$. The main improvement of the truncated basis, with respect to hierarchical B-splines, is that the overlap of consecutive levels is decreased resulting in improved stability properties and conditioning. In fact, a THB-basis is *strongly stable* with respect to the infinity norm, that is, the constants to be considered in the stability analysis of the basis do not depend on the number of levels [70]. This is an exceptional feature in the hierarchical setting. Standard hierarchical B-splines are merely weakly stable, requiring stability constants with a polynomial dependence on the number of hierarchical levels. We refer the interested reader to [69, 70] for more information.

Figure 3.30 illustrates the truncated version of the basis shown in Fig. 3.26. Note that the size of $\Omega^{(0)}$ is considerably reduced compared to the non-truncated case. The size of $\Omega^{(1)}$, on the other hand, remains the same due to the double knot at 3. Furthermore, it is pointed out that the basis functions on the last level, i.e., $\ell = 2$, are not affected by the truncation procedure. It is apparent that the truncated B-splines on level ℓ are obtained by linear combinations of inactive B-splines on the subsequent

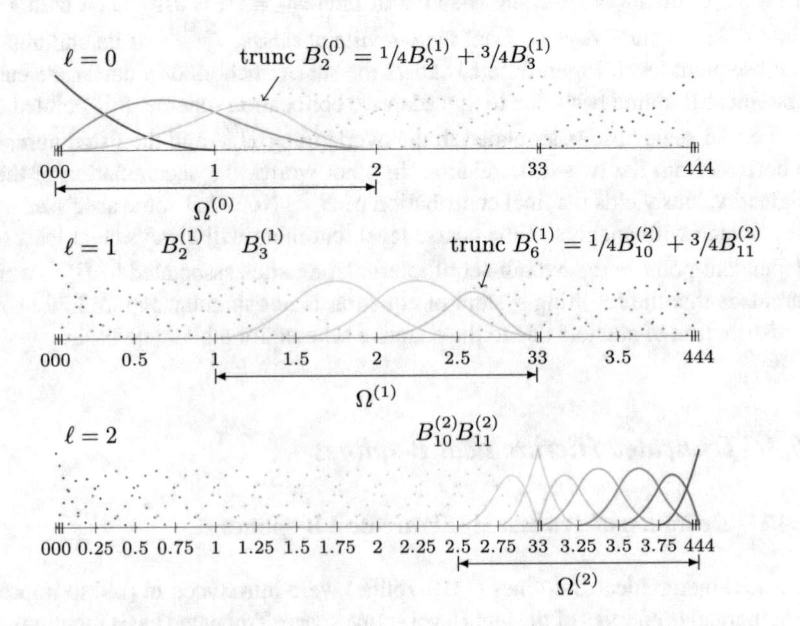

Fig. 3.30 Three-level THB-spline basis. Active functions are represented by solid lines and basis functions involved in the truncation are labelled

level $\ell + 1$. The related coefficients are derived from the two-scale relation presented in Sect. 3.6.3.2.

Using the data structure introduced in Sect. 3.6.3.4 the weights, for the linear combination that yields a truncated B-spline, can be extracted from the subdivision matrix. Consider the basis functions $\operatorname{trunc} B_2^{(0)}$ illustrated in Fig. 3.30 together with the related subdivision matrix $\mathbf{S}^{(0)}$ and vector $\mathbf{w}^{(1)}$ of the subsequent level:

$$\mathbf{w}^{(1)} = \begin{pmatrix} 0 \\ 0 \\ 0 \\ 0 \\ 1/4 \\ 3/4 \\ 1 \\ 0 \\ 0 \\ 0 \\ 0 \end{pmatrix} \quad \text{and} \quad \mathbf{S}^{(0)} = \left(\begin{array}{ccc|cccc} 1 & & & & & & \\ 1/2 & 1/2 & & & & & \\ & 3/4 & 1/4 & & & & \\ & 1/4 & 3/4 & & & & \\ \hline & & 3/4 & 1/4 & & & \\ & & 1/4 & 3/4 & & & \\ & & & 1/2 & 1/2 & & \\ \hline & & & & 1 & & \\ & & & & 1/2 & 1/2 & \\ & & & & & 1/2 & 1/2 \\ & & & & & & 1 \end{array}\right) \tag{3.46}$$

Some entries of $\mathbf{S}^{(0)}$ are not considered for the truncated basis (displayed in grey) because (i) the columns 3–6 refer to *inactive* B-splines on $\ell = 0$, and (ii) the rows 4–6 refer to *active* B-splines on $\ell = 1$. Note that the latter affects matrix entries of $B_2^{(0)}$ (column 2), which indicates that this basis function will be truncated. To be precise, this column provides the two-scale relation of $B_2^{(0)}$ to its children functions and the coefficients for the definition of $\operatorname{trunc} B_2^{(0)}$ are the remaining non-zero entries (highlighted in red) related to row 2–3, i.e., $B_2^{(1)}$ and $B_3^{(1)}$.

In order to evaluate a THB-basis efficiently, its truncated B-splines should be identified in an effective manner. This can be easily accomplished using the subdivision matrices $\mathbf{S}^{(\ell)}$. Suppose an element of a THB-basis shall be evaluated, where the element's level $\ell + 1$ is the finest level with active basis functions at an evaluation point ξ. In general, the non-zero B-splines of the element are $B_i^{(\ell+1)}$ with $i = s^{(\ell+1)} - p, \ldots, s^{(\ell+1)}$, with the knot span index $s^{(\ell+1)}$ referring to the element $\left[\xi_s^{(\ell+1)}, \xi_{s+1}^{(\ell+1)}\right)$. This element contains truncated B-splines on the previous level only if one of its $B_i^{(\ell+1)}$ is *inactive*, that is, the related subdivision weight $w_i^{(\ell+1)}$ is equal to zero. Especially for high orders it can happen that active regions of more than two levels overlap, i.e., basis functions on level $\ell - 1$ are non-zero at ξ. These situations are revealed by examining the entries of the subdivision matrix $\mathbf{S}^{(\ell-1)}$. In particular, we extract those entries that relate the inactive basis functions on level ℓ to all non-zero basis functions on level $\ell - 1$, that is, $\mathbf{S}_{jk}^{(\ell-1)}$ with $j \in \{s^{(\ell)} - p, \ldots, s^{(\ell)}\} \cap \beta_D^{(\ell)}$ and $k = s^{(\ell-1)} - p, \ldots, s^{(\ell-1)}$. If $\mathbf{S}_{jk}^{(\ell-1)}$ contains entries other than 0, there is an additional contribution from a truncated basis function on level $\ell - 1$. This check is repeated for prior levels until a null matrix is obtained or the initial level 0 is reached.

Remark The presented data structure based on subdivision matrices is only one option to implement (truncated) hierarchical B-splines. There are other concepts available and one particularly interesting for THB-splines is based on Bézier-extraction, for details see [83].

3.6.4.2 Interpolation with THB-splines

In Sect. 3.6.3.5 we introduced the concept of adaptive isogeometric collocation, which provides a very general way to associate point values to hierarchical B-splines in regions where two levels overlap. The concept, however, increases the number of anchors $\tilde{\xi}_i^{(\ell)}$.

In [117] an alternative approach for THB-splines has been presented that introduce only as many $\tilde{\xi}_i^{(\ell)}$ as basis functions. It follows the adaptive collocation idea in that Greville abscissae are used if possible, but instead of weighted collocation, the maximal value of a troublesome basis function on the coarse level determines the position of its anchor. Thus, a single point is associated to the spline rather than a set of points. This can be carried out because maximal values of THB-splines do not coincide due to the truncation procedure. However, a THB-spline may have multiple peaks in the bivariate case. In such a situation, one of the possible locations is chosen. These multiple peak cases may be avoided when the basis has a nice grading of elements, as it is obtained by the balancing criterion proposed in Sect. 3.6.3.4. In general, the determination of the location of the maximal value of a THB-spline is a relatively small overhead that has to be applied solely for the truncated functions at the overlap of two hierarchical levels. Moreover, it was demonstrated in [117] that using the maximal values lead to more accurate results and better conditioning as the adaptive scheme. Figure 3.31 compares these two setting for a cubic two-level THB-basis. The difference in the bivariate setting is shown in Fig. 3.32. Note that the number is significantly reduced compared to the adaptive approach illustrated in Fig. 3.29.

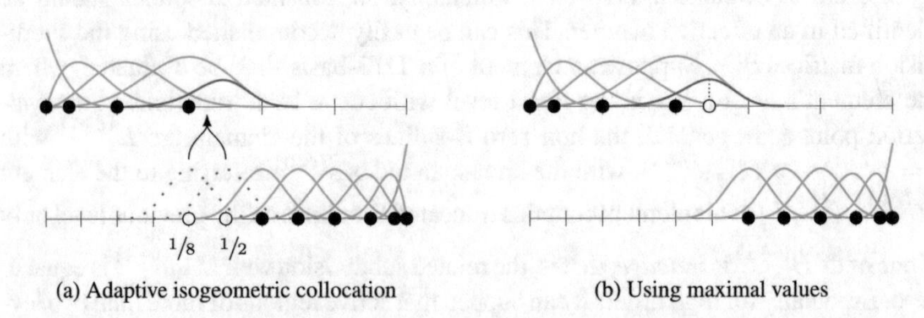

Fig. 3.31 Different interpolation sites for a THB-spline basis: (a) an adaptation of the adaptive isogeometric collocation considering only fine-scale abscissae of those B-spline that define the truncated basis function, and (b) the proposed scheme that introduces an interpolation site at the maximal value of the truncated coarse B-spline

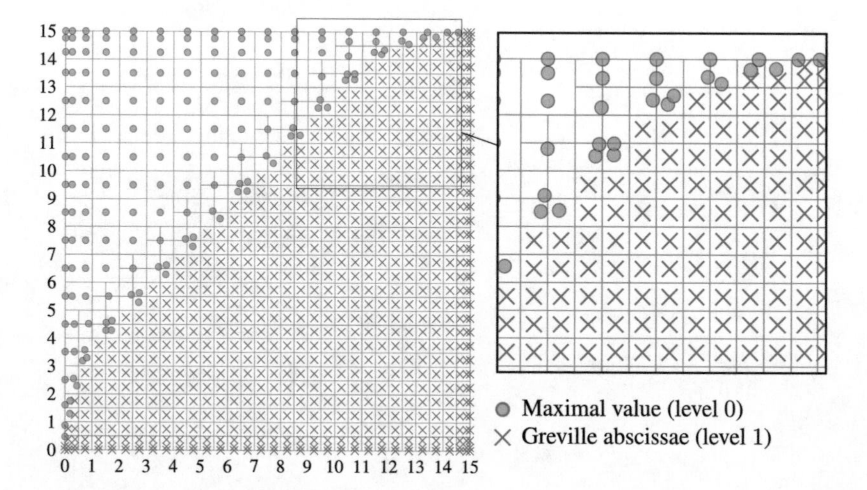

Fig. 3.32 Anchors of the bivariate THB-basis for order $p = 4$ determined by Greville abscissae and the maximal value of truncated basis functions

3.7 Summary and Conclusions

In this chapter we have introduced B-splines basis functions, which can be applied to approximate physical quantities of the problem to be solved. These function offer more options and flexibility than Lagrange polynomials. Whereas with Lagrange polynomials a subdivision into elements is required, and the basis functions are only C^0-continuous across element boundaries, B-splines allow a subdivision of the parameter space with continuous basis functions. A distinguishing feature of B-splines is that they provide control over the continuity within the parameter space. We have shown that a greater number of refinement options are available for the approximation of functions, with the k-refinement, unique to B-splines, being especially effective. The upshot of this is that the classical meshes given by an accumulation of Lagrange elements are indeed no computational necessity for numerical simulations and can be easily replaced by spline representations. Next we will show how geometries can be described with few parameters using B-splines and their generalisation called NURBS.

Chapter 4
Description of the Geometry

The previous chapter outlined the properties of B-spline as well as Lagrange basis functions. Furthermore, the use of B-spline and Lagrange polynomials to approximate a target function (see Sect. 3.1) and the different behaviour of control points and nodal points were discussed (see Sect. 3.3.3). Here we provide deeper insight into the description of geometric objects, focusing on the representation used in most CAD systems, that is, non-uniform rational B-spline (NURBS). In the context of Computer-Aided Geometric Design (CAGD), geometries are usually not used to approximate a target function but shall capture the user's design intent. There is a great number of geometric forms and modelling concepts and the reader is referred to the textbooks [44, 59, 149] for a comprehensive discussion. The following sections aim to provide a concise introduction to NURBS solid models and their representation in CAD systems in order to highlight aspects that are essential for the concepts discussed later on in this book. First, we will introduce NURBS curves to illustrate the key benefit of NURBS representations. Secondly, the effect of the various spline refinement procedures (see Sect. 3.4) on the corresponding geometric objects is investigated. Different common constructions of spline surfaces are outlined, which provides the basis for the subsequent aspect addressed, namely, the description of solid models in CAD. Finally, we discuss some aspects of NURBS models from an analysis perspective and introduce a generalised notation for spline objects used in the remainder of the book.

4.1 NURBS Curves

4.1.1 Definition

NURBS are *rational* B-splines and thus, we start with the definition of conventional B-spline curves. They are specified by B-spline basis functions $B_{i,p}$, introduced in the previous chapter, and corresponding control points $\boldsymbol{c}_i$ defined in model space

G. Beer et al., *The Isogeometric Boundary Element Method*, Lecture Notes in Applied and Computational Mechanics 90,
https://doi.org/10.1007/978-3-030-23339-6_4

(a.k.a. physical space), i.e., $\mathbb{R}^d$ with d denoting the spatial dimension. The geometrical mapping $\mathcal{X}$ from parameter space to model space is given by

$$\mathcal{X}(\xi) := \boldsymbol{C}(\xi) = \sum_{i=0}^{I-1} B_{i,p}(\xi)\ \boldsymbol{c}_i \tag{4.1}$$

and the derivative is

$$\mathbf{J}_{\mathcal{X}}(\xi) := \sum_{i=0}^{I-1} \frac{d}{d\xi} B_{i,p}(\xi)\ \boldsymbol{c}_i. \tag{4.2}$$

The parametric curve $\boldsymbol{C}(\xi)$ inherits its properties from the underlying basis. Since B-splines are piecewise polynomial functions, an object defined by (4.1) describes a piecewise polynomial curve. Furthermore, each polynomial segment is defined by a non-zero knot span of the basis' knot vector Ξ and the continuity between these polynomial segments is determined by the corresponding knot multiplicity. B-splines are very powerful when it comes to representing functions as detailed in Chap. 3 and the same holds true for the related geometric objects. However, there is a blemish from a geometric point of view – polynomials cannot exactly represent a particular group of very popular design elements, namely, conic sections. NURBS resolve this deficiency.

The distinguishing feature of NURBS is that weights w_i are associated with the control points of an object such that

$$\mathbf{c}_i^h = (w_i \boldsymbol{c}_i, w_i)^{\mathrm{T}} = (\boldsymbol{c}_i^w, w_i)^{\mathrm{T}} \in \mathbb{R}^{d+1}. \tag{4.3}$$

These *homogeneous coordinates* $\mathbf{c}_i^h$ specify a B-spline curve $\boldsymbol{C}^h(\xi)$ in a projective space $\mathbb{R}^{d+1}$. In order to obtain a curve in $\mathbb{R}^d$, the geometrical mapping (4.1) is extended by a perspective mapping $\mathcal{P}$ with the centre at the origin of $\mathbb{R}^{d+1}$. Figure 4.1 illustrates this projection which is given by

$$\boldsymbol{C}(\xi) = \mathcal{P}\left(\boldsymbol{C}^h(\xi)\right) = \frac{\boldsymbol{C}^w(\xi)}{w(\xi)} \quad \text{with} \quad w(\xi) = \sum_{i=0}^{I-1} B_{i,p}(\xi)\, w_i \tag{4.4}$$

where $\boldsymbol{C}^w = \left(x_1^h, \ldots, x_d^h\right)^{\mathrm{T}}$ are the homogeneous vector components of the curve. The resulting $\boldsymbol{C}(\xi)$ describes a NURBS curve, i.e., a piecewise rational polynomial. The effect of modifying weights is indicated in Fig. 4.2. Note that decreasing the value of a weight leads to an increased distance between the corresponding control point and the NURBS object.

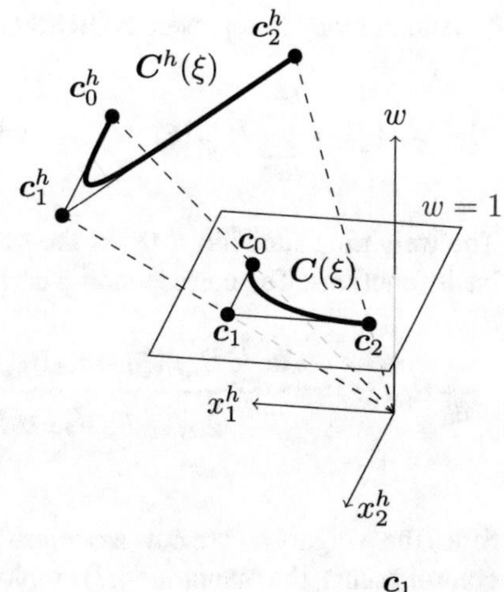

Fig. 4.1 Perspective mapping $\mathcal{P}$ of a quadratic B-spline curve $C^h(\xi)$ in homogeneous form $\mathbb{R}^3$ to a circular arc $C(\xi)$ in model space $\mathbb{R}^2$. The mapping is indicated by dashed lines

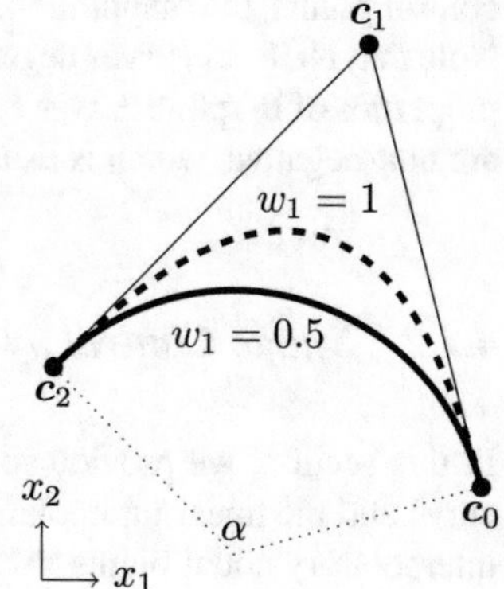

Fig. 4.2 Effect of modified weights w_i in model space: a B-spline curve (dashed) where all weights are 1 and a NURBS curve (solid) with $w_1 = \cos(\alpha/2)$ to define a circular arc with opening angle $\alpha = 120°$

The derivative of the NURBS geometrical mapping is defined by

$$\mathbf{J}_{\mathcal{X}}(\xi) := \frac{w(\xi)\frac{\partial C^w(\xi)}{\partial \xi} - \frac{\partial w(\xi)}{\partial \xi} C^w(\xi)}{(w(\xi))^2} \tag{4.5}$$

with

$$\frac{\partial w(\xi)}{\partial \xi} = \sum_{i=0}^{I-1} \frac{d}{d\xi} B_{i,p}(\xi)\, w_i \quad \text{and} \quad \frac{\partial C^w(\xi)}{\partial \xi} = \sum_{i=0}^{I-1} \frac{d}{d\xi} B_{i,p}(\xi)\, c_i^w\,. \tag{4.6}$$

Another way to represent NURBS curves is

$$\boldsymbol{C}(\xi) = \sum_{i=0}^{I-1} R_{i,p}(\xi)\ \boldsymbol{c}_i \qquad \text{with} \qquad R_{i,p}(\xi) = \frac{w_i B_{i,p}(\xi)}{w(\xi)}. \tag{4.7}$$

The weighting function $w(\xi)$ is the same as in Eq. (4.4) and $R_{i,p}$ denote NURBS basis functions. The corresponding derivatives are given by

$$\frac{d}{d\xi} R_{i,p}(\xi) = \frac{w_i \frac{D}{d\xi} B_{i,p}(\xi) - w_i B_{i,p}(\xi)\,\beta}{\sum_{j=0}^{J-1} w_j B_{j,p}(\xi)} \quad \text{with} \quad \beta = \frac{\sum_{j=0}^{J-1} w_j \frac{d}{d\xi} B_{j,p}(\xi)}{\sum_{j=0}^{J-1} w_j B_{j,p}(\xi)}. \tag{4.8}$$

Since the weights w_i are now associated with the basis functions $B_{i,p}$ rather than the control points, the mapping (4.7) employs control points $\boldsymbol{c}_i$ defined in model space. Note that NURBS curves degenerate to B-spline curves if all weights are equal. The properties of B-spline curves apply to their rational counterpart as well, if the weights are non-negative, which is usually the case.

4.1.2 Shape Control by the Control Polygon

In this section, we provide some information on the interplay between a NURBS curve and the linear interpolation of the control points – the control polygon. Using interpolatory nodal points to define a shape can lead to unintended oscillations (as shown in Sect. 3.3.3), whereas control points implement an intuitive way to manipulate curves.

First of all, a B-spline curve is contained within the *convex hull* of its control polygon. To be precise, a polynomial segment related to a knot span s, i.e., $\xi \in [\xi_s, \xi_{s+1})$, is in the convex hull of the control points $\boldsymbol{c}_{s-p}, \ldots, \boldsymbol{c}_s$. Figure 4.3 displays convex hulls for different curves. Note that they become broader with increasing continuity. The convex hull property results from the fact that B-splines form a positive partition of unity (3.16). Consequently, it holds true also for NURBS curves, if their weights are non-negative. This property allows applying geometric operations to the control polygon rather than the actual curve, leading to efficient implementations.

Another advantage of the control polygon is that it provides a means to specify continuity relationships between curves. A very helpful feature in this regard is that a segment of a control polygon, that emanates from an interpolatory control point such as an endpoint of a curve, specifies the *tangent* to the curve at this point. Hence, connecting two curves so that their tangents share a comment direction is trivial. We demonstrate this by joining two Bézier curves with $\xi \in [0, 1]$ for both. A Bézier curve is a special case of a B-spline, where the basis is defined by Bernstein polynomials. Hence, the first and last control points are interpolatory and there is only a single

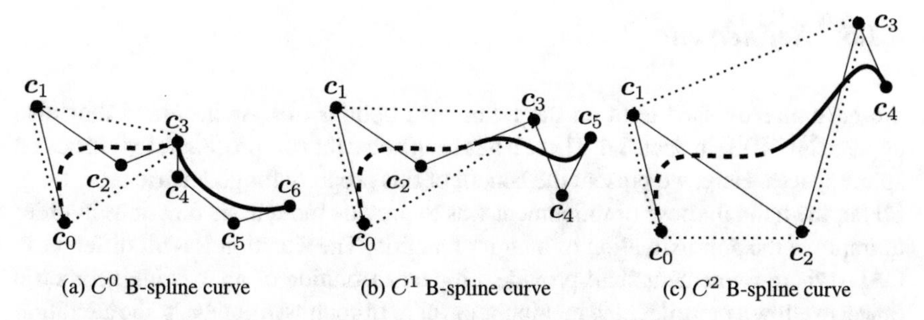

(a) C^0 B-spline curve (b) C^1 B-spline curve (c) C^2 B-spline curve

Fig. 4.3 The convex hull of a cubic B-spline curve segment with different continuity C^i. Each curve consists of two elements displayed as dashed and solid lines, respectively. The left element is the same for all examples and dotted boxes mark its convex hull

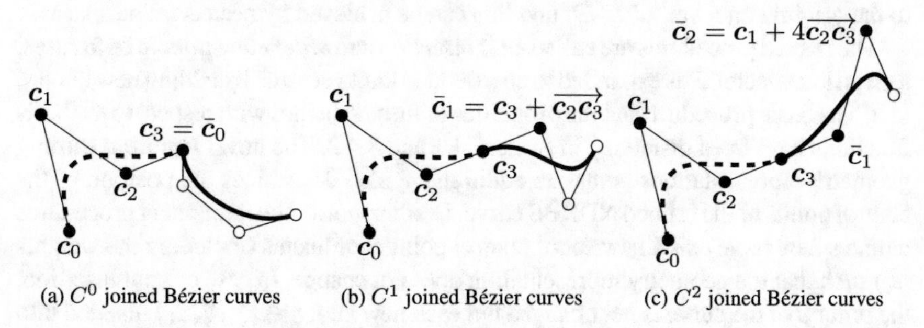

(a) C^0 joined Bézier curves (b) C^1 joined Bézier curves (c) C^2 joined Bézier curves

Fig. 4.4 Enforcing continuity using control points: the control points c_i of the left curve are fixed and the control points $\bar{c}_i$ are adjusted to obtain the continuity sought. The vector $\overrightarrow{c_2 c_3}$ represents the last segment of the control polygon of the first curve. The white control points mark those $\bar{c}_j$ that are not affected by the continuity condition

element. Figure 4.4 shows the construction of the Bézier version of the B-splines curves of Fig. 4.3, where the continuity is not determined by the basis functions but the positioning of the control points near the joint of connected Bézier curves. Note that increasing the desired continuity at the joint, increases the number of control points $\bar{c}_j$ that have to be adjusted based on the c_i of the first element. These relations between the control points are generally derived from the conditions

$$\frac{d^k}{d\xi^k} C_0(1) = \frac{d^k}{d\xi^k} C_1(0) \quad \Leftrightarrow \quad \sum_{i=0}^{3} \frac{d^k}{d\xi^k} B_{i,3}(1) c_i = \sum_{j=0}^{3} \frac{d^k}{d\xi^k} B_{j,3}(0) \bar{c}_j \tag{4.9}$$

for $k = 0, 1, 2$. Finally, it is noted that the specifications of $\bar{c}_0$, $\bar{c}_1$, and $\bar{c}_3$ in Fig. 4.4 fulfil the continuity conditions (4.9) only if the curves' parameterisation is the same. Otherwise, geometric continuity G^k is obtained, that is, the curves share a common tangent direction (G^1) and centre of curvature (G^2) at the joint.

4.1.3 Refinement

We have already shed light on the refinement options of B-splines (and they also apply to NURBS) in Sect. 3.4. There it has been noted that the procedures yield nested spline spaces. Here, we present the benefit of this property for geometric modelling. So far, the main purpose of refinement was to provide more basis functions in order to improve the approximation of a target function. The situation is a bit different in CAD, where refinement shall provide a higher resolution of an existing geometric object to allow a more detailed manipulation of it. In contrast to analysis, the alteration of the represented model should not be part of the refinement scheme but defined by the designer once the refinement has been applied. Therefore it is essential that a given NURBS curve $\boldsymbol{C}$ can be refined such that the resulting object $\hat{\boldsymbol{C}}$ is equivalent to the original one, i.e., $\boldsymbol{C} \equiv \hat{\boldsymbol{C}}$, and this can be achieved by nested spline spaces.

The related procedures are called *knot insertion* and *order elevation*. In both cases, a given knot vector Ξ is expanded to an extended knot vector $\hat{\Xi}$ by adding new knots $\hat{\xi}$ to it. The basic procedure and the properties of these schemes with respect to the basis functions have been discussed in Sects. 3.4.1 and 3.4.2. The novel feature regarding geometry representations is that an additional $\hat{\xi}$ also determines the position of the control points of the refined NURBS curve. In other words, the refinement procedures define a new basis *and* a new set of control point coordinates (including the weights w_i) such that the geometry representation does not change. In case of knot insertion, the order p of the curve is not changed but each new knot $\hat{\xi} \in [\xi_s, \xi_{s+1})$ inserted into the sth knot span introduces a new control point and the coordinates of neighbouring control points are adjusted. These new coordinates $\hat{\boldsymbol{c}}_i^h$ are determined by

$$\hat{\boldsymbol{c}}_k^h = \alpha_k \boldsymbol{c}_k^h + (1 - \alpha_k)\, \boldsymbol{c}_{k-1}^h \quad \text{with} \quad \alpha_k = \begin{cases} 1 & k \leqslant s - p \\ \frac{\hat{\xi} - \xi_k}{\xi_{k+p} - \xi_k} & s - p + 1 \leqslant k \leqslant s \\ 0 & k \geqslant s + 1 \end{cases} . \tag{4.10}$$

Attentive readers may recognise that Eq. (4.10) is identical with the formula for the coefficients of a subdivision matrix (3.34) which led to the two-scale relation of the nested spaces used in hierarchical refinement (see Sect. 3.6.3). Following Piegl and Tiller [132], this relation is also crucial for the update of the control points due to order elevation which is performed by:

1. Subdividing of the curve into Bézier segments by knot insertion (4.10) at the initial knot values so that the multiplicity of all knots is $m = p$.
2. Order elevation of the resulting Bézier segments given by

$$\hat{\boldsymbol{c}}_i^h = (1 - \alpha_i)\, \boldsymbol{c}_i^h + \alpha_i \boldsymbol{c}_{i-1}^h \quad \text{with} \quad \alpha_i = \frac{i}{p+1}, \quad i = 0, \ldots, p+1. \tag{4.11}$$

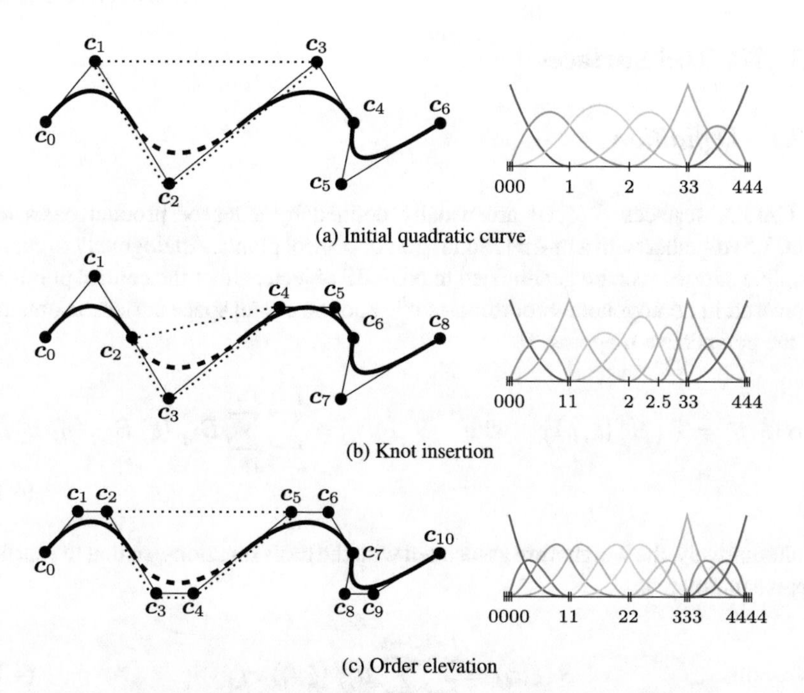

Fig. 4.5 Example of a refinement of a curve (left) and its basis (right): (a) the initial B-spline basis given by $\Xi = (0, 0, 0, 1, 2, 3, 3, 4, 4, 4)$ and the corresponding piecewise polynomial curve, (b) the refined version due to the insertion of the knots $\hat{\xi}_i = \{1, 2.5\}$, and (c) the representation of the initial object by cubic basis functions. On the left, circles denote the control points and dotted lines indicate the convex hull of the dashed curve segment defined in $\xi \in [1, 2)$

3. Removing of superfluous knots which separates the elevated segments by the reverse process of knot insertion to obtain a NURBS curve with the initial continuity again.

The effect of these refinement schemes on the basis functions and the corresponding geometry is illustrated in Fig. 4.5. Note that the geometry itself remains unchanged, despite the fact that the set of control points and basis functions is updated accordingly to the refinement scheme applied. Furthermore, it is worth noting that the convex hull spawn by the control polygon becomes a better estimation of the curve and its individual segments. This is indicated in Fig. 4.5 by the dotted lines that enclose the second polynomial segment.

4.2 NURBS Surfaces

4.2.1 *Definition*

In CAGD, surfaces $\boldsymbol{S}(\xi,\eta)$ are usually defined by a tensor product basis (cf., Sect. 3.5) together with a bi-directional grid of control points. Analogously to curves, B-spline surfaces can be generalised to NURBS objects; either the control points are expressed in homogeneous coordinates $\boldsymbol{c}^h_{i,j}$ and the model space surface is obtained by the perspective mapping $\mathcal{P}$

$$\boldsymbol{S}(\xi,\eta) = \mathcal{P}\left(\boldsymbol{S}^h(\xi,\eta)\right) \quad \text{with} \quad \boldsymbol{S}^h(\xi,\eta) = \sum_{i=0}^{I-1}\sum_{j=0}^{J-1} B_{i,p}(\xi)\, B_{j,q}(\eta)\; \boldsymbol{c}^h_{i,j}, \tag{4.12}$$

or alternatively, the weights are associated with the basis functions, so that the surface representation reads

$$\boldsymbol{S}(\xi,\eta) = \sum_{i=0}^{I-1}\sum_{j=0}^{J-1} R^{pq}_{ij}(\xi,\eta)\; \boldsymbol{c}_{i,j} \tag{4.13}$$

with

$$R^{pq}_{ij}(\xi,\eta) = \frac{B_{i,p}(\xi)\, B_{j,q}(\eta)\, w_{ij}}{\sum_{m=0}^{M-1}\sum_{n=0}^{N-1} B_{m,p}(\xi)\, B_{n,q}(\eta)\, w_{mn}} = \frac{B^{pq}_{ij}(\xi,\eta)\, w_{ij}}{w(\xi,\eta)} = \frac{\boldsymbol{A}(\xi,\eta)}{w(\xi,\eta)}. \tag{4.14}$$

There is a subtle difference between $R^{pq}_{ij}(\xi,\eta)$ and its non-rational counterpart $B^{pq}_{ij}(\xi,\eta)$: strictly speaking, the basis functions (4.14) do not form a tensor product basis, since they are only products of univariate basis functions if the weights can be expressed as $w_{ij} = w_i w_j$. Nonetheless, multivariate NURBS have the same benefits (e.g., efficiency) and disadvantages (e.g., local refinement) as multivariate B-splines. All B-spline algorithms can be applied to $\boldsymbol{S}^h(\xi,\eta)$ by simply allowing homogeneous control points and the final evaluation of a point on the rational surface is performed by the trivial projection $\mathcal{P}$. Derivatives of NURBS surfaces, however, are more involved. To be precise, for B-spline patches derivatives are given by

$$\frac{\partial^{k+l}}{\partial^k\xi\,\partial^l\eta}\boldsymbol{S}^h(\xi,\eta) = \sum_{i=0}^{I-1}\sum_{j=0}^{J-1} \frac{d^k}{d^k\xi}B_{i,p}(\xi)\,\frac{d^l}{d^l\eta}B_{j,q}(\eta)\; \boldsymbol{c}^h_{ij} \tag{4.15}$$

whereas derivatives of NURBS patches become

$$
\begin{aligned}
\frac{\partial^{k+l}}{\partial^k \xi \partial^l \eta} \boldsymbol{S}(\xi, \eta) = \frac{1}{w(\xi, \eta)} \Bigg[& \frac{\partial^{k+l}}{\partial^k \xi \partial^l \eta} \boldsymbol{A}(\xi, \eta) \\
& - \sum_{i=1}^{k} \binom{k}{i} \frac{\partial^i}{\partial^i \xi} w(\xi, \eta) \frac{\partial^{k-i+l}}{\partial^{k-i} \xi \partial^l \eta} \boldsymbol{S}(\xi, \eta) \\
& - \sum_{j=1}^{l} \binom{l}{j} \frac{\partial^j}{\partial^j \eta} w(\xi, \eta) \frac{\partial^{k+l-j}}{\partial^k \xi \partial^{l-j} \eta} \boldsymbol{S}(\xi, \eta) \\
& - \sum_{i=1}^{k} \binom{k}{i} \sum_{j=1}^{l} \binom{l}{j} \frac{\partial^{i+j}}{\partial^i \xi \partial^j \eta} w(\xi, \eta) \frac{\partial^{k-i+l-j}}{\partial^{k-i} \xi \partial^{l-j} \eta} \boldsymbol{S}(\xi, \eta) \Bigg] .
\end{aligned}
\tag{4.16}
$$

The coefficients related to $\boldsymbol{S}(\xi, \eta)$ on the right-hand side of Eq. (4.16) can be successively evaluated based on the derivatives of the nominator $\boldsymbol{A}(\xi, \eta)$ and denominator $w(\xi, \eta)$ of the rational patch which can be computed for $m = 0, \ldots, k$ and $n = 0, \ldots, l$ by

$$
\frac{\partial^{m+n}}{\partial^m \xi \partial^n \eta} \boldsymbol{A}(\xi, \eta) = \sum_{i=0}^{I-1} \sum_{j=0}^{J-1} \frac{d^m}{d^m \xi} B_{i,p}(\xi) \frac{d^n}{d^n \eta} B_{j,q}(\eta)\ w_{ij}\ \boldsymbol{c}_{ij}, \tag{4.17}
$$

$$
\frac{\partial^{m+n}}{\partial^m \xi \partial^n \eta} w(\xi, \eta) = \sum_{i=0}^{I-1} \sum_{j=0}^{J-1} \frac{d^m}{d^m \xi} B_{i,p}(\xi) \frac{d^n}{d^n \eta} B_{j,q}(\eta)\ w_{ij}. \tag{4.18}
$$

Suppose we want to compute the first derivative in ξ-direction, i.e., $k = 1$ and $l = 0$, the general form (4.16) boils down to

$$
\frac{\partial}{\partial \xi} \boldsymbol{S}(\xi, \eta) = \frac{1}{w(\xi, \eta)} \left(\frac{\partial}{\partial \xi} \boldsymbol{A}(\xi, \eta) - 1 \frac{\partial}{\partial \xi} w(\xi, \eta)\, \boldsymbol{S}(\xi, \eta) \right). \tag{4.19}
$$

For a NURBS basis function (4.14), the same derivative is given by

$$
\frac{\partial R_{ij}^{pq}(\xi, \eta)}{\partial \xi} = \frac{1}{w(\xi, \eta)} \left(\frac{d}{d\xi} B_{i,p}(\xi) B_{j,q}(\eta)\ w_{ij} - \beta B_{i,p}(\xi) B_{j,q}(\eta)\ w_{ij} \right) \tag{4.20}
$$

with

$$
\beta = \frac{\sum_{m=0}^{M-1} \sum_{n=0}^{N-1} \frac{d}{d\xi} B_{m,p}(\xi) B_{n,q}(\eta) w_{mn}}{w(\xi, \eta)}. \tag{4.21}
$$

For further details on computing derivatives of NURBS patches the interested reader is referred to Piegl and Tiller [132, Sect. 4.5].

4.2.2 Constructing Surfaces by Boundary Curves

Spline surfaces may be used to approximate given data (as discussed in Sect. 3.1) or they are constructed by defining the location of their control points. Often, a spline surface shall blend an area defined by its boundary curves. In fact, the most basic surface construction scheme is to connect two curves $\boldsymbol{C}_i$ with $i = \{1, 2\}$ by linear interpolation. The resulting *ruled surfaces* are defined as

$$\begin{aligned} \boldsymbol{S}^r(\xi,\eta) &= (1-\eta)\,\boldsymbol{C}_1(\xi) + \eta\,\boldsymbol{C}_2(\xi) \\ &= (1-\eta)\,\boldsymbol{S}^r(\xi,0) + \eta\,\boldsymbol{S}^r(\xi,1) \end{aligned} \tag{4.22}$$

where $\xi, \eta \in [0, 1]$. If both $\boldsymbol{C}_i$ have the same order and knot vector, it is straightforward to represent $\boldsymbol{S}^r$ as a single tensor product surface. In this case the connection lines on $\boldsymbol{S}^r$ associate points of equal parameter value. Alternatively, the ruling (4.22) could also be performed according to relative arc length. This yields, however, a different geometry which cannot be converted to a NURBS surface [132].

The construction of *Coons patches* is another very common procedure. Thereby, a surface $\boldsymbol{S}^c$ is sought to fit four boundary curves $\boldsymbol{C}_i(\xi)$ and $\boldsymbol{C}_j(\eta)$ with $i = \{1, 2\}$ and $j = \{3, 4\}$. The parameter range is again $\xi, \eta \in [0, 1]$. The curves have to satisfy the following compatibility conditions at the corners of the surface

$$\begin{aligned} \boldsymbol{S}^c(0,0) &= \boldsymbol{C}_1(\xi=0) = \boldsymbol{C}_3(\eta=0) \\ \boldsymbol{S}^c(1,0) &= \boldsymbol{C}_1(\xi=1) = \boldsymbol{C}_4(\eta=0), \\ \boldsymbol{S}^c(0,1) &= \boldsymbol{C}_2(\xi=0) = \boldsymbol{C}_3(\eta=1), \\ \boldsymbol{S}^c(1,1) &= \boldsymbol{C}_2(\xi=1) = \boldsymbol{C}_4(\eta=1). \end{aligned} \tag{4.23}$$

Using a bilinear interpolation, a Coons patch is given by

$$\boldsymbol{S}^c(\xi,\eta) = \boldsymbol{S}^r_\xi(\xi,\eta) + \boldsymbol{S}^r_\eta(\xi,\eta) - \boldsymbol{S}^r_c(\xi,\eta) \tag{4.24}$$

where $\boldsymbol{S}^r_\xi$ and $\boldsymbol{S}^r_\eta$ are ruled surfaces based on $\boldsymbol{C}_i(\xi)$ and $\boldsymbol{C}_j(\eta)$, respectively, and $\boldsymbol{S}^r_c$ is the bilinear interpolant to the four corner points

$$\boldsymbol{S}^r_c(\xi,\eta) = \begin{bmatrix} 1 \\ \xi \end{bmatrix}^{\mathrm{T}} \begin{bmatrix} \boldsymbol{S}^c(0,0) & \boldsymbol{S}^c(0,1) \\ \boldsymbol{S}^c(1,0) & \boldsymbol{S}^c(1,1) \end{bmatrix} \begin{bmatrix} 1 \\ \eta \end{bmatrix}. \tag{4.25}$$

Figure 4.6 displays these different components of a Coons patch. Equation (4.24) can be generalised by using two arbitrary smooth interpolation functions $f_0(s)$ and $f_1(s)$ fulfilling

$$f_k(\ell) = \delta_{k\ell}, \;\; k,\ell = 0,1 \;\; \text{and} \;\; f_0(s) + f_1(s) = 1, \;\; s \in [0,1], \;\; s = \xi, \eta. \tag{4.26}$$

The corresponding Coons patch can be expressed in matrix form [139] as

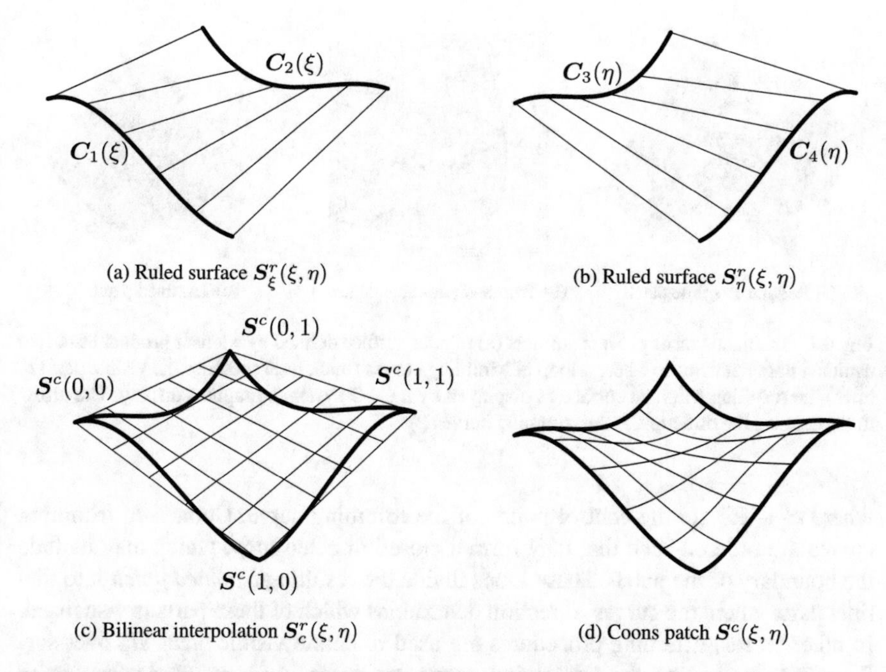

(a) Ruled surface $S^r_\xi(\xi,\eta)$

(b) Ruled surface $S^r_\eta(\xi,\eta)$

(c) Bilinear interpolation $S^r_c(\xi,\eta)$

(d) Coons patch $S^c(\xi,\eta)$

Fig. 4.6 Components of a bilinear Coons patch defined by the boundary curves $C_i(\xi)$ and $C_j(\eta)$ highlighted by thick lines

$$S^c(\xi,\eta) = -\begin{bmatrix} -1 \\ f_0(\xi) \\ f_1(\xi) \end{bmatrix}^{\mathrm{T}} \begin{bmatrix} \mathbf{0} & S^c(\xi,0) & S^c(\xi,1) \\ S^c(0,\eta) & S^c(0,0) & S^c(0,1) \\ S^c(1,\eta) & S^c(1,0) & S^c(1,1) \end{bmatrix} \begin{bmatrix} -1 \\ f_0(\eta) \\ f_1(\eta) \end{bmatrix} \tag{4.27}$$

with $\mathbf{0} \in \mathbb{R}^d$ denoting the zero vector. Various functions may be used to specify f_k such as Hermite polynomials or trigonometric functions. In the case of Bernstein polynomials, the surfaces S^r_ξ, S^r_η, and S^r_c are in Bézier or B-spline form and the resulting Coons patch can be represented as a single NURBS surface.

4.2.3 Trimmed Surfaces

In order to employ tensor product surfaces for arbitrary shapes (e.g., ones that are not defined by four boundaries) trimming procedures may be used. For this purpose, curves are defined within the parameter space of a surface $S(\xi,\eta)$. These *trimming curves* $C^t(\zeta)$ are usually B-spline or NURBS curves. They are given by

$$C^t(\zeta) = \begin{bmatrix} \xi(\zeta) \\ \eta(\zeta) \end{bmatrix} = \sum_{i=0}^{I-1} R_{i,p}(\zeta)\, c^t_i \tag{4.28}$$

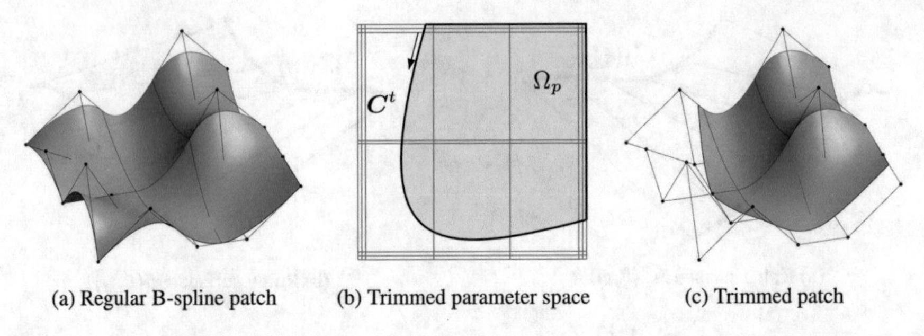

Fig. 4.7 Trimmed tensor product surface: (a) regular surface defined by a tensor product basis, (b) trimmed parameter space where a loop of trimming curves (thick line) specifies the visible part Ω_p of (c) the resulting trimmed surface as displayed by a CAD system's graphics display. The arrow in (b) denotes the direction of the trimming curves

where $\boldsymbol{c}_i^t \in \mathbb{R}^2$ are the control points of the trimming curve. Connected trimming curves are ordered such that they form a closed directed *loop*, which may include the boundary of the patch. These loops divide the resulting *trimmed patch* into distinct parts where the curves' direction determines which of these parts is visualised. In other words, trimming procedures are used to define visible areas Ω_p over surfaces independent of the underlying parameter space. As a result, surfaces with non-rectangular topologies can be represented in a very simple way. An example of a trimmed patch is shown in Fig. 4.7. It is emphasised that the mathematical description of the original patch, i.e., the tensor product basis and the related control grid, does *not* change and is *never* updated to reflect the trimmed boundary represented by the independent trimming curves. Trimmed surfaces should be considered as an "engineering" extension of tensor product patches [59]. On the one hand, they permit a pragmatic way to define arbitrary surface topologies and provide a means for visually displaying them in graphics systems. On the other hand, they do not offer a canonical solution to related problems such as a smooth connection of two adjacent patches along a trimming curve, although the graphics system leads the user to believe so. In fact, enormous effort has been and is still devoted to resolve the shortcomings of trimming procedures, see e.g., [114].

Often, trimmed surfaces result from surface-to-surface intersection (SSI) operations [85, 128]. Most issues related to models defined by trimmed surfaces originate from the problem of resolving the high complexity of the exact solution of intersections between arbitrary freeform surfaces. For instance, the intersection of two bi-cubic surfaces has an order of 324 [163]. In fact, the SSI problem has been shown to be mathematically intractable [161]. Hence, a large number of related approximation schemes has been developed, which try to find a good balance between accuracy, efficiency, and robustness. Figure 4.8 shows some inaccuracies of a trimmed model defined by a torus intersected by a plane. Note that the discrepancy between the computed intersection $\boldsymbol{C}_{Trimmed}$ and the related exact solution $\boldsymbol{C}_{Torus}$ is scarcely visible. The imperfections of trimmed geometries are usually very well hidden from the user, but they surface as soon as a design model is applied to downstream applications.

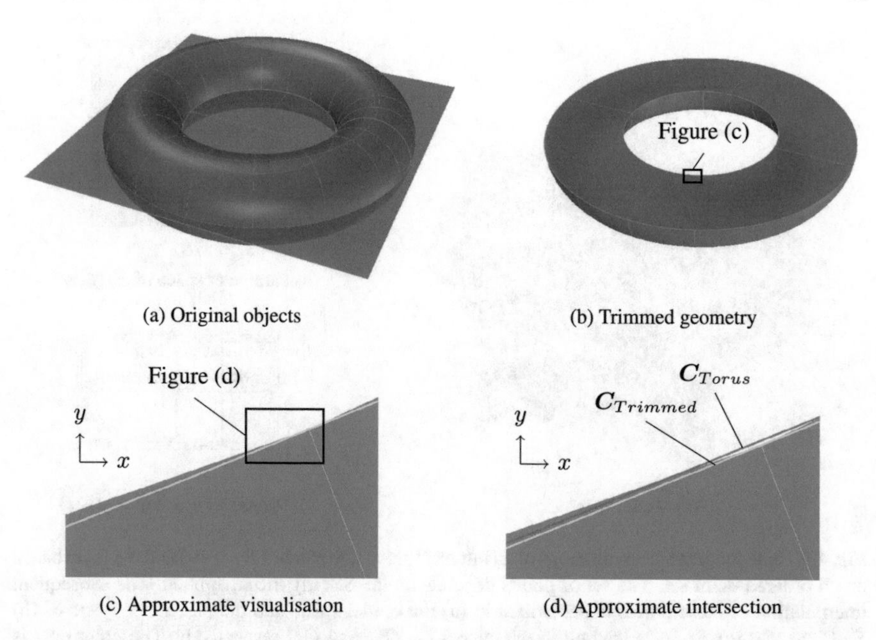

Fig. 4.8 Model of a half of a torus: (a) initial non-trimmed torus and a surface defined in the xy-plane, (b) the resulting trimmed object, (c) a close-up showing the deviation from the visualisation mesh (blue background) and the computed intersection curve $C_{Trimmed}$ (yellow), and (d) a close-up illustrating the difference of the original inner circle of the torus C_{Torus} (red) to the SSI result. The images (c) and (d) are captured in top view

To use the words of Piegl [133]: "While one can cheat the eye in computer graphics and animation, the milling machine is not as forgiving". Without going into more detail on SSI operations, we want to outline the usual representation of their result.

In general, three *distinct* representations of an intersection are obtained, as illustrated in Fig. 4.9. On the one hand, the intersection curve in the model space is computed. This may seem to be the main objective of SSI at first glance, yet it is just a part of the overall solution process. The intersection curve has to be represented in each parameter space by trimming curves to define the visible area of the patch as described above. Intersection and trimming curves can be defined by any kind of representation, but usually, low-order B-splines are used. They are constructed based on a set of sampling points that result from the applied SSI algorithm [120]. Subsequently, an interpolation scheme or another curve-fitting technique is used to generate a continuous approximation of the intersection in model space $\hat{\boldsymbol{C}}$. This curve does not lie on either of the intersecting surfaces. A trimming curve $\boldsymbol{C}^t$ is obtained based on the sampling points given in the corresponding parameter space [141]. The related curve $\hat{\boldsymbol{C}}^t$ in the model space is obtained by evaluating the equation of the surface $\boldsymbol{S}$ along $\boldsymbol{C}^t$. Alternatively, $\hat{\boldsymbol{C}}^t$ may be represented explicitly. Such an expansion of $\boldsymbol{S} \circ \boldsymbol{C}^t$ into an explicit representation $\hat{\boldsymbol{C}}^t$ may be used to join another patch to a trimmed surface, but there is no computational benefit [107]. In general,

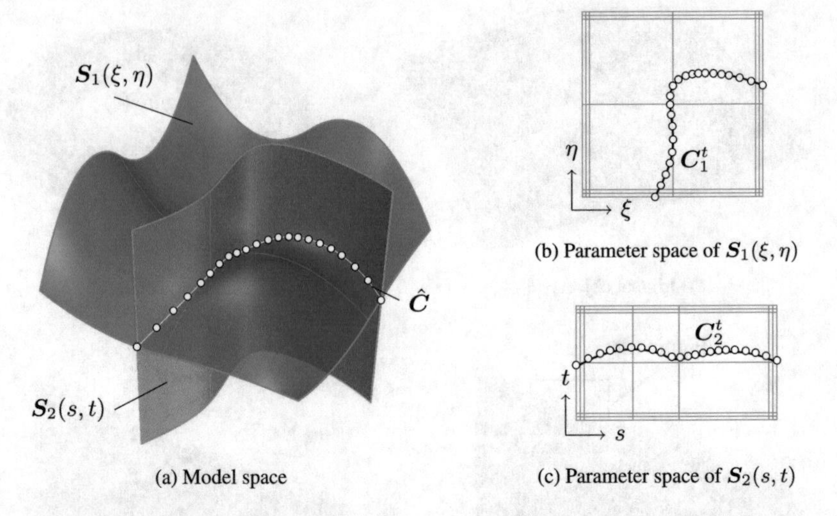

(a) Model space

(b) Parameter space of $S_1(\xi, \eta)$

(c) Parameter space of $S_2(s,t)$

Fig. 4.9 Independent approximations of an intersection of two patches $S_1(\xi, \eta)$ and $S_2(s,t)$ based on an ordered point set. The set of points depends on the SSI algorithm applied. The subsequent interpolation of these points is performed in (a) the model space and the parameter space of (b) $S_1(\xi, \eta)$ and (c) $S_2(s,t)$ leading to the curves $\hat{C}$, C_1^t, and C_2^t, respectively. The point data is usually discarded once the curves are constructed

curves on surfaces have an order of $n(p+q)$ with n denoting the order of the trimming curve and p and q correspond to the orders of the trimmed surface [59]. Renner and Weiß [141] compared exact and approximate representations of $\hat{C}^t$ and concluded that avoiding high orders is the main reason for preferring an approximation scheme. Regardless of its representation, $\hat{C}^t$ does *not* coincide with the intersection curve in model space $\hat{C}$. In addition, all procedures related to trimming curves are performed for each patch separately. Hence, the images of these curves $\hat{C}_i^t$ do not coincide, neither with each other, nor with $\hat{C}$. As a consequence, gaps and overlaps occur between intersecting patches. Furthermore, there is *no connection* between these three representations of the intersection; although the sample points provide some information during the construction, this data is only stored temporarily during the SSI procedure and never retained in memory for further use.

Currently, the most common geometric modelling kernels are `ACIS`, `C3D`, and `Parasolid`. They provide software components for the representation and manipulation of objects, and form the geometric core of many CAD applications. All of them use splines for the description of trimming curves [40, 46, 166]. Yet, the representation of the intersection curve in model space varies: `ACIS` defines it by a three-dimensional B-spline curve, `Parasolid` uses a set of sorted intersection points that can be interpreted as a linear approximation, and in `C3D` the intersection curve is not stored at all. In `C3D`, trimming curves are computed such that they have the same radius and derivatives at the same parametric values. However, this is only satisfied at the intersection point used for the construction, for details see [74].

4.3 Solid Models

There are different concepts to represent solid models; an overview is provided in [142] and a historical summary can be found in [143]. The majority of commercial CAD systems use boundary representations (B-rep) where a solid is implicitly defined by its boundary. Usually, this involves a collection of several surfaces. A single regular NURBS patch may be closed to represent a cylinder or a torus. Spherical objects may be represented as well, if degenerated edges (i.e., edges that collapse to a point) are introduced. Yet, more complicated objects such as a double torus require a partition into *multiple* patches. The connection of two adjacent surfaces is complicated, especially if a certain continuity is desired. In fact, there is no known method to ensure exact continuity across the intersection of two trimmed surfaces [59]. In general, a *non-conforming* parameterisation along surface boundaries is to be expected.

Besides the geometric representation of boundary patches, B-reps are characterised by explicitly defining the *topology* of an object. This addresses the connectivity of the various components, and the corresponding entities are termed

- *vertices* relating to points,
- *edges* relating to curves,
- *faces* relating to surfaces.

A solid is then defined by a face-edge-vertex graph, a concept that goes back to the seminal work of Baumgart [7]. It is emphasised that the descriptions of an object's shape, i.e., geometry, and its structure, i.e., topology, are separated [175]. By definition, a B-rep model always consists of a data structure of both topological and geometric objects. Regarding isogeometric analysis, the parameterisation of the geometry is a further important issue that should be taken into account. Figure 4.10 summarises these various perspectives of a model representing a simple solid. The corresponding object consists of three trimmed surfaces and a regular one. It is apparent that even simple models rely on multiple surfaces with non-conforming parameterisation along their intersections.

Finally, it should be mentioned that B-reps are also used to describe dimensionally reduced objects, i.e., shell structures. In this case, the boundary patches specify the object itself rather than its outer skin. In CAGD, the terms *surface* model and *solid* model do not refer to the dimension of an object; they rather indicate if a model contains topology information (solid model) or not (surface model).

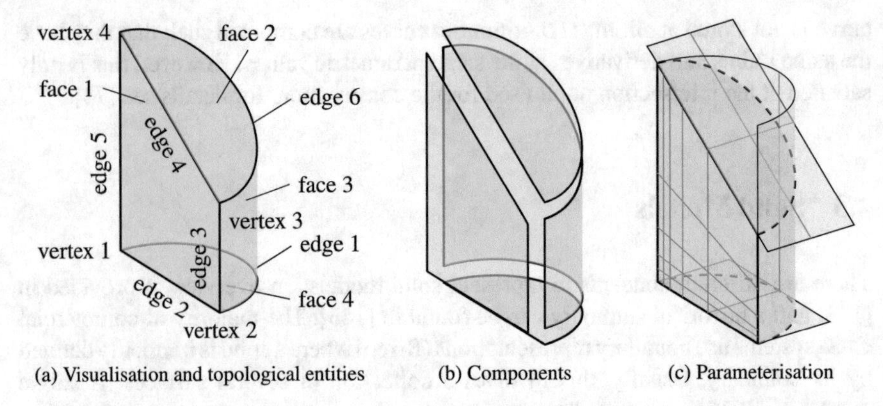

Fig. 4.10 Different perspectives of a solid model: (a) visible part of the object and its topological entities (Strictly speaking the geometric objects, i.e., points, curves, and surfaces, that are related to the topological entities are displayed), (b) the geometric segments of the B-rep and (c) the underlying mathematical parameterisation of each surface. In (c), dashed lines mark the boundary of the visible area and grey lines indicate the underlying tensor product basis. Note that the parameterisation along common edges does not match

4.4 NURBS in the Context of Approximation

NURBS are the de facto standard in CAGD for defining geometric objects due to their rich representational power, including conic sections and their generalisation, i.e., quadrics. In the following, we take a look at NURBS geometries from an analysis perspective and emphasise certain advantages and disadvantages.

Representation of a solid. NURBS solids are B-reps defining the geometry and topology of the object's boundary. In analysis, on the other hand, the notion of a "solid" is often associated with a volume representation. It is worth noting that there are attempts in CAGD to introduce volumetric spline representations, see e.g., [24, 111, 118], but they are not common in practice yet. Deriving volumetric analysis models from B-reps provides an active research area in isogeometric analysis, see e.g., [194, 197], which is quite challenging. The BEM provides an elegant and straightforward solution to this discrepancy between the representation of solids and thus, it is a perfect companion for the isogeometric paradigm.

Trimmed basis functions. In conventional meshes, it is usually not necessary to deal with trimmed basis functions, since the entire element is part of the domain of interest. However, there are exceptions such as immersed or fictitious finite element methods [156]. These schemes require special techniques to deal with elements that are cut by the boundary of the domain. The same holds true for trimmed NURBS patches as we will discuss in Chap. 10.

Mesh refinement. An analysis model usually requires a finer representation than a CAD model, since it has to represent the geometry *and* the variation of physical quantities on the geometry. Nevertheless, conventional meshes for simulations only

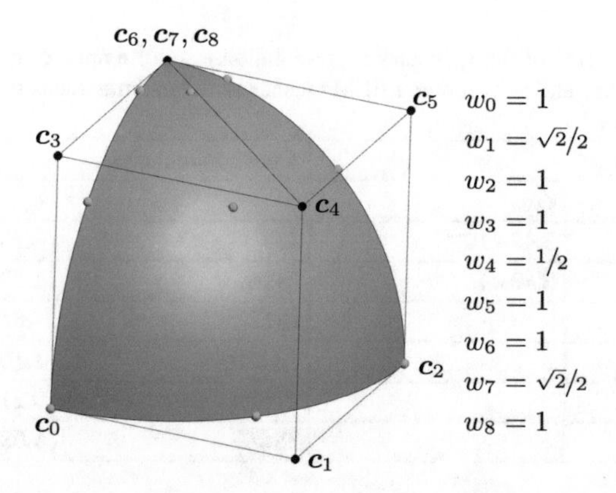

Fig. 4.11 NURBS representation of a quarter of a hemisphere. The weights of the control points are determined based on the rule given in Fig. 4.2: the angle $\alpha = 90°$ for both parametric directions and the bivariate weights are computed by the tensor product of the univariate ones (i.e., $w_0 = 1, w_1 = \sqrt{2}/2, w_2 = 1$) . The orange points indicate the abscissae of the basis functions

approximate the object considered, and several refinement steps may be required to reduce the geometric error to an acceptable level. This is not the case when NURBS models are used. In fact, already the initial model, extracted from the CAD system, is the best representation of the design intent available; refinement by knot insertion or order elevation does not alter the geometry but enriches its underlying basis. This is indeed a great benefit since it elegantly eliminates an error source of the simulation.

We demonstrate this advantage by the representation of a sphere using (a) conventional BEM meshes and (b) NURBS models. Both representations use quadratic basis function. The simplest NURBS representation of a sphere is derived from a quarter circle that is rotated around one of its endpoints. This construction is illustrated in Fig. 4.11 where the curve defined by the control points c_0, c_3, and c_6 is rotated around c_6. The related weights are determined by the tensor product of the weights of a circular arc (see Fig. 4.2) with an opening angle of $\alpha = 90°$, which are defined in both parametric directions. Note that the resulting quarter of a hemisphere has a degenerated edge at the top. We can still apply this representation to interpolation and collocation problems when the abscissae of the basis functions along the degenerated edge are shifted into the domain. In Fig. 4.11 this is indicated by the orange points.

For this example, we exported the sphere model from the CAGD software Rhinoceros, which uses the representation with 8 quarter hemispheres with a degenerated edge. The precision of the exported data was set to $\epsilon_e = 10^{-8}$. The relative error of the geometry representation to an analytic sphere ϵ_{geo} is given in Table 4.1. Further, three discretisations are illustrated in Fig. 4.12. The error ϵ_{geo} demonstrates clearly the superiority of the isogeometric concept for accurate geometry representations. The unrefined NURBS patch provides already a precise geometric model, while a large number of conventional Lagrange elements is required for an adequate approximation. This is particularly important for simulation processes which employ

Table 4.1 Relative error of the geometry representation ϵ_{geo} of a sphere measured in L_2-norm due to an isogeometric and conventional BEM meshes with quadratic basis functions and various degrees of freedom n

Isogeometric mesh		Conventional mesh	
n	ϵ_{geo}	n	ϵ_{geo}
216	3.42×10^{-8}	483	4.76×10^{-1}
288	3.42×10^{-8}	2436	1.53×10^{-2}
		9312	1.81×10^{-3}
		12 120	1.10×10^{-3}
		21 858	3.21×10^{-4}
		38 886	1.08×10^{-4}

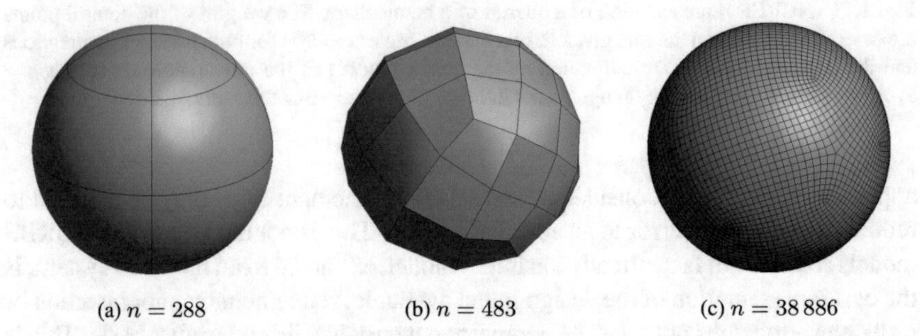

(a) $n = 288$ (b) $n = 483$ (c) $n = 38\,886$

Fig. 4.12 Various discretisations of a boundary of a sphere with different degrees of freedom n: (a) finest NURBS model, (b) coarsest, and (c) finest conventional BEM mesh listed in Table 4.1

adaptive refinement during the analysis. Note that ϵ_{geo} correlates to the accuracy of the input data ϵ_e in the isogeometric case, hence it can be controlled by the user.

General approximation quality. The previous example highlights the power of NURBS and raises the question if these rational basis functions are in general better than their polynomial counterparts. In order to assess this assumption, we repeat the first example of Sect. 3.4.4, where the approximation of a smooth function $f(\xi)$ is considered and the quality of various refinement schemes is compared by the L_2 approximation error (3.25). In particular, we compare the performance of knot insertion and k-refinement using B-splines and NURBS basis functions. For the latter, the weights are determined by setting the initial weights of a quadratic NURBS basis with a single element to 0.4, 1, and 1.6; these weights are updated during refinement based on the rules presented in Sect. 4.1.3. The approximation errors of all discretisations are summarised in Fig. 4.13.

It is apparent that the B-spline and NURBS approximations behave similar. If the best values for the weights are not known, using B-splines is perhaps the better choice, since their evaluation is more efficient. Still, conic sections and their three-

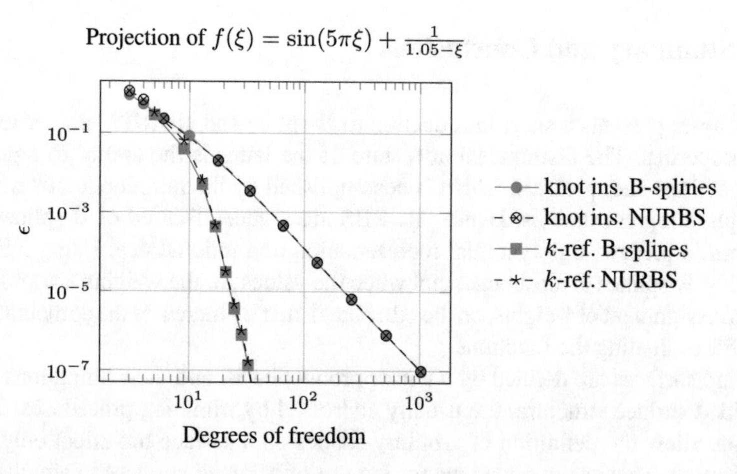

Fig. 4.13 Approximation of a smooth function with B-splines and NURBS

dimensional counterparts, e.g., spheres and cylinders, are very common design elements for which the proper weights are well-known, and in such situations NURBS surpass the approximation quality of other basis functions.

4.5 Unified Notation for the Remainder of This Book

In the subsequent chapters it is not necessary to distinguish between curves and spline surfaces or B-spline and NURBS objects, in most situations. Hence, we introduce a generalised notation. First, the term "patch" is used for spline curves and surfaces based on B-splines and NURBS. A point $\boldsymbol{x}$ on a patch is given by

$$\boldsymbol{x}(\cdot) = \sum_{i=1}^{I} R_i(\cdot)\, \boldsymbol{x}_i, \tag{4.29}$$

with $(\cdot)$ being a placeholder for the local coordinate(s) of the object, R_i are B-spline or NURBS basis functions, and $\boldsymbol{x}_i$ refer to the coordinates of the corresponding control points. The index i denotes the *global* index within the patch sorted first by the ξ-direction and then by the η-direction, if applicable. I is the total number of basis functions within the basis.

4.6 Summary and Conclusions

This chapter presents a short introduction to B-spline and NURBS geometries and their properties. The distinguishing feature of the latter is the ability to represent conic sections and quadrics, which is accomplished by the introduction of weights to B-spline representations. Hence, NURBS are a generalisation of B-splines that transform a piecewise polynomial representation to a rational one. Using NURBS instead of B-splines is most beneficial when the values for the weights are known; a random assignment of weights, on the other hand, merely increases the computational effort for evaluating the functions.

Spline surfaces are defined by a tensor product basis, and their limitations (e.g., four-sided surface structure) are usually addressed by trimming procedures. These concepts allow the definition of arbitrary shapes on a surface but affect only their visualisation. This has implications for the use of trimmed surfaces to simulations, which we will address in Chap. 10.

Often, a collection of trimmed surfaces defines a CAD model. The model's B-rep data structure stores the connectivity of these surfaces and the topology of the geometry. The main differences to conventional meshes are that (i) refinement does not alter the definition of a geometry model but merely enriches its underlying basis, and that (ii) there is usually a non-conforming parameterisation between adjacent patches. The latter is a problem in simulations, but it can be readily addressed in the context of BEM as we will show in Chap. 6. The former is a huge benefit that we will utilise to derive efficient BEM discretisations and refinement strategies.

Chapter 5
Getting Geometry Information from CAD Programs

The long-term goal of isogeometric analysis is to integrate design and analysis. Hence, it is worth to take a brief look at how information can be extracted from CAD systems. The following discussion is by no means complete, but we aim to highlight some aspects of this topic. The chapter starts with general considerations for exchanging data between different computer software systems. Next, we provide a brief introduction to the Standard for the Exchange of Product Model Data (STEP). Finally, the idea of Bézier extraction is presented.

5.1 General Considerations

In CAD systems, parameters and constraints govern the design of a model, rather than the definition of specific control points. Further essential components are local features and the construction history. All these various factors form the *design intent* [96]. Each software has its native data structure to keep track of the geometry, the topology, and the design intent of its models. Thus, the exchange of information between systems with different native structures requires a translation process. The conversion of formats may seem like an easy task, but it is indeed very complicated. Usually, there is no direct mapping from one data structure to another; especially for information related to the design intent since there are no canonical guidelines for its representation. Consequently, the exchange of the complete data of a CAD model between different systems is hardly possible. When the systems have different purposes (e.g., design and analysis), the situation gets even more delicate. In the end, most exchange procedures transfer only the geometric data of the final CAD model.

This interoperability issue was investigated in a study focusing on the U.S. automotive supply chain [176]. The solutions considered were: (i) standardisation on a single system, (ii) point-to-point translation, and (iii) neutral format translation.

G. Beer et al., *The Isogeometric Boundary Element Method*, Lecture Notes in Applied and Computational Mechanics 90,
https://doi.org/10.1007/978-3-030-23339-6_5

In the case of a single system standardisation, the same native format is used for all processes, e.g., design and analysis. The main advantage is that the compatibility of the model data is assured since no translation is required. However, this approach implies the restriction to a single system. Consequently, every part depends on the dominant application of the software. Most importantly, translation problems can arise even within one system, for instance, due to different software versions. In fact, this is a common problem – imagine the result when you open a 10-year-old PowerPoint presentation.

The basic idea of point-to-point translation is to convert a native format of one system directly to a native format of another one. This concept works reasonably well for unambiguous data exchange tasks. However, it is like translating a text from one language – let's say Chinese – to another one – like Spanish; often the proper interpretation of a given piece of information in the other format is not clear. Furthermore, a high degree of vendor cooperation is necessary to develop a direct translator. Similar to the previous strategy, direct translators have to be rewritten for each new system or perhaps even for new versions of the same software.

Neutral format translation is based on a standardised neutral format for the exchange of (geometric) data. This approach enables independent development of various tools working on the same model. The minimisation of dependencies simplifies the maintenance of each software and eventually leads to robust implementations since a clean code is designed to do one thing well, as noted by the inventor of C++ Bjarne Stroustrup [110]. Furthermore, vendors are more willing to develop translators for neutral formats since it does not require the disclosure of proprietary code. This is beneficial since interpretation errors of a native format are most likely minimised when the vendors provide the conversion themselves. An additional advantage of neutral formats is that they are ideally suited for long term storage of data. However, there are a number of weaknesses. Currently, it is not possible to capture the design intent and thus, translation to a neutral format provides only a snapshot of the current geometric model. The amount of translations is higher than in the other data exchange concepts, and every translation can lead to a loss of information in general. Hence, the quality of an exchanged model depends on the capability of the neutral format used.

In the context of isogeometric analysis, the minimal requirement is the accurate exchange of geometrical information of the final model. Topology is also essential in order to assess the connectivity between patches. The reconstruction of topological data based on edge comparison or related strategies is very cumbersome and extremely error-prone, especially in cases of trimmed geometries where edges only coincide within a certain tolerance as discussed in Sect. 4.2.3. The following two approaches are suggested: (i) direct extraction of topological data from CAD software by a point-to-point translation or (ii) using a neutral format that is able to cope with topological data. The former may be preferred if there is a cooperation with a CAD vendor and the developments focus on the specific product. However, neutral exchange formats will be discussed in the following because they are the most general and independent approaches.

5.2 Neutral Format Translators

Concepts for a neutral data exchange format emerged in the 1970s. These attempts were borne by a variety of partners from industry, academia, and government [73]. Based on the initiative of the CAD user community, in particular, General Electric and Boeing, vendors agreed to create an American national standard for CAD data exchange. The final result was the first version of the Initial Graphics Exchange Specification (IGES) [122]. IGES continues to be deployed in industry, but it has several shortcomings. Most importantly, IGES is a *stagnant* exchange format and the last official version of IGES, i.e., version 5.3 [93], was published more than 20 years ago in 1996. The main legacy of IGES is the disclosure of several weaknesses of the neutral exchange concept, thereby enhancing new emerging standards. The most notable one is the Standard for the Exchange of Product Model Data (STEP), which provides a broader, more robust standard for the exchange of data [73]. However, the broad scope of STEP may seem overwhelming, and hence, the following subsection aims to provide a short overview of this specific neutral exchange format.

5.2.1 STEP

Since 1984 the International Organisation for Standardisation (ISO) has been working on a standard for the exchange of *product data* and its first parts were published in 1994 [135]. The objective of this development effort – one of the largest ever undertaken by ISO – is the complete and unambiguous definition of a product throughout its entire life cycle, that is independent of any computer system [154]. Hence, the corresponding standard includes the exchange of CAD data, yet its scope is much broader. According to Tassey et al. [176], STEP is superior to other translators because it

- addresses many types of data,
- incorporates a superset of elements common to all systems,
- supports special application needs, and
- provides for international exchanges.

In their paper, several studies are discussed in which STEP excels with respect to the quality of exchanging data of industrial examples. In addition, this standard is constantly developed and improved, e.g., by its enhancements for isogeometric analysis [170].

STEP is the informal term for the standard officially denoted as ISO 10303. It is organised by an accumulation of various *parts* unified by a set of fundamental principles. These parts are referred to as ISO 10303-xxx, where xxx specifies the part number. Each of them is separately published and has to pass several development phases. Each part is associated with one of the following *series*: (i) description methods, (ii) implementation specifications, (iii) conformance testing, (iv) generic

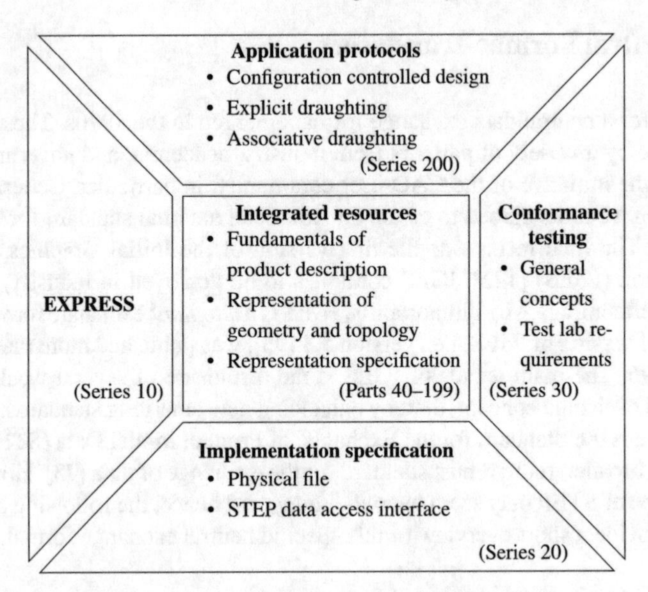

Fig. 5.1 Structure of STEP (re-execution of the original diagram [102])

integrated resources, (v) application integrated resources, (vi) application protocols. Figure 5.1 gives an overview of these various components of STEP.

The description is provided by the common specification language EXPRESS (Series 10) specifying data types, entities, rules, functions, and so on [165]. It is not a programming language but has an object-oriented flavour. The implementation specifications (Series 20) defines the transfer of data. The exchange by a neutral ASCII file is addressed in Part 21, "clear text encoding the exchange structure". This STEP-file transfer is the most widely used data exchange form of STEP [154]. However, other approaches, like shared memory access, are covered by the series as well; see Part 22, "standard data access interface". Conformance tests provide the verification requirements (Series 30).

The most fundamental components of STEP are the integrated resources. They contain generic data such as geometric information and display attributes (Series 40) as well as further elementary units that are specialised for certain application areas (Series 100). For example, Part 42, "geometric and topological representations", focuses on the definition of geometric models in general, while Part 104, "finite element analysis", is devoted to applications in the context of FEM. These parts provide the entities needed to build application protocols denoted as APs (Series 200). They are the link to the needs of industry and other users. Their purpose is to interpret the STEP data in the context of a specific application which may be part of one or more stages of a life cycle of a particular product. Part 209, "multidisciplinary analysis and design", addresses engineering analysis. Each STEP application protocol is further subdivided into a set of conformation classes (CCs). These subsets must be completely implemented if a translator claims to be conformable with the standard

[136]. Hence, it is important to know what conformance classes are supported by a software system. This modular structure, with several APs and their CCs, may seem complex and daunting, but it gives users the necessary transparency of what can be expected of the data exchanged. Moreover, the complexity of the overall concept of STEP does not imply that it is difficult to use. The following paragraphs provide a glimpse of how STEP-files specify surfaces and their connectivity.

File structure. STEP files are easy to read since the language used is based on an English-like syntax [165]. In general, an accumulation of entities pointing to each other shapes the structure of the exchange data is used. Lines specifying entities begin with the symbol #, followed by the unique identifier of the corresponding object. This identifier is used to connect various entities with each other as shown in Fig. 5.2(a). In addition, an entity may consist of integers, real numbers, Booleans (.F./.T.), and enumeration flags (e.g., .UNSPECIFIED.).

Surfaces representation. STEP provides the fundamental information of B-spline patches, i.e., orders, knot vectors, and control points, together with auxiliary information. Thereby, the distinct components are grouped in the surface STEP entity using brackets. Figure 5.2(b) shows a representation of a regular square patch. Note

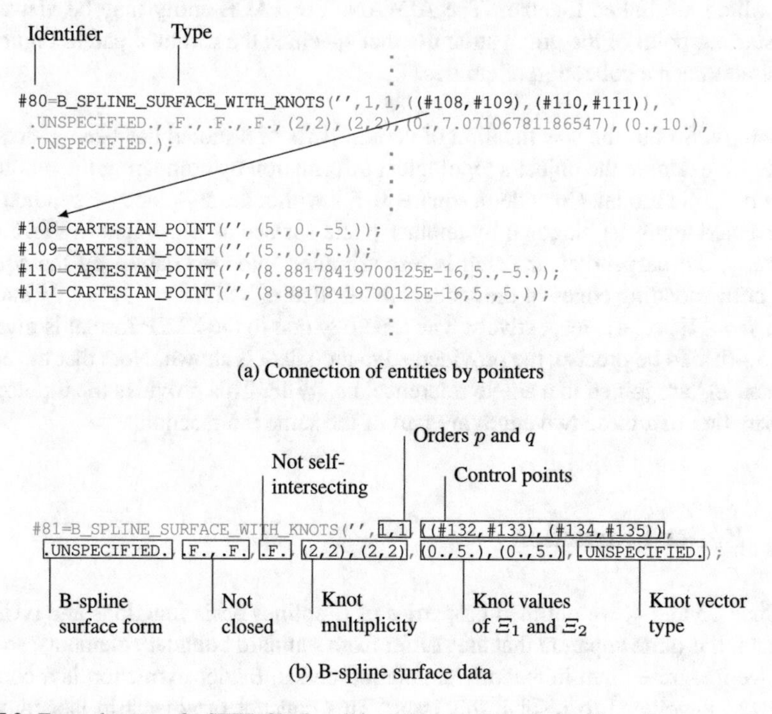

Fig. 5.2 Example parts of a STEP-file describing a planar square defined by a linear non-trimmed B-spline surface

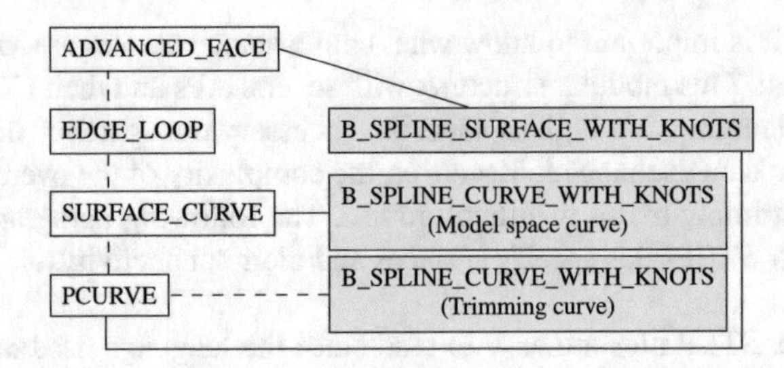

Fig. 5.3 Graph representation of a trimmed surface. Entities that provide geometrical information are highlighted in grey. For the sake of clarity, intermediate nodes may be skipped which is indicated by dashed lines

that knot vectors are specified by knot values with their multiplicity and that coordinates of control points are stored within its own entity. Hence, the interpretation of the knot vector does not involve error-prone comparisons of floating point numbers. Regarding trimmed surfaces, the following information is provided: (i) the surface, (ii) the related loops of trimming curves, and (iii) their counterparts in model space. The trimmed surface data is not coalesced in a single object, but it is embedded in a graph structure. Hence, the information is represented by various different entities which are linked together. The ADVANCED_FACE entity may be viewed as the starting point of the graph structure that specifies the trimmed patch. Figure 5.3 illustrates such a collection of entities.

Topology. So far, the specification of certain parts of a model has been addressed. Here, we examine the object's topological information by comparing the output for a simple solid model. Consider a square $[0,5]^2$ within the xy-plane perpendicularly intersected along its diagonal by another plane surface as illustrated in Fig. 5.4(a). Thereby, the perpendicular patch is also trimmed into two halves by the square. The corresponding edges of the model are labelled $e_i^{\triangle}$ with $i = \{1, \ldots, 3\}$ and $e_j^{\square}$ with $j = \{1, \ldots, 4\}$, respectively. The topology due to the STEP format is given in Fig. 5.4(b). To be precise, the provided edge loop data is shown. Note that the edges $e_1^{\square}$ and $e_3^{\triangle}$ are joined in a single reference, i.e., $\#48$. This provides the topological information that these two edges are part of the same intersection.

5.3 Bézier Extraction

In Chaps. 3 and 4, we outlined properties of B-splines basis functions and NURBS objects. It is quite apparent that they differ from standard boundary elements, so how can we integrate them in existing simulation codes. Bézier extraction is a concept that was developed to resolve this issue. This concept originates in isogeometric analysis [30] and provides a means to represent the elements of a patch such that they can be treated as conventional (Lagrange) elements. The basic idea is to express

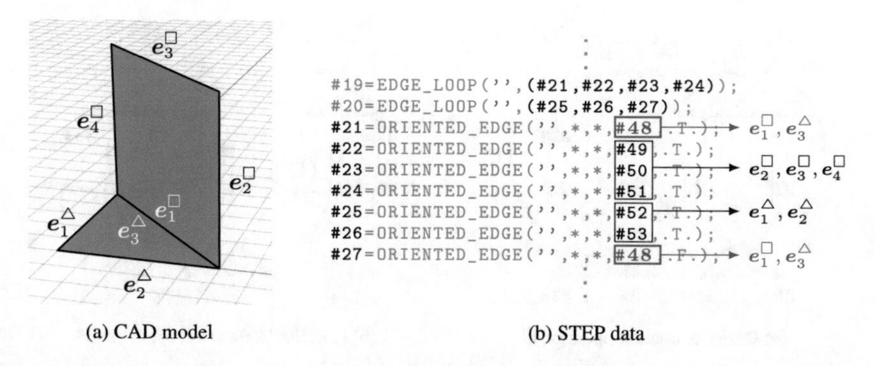

Fig. 5.4 An example of (a) a simple solid model and (b) the representation of its topology in STEP

the original spline object by C^0-continuous Bézier elements – one for each non-zero knot span. As a result, each element is defined by a Bernstein basis and the only difference between elements is their polynomial order. Thus, isogeometric elements can be implements like conventional element-types where the main changes only affect the basis function subroutine. The compelling feature of Bézier extraction is that the number of degrees of freedom as well as the smoothness of the original basis is not affected by the established element-type representation.

The technical foundation of Bézier extraction has already been introduced in Sect. 3.6.3.2 where we presented the subdivision matrix (3.34) as a concise representation of the knot insertion coefficients. To be precise, in this former section, the subdivision matrix encodes the relation between two different scales of a hierarchical B-spline basis. However, we can use the same strategy to link a smooth spline basis to its C^0-continuous complement. The main difference is that the inserted knots do not add new elements but increase the multiplicity of existing knots as illustrated in Fig. 5.5(a). Following Eq. (3.34), the corresponding subdivision matrix is given by

$$\mathbf{S} = \begin{pmatrix} 1 & & & & \\ & 1 & & & \\ & \tfrac{1}{2} & \tfrac{1}{2} & & \\ & & 1 & & \\ & & \tfrac{1}{2} & \tfrac{1}{2} & \\ & & & 1 & \\ & & & & 1 \end{pmatrix} \tag{5.1}$$

and links, for instance, the control points $\boldsymbol{x}_j^b$ of the Bézier elements to the control points of the original spline

$$\{\boldsymbol{x}_j^b\} = \mathbf{S}\,\{\boldsymbol{x}_i\} \qquad \text{where} \qquad \mathbf{S} \in \mathbb{R}^{J\times I}, \{\boldsymbol{x}_j^b\} \in \mathbb{R}^{J\times d}, \{\boldsymbol{x}_i\} \in \mathbb{R}^{I\times d}. \tag{5.2}$$

The Bézier extraction operator, on the other hand, describes the mapping from the Bernstein basis functions to the original B-splines, i.e.,

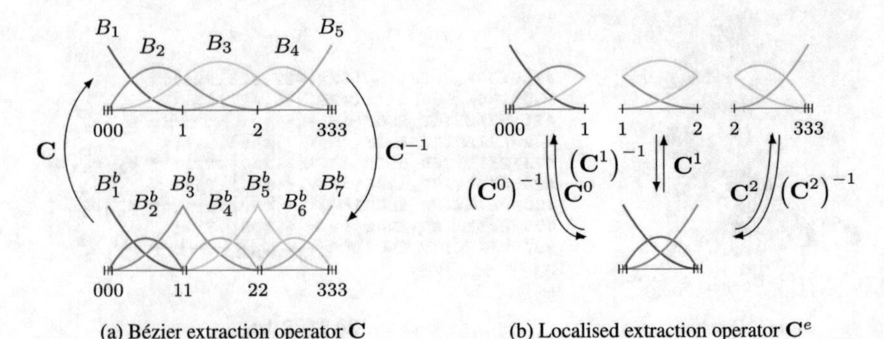

(a) Bézier extraction operator **C** (b) Localised extraction operator $\mathbf{C}^e$

Fig. 5.5 Illustration of the Bézier extraction operator given in Eq. (5.4): (a) global relation of the overall basis and (b) localised version for each individual element

$$\{B_i(\xi)\} = \mathbf{C}\{B_j^b(\xi)\} \tag{5.3}$$

with the operator $\mathbf{C}$ given by

$$\mathbf{C} = \mathbf{S}^{\mathrm{T}} = \begin{pmatrix} 1 & & & & & & \\ & 1 & \tfrac{1}{2} & & & & \\ & & \tfrac{1}{2} & 1 & \tfrac{1}{2} & & \\ & & & & \tfrac{1}{2} & 1 & \\ & & & & & & 1 \end{pmatrix}. \tag{5.4}$$

The highlighted matrix blocks are the localised operators $\mathbf{C}^e$. They define the relation between the basis functions for each element as illustrated in Fig. 5.5(b).

The operators can also be applied to NURBS. Therefore, we define the homogeneous control points of the Bézier elements by

$$\{\boldsymbol{x}_j^{bh}\} = \mathbf{S}\{\boldsymbol{x}_i^h\} \qquad \text{with} \qquad \boldsymbol{x}_j^{bh} = (w_j^b \boldsymbol{x}_j^b, w_j^b) \qquad \text{and} \qquad \boldsymbol{x}_i^h = (w_i \boldsymbol{x}_i, w_i) \tag{5.5}$$

where w_i are the known NURBS weights of the original basis and w_j^b are the Bézier counterparts. From $\boldsymbol{x}_j^{bh}$, it is straightforward to compute the Bézier control points in model space $\boldsymbol{x}_j^b \in \mathbb{R}^d$. Finally, a NURBS curve can be represented as

$$\boldsymbol{x}(\xi) = \sum_{i=1}^{I} \frac{B_i(\xi)\, w_i}{w(\xi)}\, \boldsymbol{x}_i = \sum_{j=1}^{J} \frac{B_j^b(\xi)\, w_j^b}{w_j^b(\xi)}\, \boldsymbol{x}_j^b \tag{5.6}$$

with

$$w^b(\xi) = \sum_{k=0}^{K} B_k^b(\xi)\, w_k^b. \tag{5.7}$$

Bézier extraction can also be extended to geometry independent field approximation which will be introduced in Sect. 6.5.3. In this case, we define several extraction operators; one for the geometry and each physical field.

The main advantage of Bézier extraction is that the resulting Bézier elements can be treated during the integration like conventional Lagrange elements. Although computations are now performed over the Bernstein basis, the extraction operator constrains the degrees of freedom of the Bézier elements to maintain the continuity of the original NURBS object. The original basis is also the one that determines the location of collocation points used and the assembly of multiple patches. Hence, the concept facilitates the integration of isogeometric analysis into existing simulations software without any restriction on the capabilities of splines. In case an isogeometric analysis code is developed from scratch, Bézier extraction is not necessary.

5.4 Summary and Conclusions

At the beginning of this chapter, the problem of exchanging data between different systems was outlined from a general point of view. It is argued that the use of neutral exchange standards is the most comprehensive approach for this task. Nevertheless, every mapping from one system to another may cause problems, especially with respect to the design intent of a model where no canonical representation exists. There is often no one-to-one translation from one format to another, which leaves room for (mis)interpretation. As a result, exchange is usually restricted to snapshots of an object's geometry. A proper transfer of topology data requires that (i) the design model is properly constructed as a coherent solid model and (ii) the neutral format is able to capture this information. It is apparent that the capability of the neutral standard applied is essential for the quality and success of the exchange.

With this in mind, STEP should generally be preferred as a neutral exchange format. It is one of the most advanced formats and excels with respect to the quality of exchanging data of industrial examples [176]. The most important feature of STEP is its extensibility. Particularly interesting for isogeometric analysis is the specification of volumetric NURBS and local refinement in the next versions of Part 42 and other parts [170]. Theoretically, the broad scope and modular structure of STEP provides coverage of various application domains which are indicated by the application protocols and their conformance classes. However, this functionality has to be supported by the CAD vendors. Most vendors have chosen to implement only certain parts of STEP, i.e., some conformance classes of AP 203 and AP 214 [154]. It is not surprising that vendors seem to show little interest in neutral exchange formats, since their implementation slows down the development of the actual software and users become more independent from their products. Hence, it is likely that neutral file formats will always provide less information than the original model. Translation errors may be avoided if the required data is extracted directly from the native format, but this requires vendor interaction and the restriction to a single software. This alternative is not very sustainable since a native format may become obsolete after a new software version is released.

A concept that simplifies the integration of CAD data to existing simulation codes is Bézier extraction. It is not an exchange standard but an implementation strategy for isogeometric analysis that allows the representation of smooth spline patches by an element-like structure. Bernstein polynomials form the basis of these Bézier elements, and the so-called extraction operator encodes the relation of the Bernstein basis to the original one. Some CAD software tools even provide options to export Bézier elements and the corresponding operators. However, the capabilities of this exchange option are somewhat limited compared to STEP, but the approach is tailored to the needs of most isogeometric analysis users – at least in academia. Finally, we want to note that Bézier extraction is not required when an isogeometric solver is developed from scratch; for instance, all results in this book have been computed without this concept.

Chapter 6
Numerical Treatment of Integral Equations

In this chapter, methods are presented for the solution of the integral equations. For practical problems they cannot be solved analytically and we have to revert to a numerical solution. It should be noted here that the implementation of the BEM for different physical problems involves a change in the fundamental solutions, so the procedures outlined here are valid for the wide range of problems discussed in this book. We use the vectors $\mathbf{u}$ and $\mathbf{t}$ for the explanation but the number of components can range from one for potential problems to three for three-dimensional elasticity.

6.1 Introduction

The definition of the boundary geometry alone is not sufficient to define the problem domain. It is also necessary to specify where the material is located, in other words is it the boundary of a hole in an infinite domain or of a finite domain. This is determined by the orientation of the normal vector, $\mathbf{n}$, which influences the fundamental solution T. The definition of finite and infinite domains is shown in Fig. 6.1. The normal is also known as "outward normal" as it always points away from the material.

In a boundary value problem, either the primary variable $\mathbf{u}$ or the secondary variable $\mathbf{t}$ is specified on the boundary and a solution for the other is sought. The following boundary conditions (BC) are possible:

- Dirichlet BC. The value of $\mathbf{u}$ is specified.
- Neumann BC. The value of $\mathbf{t}$ is specified.
- Robin BC. The value of $\mathbf{t}$ is specified as a function of $\mathbf{u}$

Two approaches are available to satisfy the integral equations on the boundary: one where the integral equation is satisfied only at certain boundary points and one where the error in satisfaction of the integral equation is minimised. The first is known as *Collocation* and second as the *Galerkin* method. Since *Collocation* is

G. Beer et al., *The Isogeometric Boundary Element Method*, Lecture Notes in Applied and Computational Mechanics 90,
https://doi.org/10.1007/978-3-030-23339-6_6

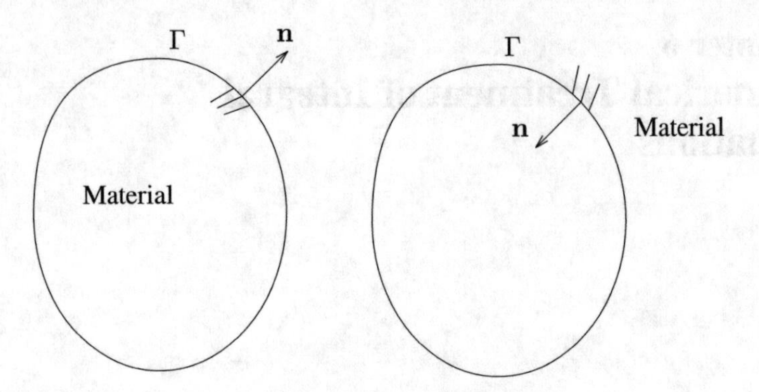

Fig. 6.1 Definition of a finite (left) and infinite (right) domain

much more widely used in engineering practice we will concentrate on this method. For extracting the unknowns from the integrals we approximate them with the basis functions introduced earlier. To solve the problem with the Cauchy principal value we apply a clever solution that also computes the free term.

6.2 Satisfaction of Integral Equations

We need to satisfy the following integral equation on the entire boundary, i.e. at an infinite number of points $\tilde{\boldsymbol{x}}$

$$\mathbf{c}(\tilde{\boldsymbol{x}})\,\mathbf{u}(\tilde{\boldsymbol{x}}) = \int_{\Gamma} (\mathsf{U}(\tilde{\boldsymbol{x}}, \hat{\boldsymbol{x}})\,\mathbf{t}(\hat{\boldsymbol{x}}) - \mathsf{T}(\tilde{\boldsymbol{x}}, \hat{\boldsymbol{x}})\,\mathbf{u}(\hat{\boldsymbol{x}}))\,d\Gamma(\hat{\boldsymbol{x}}). \tag{6.1}$$

Of course, this is not possible, so one approach that can be taken is to minimise the error in satisfying the integral equation. Using the *Galerkin* method we obtain

$$\begin{aligned} 0 = &\int_{\Gamma} W_n\,\mathbf{c}(\tilde{\boldsymbol{x}})\,\mathbf{u}(\tilde{\boldsymbol{x}})\,d\Gamma(\hat{\boldsymbol{x}}) \\ &- \int_{\Gamma} W_n \int_{\Gamma} (\mathsf{U}(\tilde{\boldsymbol{x}}, \hat{\boldsymbol{x}})\,\mathbf{t}(\hat{\boldsymbol{x}}) - \mathsf{T}(\tilde{\boldsymbol{x}}, \hat{\boldsymbol{x}})\,\mathbf{u}(\hat{\boldsymbol{x}}))\,d\Gamma(\hat{\boldsymbol{x}}) \end{aligned} \tag{6.2}$$

where W_n are suitable weight functions. This way a square system of N equations, where N is the number of unknowns, can be obtained which can be solved. This approach is favoured by mathematicians as it allows theoretical error analyses to be carried out. However, it involves a double integration and is computationally expensive.

Another, simpler approach is to enforce the integral equations only at certain points $\tilde{\boldsymbol{x}}_n$, the *Collocation points*

$$\mathbf{c}(\tilde{\boldsymbol{x}}_n)\,\mathbf{u}(\tilde{\boldsymbol{x}}_n) = \int\limits_{\Gamma} \left(\mathsf{U}(\tilde{\boldsymbol{x}}_n, \hat{\boldsymbol{x}})\,\mathbf{t}(\hat{\boldsymbol{x}}) - \mathsf{T}(\tilde{\boldsymbol{x}}_n, \hat{\boldsymbol{x}})\,\mathbf{u}(\hat{\boldsymbol{x}})\right)\,d\Gamma(\hat{\boldsymbol{x}}). \tag{6.3}$$

A sufficient number of collocation points has to be provided to allow the solution, i.e. the number of collocation points times the dimensionality of the problem has to equal the number of unknown parameters. Collocation is favoured by engineers since the programming is not only simpler, but practical applications have shown [21] that the increase in computational effort is not rewarded by a significantly better quality of the results. Therefore it will be used in this book. In fact the majority of the papers on the IGABEM favour this method.

6.3 Singular Integrals

Fundamental solutions tend to infinity as the field point $\hat{\boldsymbol{x}}$ approaches the source point $\tilde{\boldsymbol{x}}$. This happens with various degrees and we distinguish between weakly and strongly singular Kernels. Weakly singular Kernels are the ones that tend to infinity with $\mathcal{O}(\ln\frac{1}{r})$ in 2-D and $\mathcal{O}(\frac{1}{r})$ in 3-D. Strongly singular Kernels tend to infinity with $\mathcal{O}(\frac{1}{r})$ in 2-D and $\mathcal{O}(\frac{1}{r^2})$ in 3-D. Weakly singular Kernels can be integrated, but special integration techniques must be applied as will be shown later. Integrals of strongly singular Kernels do not exist, but a Cauchy principal value can be determined.

6.4 Solving the Cauchy Principal Value Problem

The integral involving the Kernel T is strongly singular and only a Cauchy principal value can be calculated. The integral can be evaluated numerically by surrounding the source point with an artificial boundary as explained in [21] but the procedure is rather complicated. Fortunately, one can avoid the computation of the Cauchy principal value, using a simple solution of the integral equation. For a finite domain problem, if we solve the integral equation for a Dirichlet BC $\mathbf{u}$ that is the same everywhere on the boundary, then all values of the unknown $\mathbf{t}$ must be zero. We explain the method in solid mechanics and this corresponds to a rigid body motion and is also known as the "rigid body motion trick" (see Fig. 6.2 on the left).[1] By applying the constant Dirichlet BC $\mathbf{u}(\hat{\boldsymbol{x}}) = \mathbf{u}_R$ we can write Eq. (6.1), knowing that the solution $\mathbf{t}$ is zero everywhere

$$\mathbf{c}(\tilde{\boldsymbol{x}})\,\mathbf{u}_R = -\mathbf{u}_R \int\limits_{\Gamma} \mathsf{T}(\tilde{\boldsymbol{x}}, \hat{\boldsymbol{x}})\,d\Gamma(\hat{\boldsymbol{x}}) \tag{6.4}$$

[1]This can be also applied to potential problems with the rigid body motion being replaced by a potential that is constant over the whole boundary and where the flow must be zero.

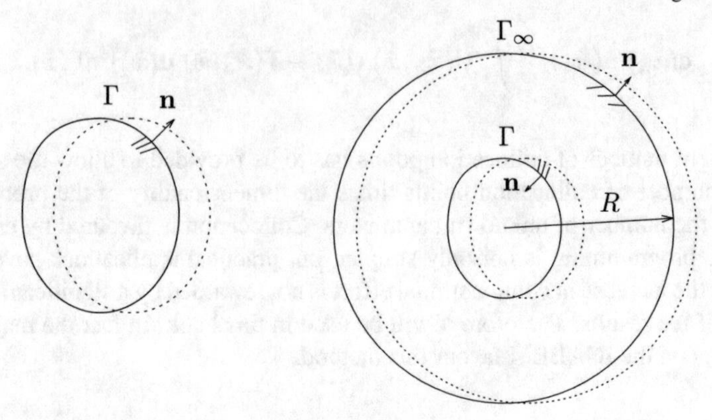

Fig. 6.2 "Rigid body trick" applied to a finite and infinite domain

and solve for the free term

$$\mathbf{c}(\tilde{\boldsymbol{x}}) = -\int_{\Gamma} \mathsf{T}(\tilde{\boldsymbol{x}}, \hat{\boldsymbol{x}})\, d\Gamma(\hat{\boldsymbol{x}}). \tag{6.5}$$

Substituting this into Eq. (6.3) we obtain

$$-\mathbf{u}(\tilde{\boldsymbol{x}}_n)\int_{\Gamma} \mathsf{T}(\tilde{\boldsymbol{x}}, \hat{\boldsymbol{x}}) d\Gamma(\hat{\boldsymbol{x}}) = \int_{\Gamma} \left(\mathsf{U}(\tilde{\boldsymbol{x}}_n, \hat{\boldsymbol{x}})\mathbf{t}(\hat{\boldsymbol{x}}) - \mathsf{T}(\tilde{\boldsymbol{x}}_n, \hat{\boldsymbol{x}})\mathbf{u}(\hat{\boldsymbol{x}})\right) d\Gamma(\hat{\boldsymbol{x}}) \tag{6.6}$$

or

$$\int_{\Gamma} \mathsf{U}(\tilde{\boldsymbol{x}}_n, \hat{\boldsymbol{x}})\, \mathbf{t}(\hat{\boldsymbol{x}})\, d\Gamma(\hat{\boldsymbol{x}}) - \int_{\Gamma} \mathsf{T}(\tilde{\boldsymbol{x}}_n, \hat{\boldsymbol{x}}) \left[\mathbf{u}(\hat{\boldsymbol{x}}) - \mathbf{u}(\tilde{\boldsymbol{x}}_n)\right]\, d\Gamma(\hat{\boldsymbol{x}}) = 0. \tag{6.7}$$

We can see that since the term inside the square parentheses tends to zero as $\hat{\boldsymbol{x}}$ approaches the source point $\tilde{\boldsymbol{x}}_n$ the integrand becomes weakly singular. We have thus not only elegantly solved the Cauchy principal value problem but it is also no longer required to compute the free term.

Obviously, this cannot be applied directly to an infinite domain problem. However, we can temporarily create a finite domain by surrounding Γ with a circular or spherical auxiliary surface Γ_∞ with a radius R (see Fig. 6.2 on the right). Equation (6.5) has to be modified to include the auxiliary surface

$$\mathbf{c}(\tilde{\boldsymbol{x}}) = -\left(\int_{\Gamma} \mathsf{T}(\tilde{\boldsymbol{x}}, \hat{\boldsymbol{x}})\, d\Gamma(\hat{\boldsymbol{x}}) + \int_{\Gamma_\infty} \mathsf{T}(\tilde{\boldsymbol{x}}, \hat{\boldsymbol{x}})\, d\Gamma_\infty(\hat{\boldsymbol{x}})\right). \tag{6.8}$$

The second integral is also known as *Azimuthal integral* and can be evaluated analytically using polar coordinates. For a plane problem, we have

$$\int_{\Gamma_\infty} \mathsf{T}(\tilde{\boldsymbol{x}}, \hat{\boldsymbol{x}})\, d\Gamma_\infty(\hat{\boldsymbol{x}}) = \int_{0}^{2\pi} \frac{1}{R}\bar{\mathsf{T}} R d\varphi = -\mathbf{I} \tag{6.9}$$

where $\bar{\mathsf{T}}$ is the part of the fundamental solution that does not contain R and $\mathbf{I}$ is the identity matrix. We can see that this result is independent of R and therefore also valid for $R = \infty$. Substituting this into Eq. (6.3) we obtain

$$-\left(\int_{\Gamma} \mathsf{T}(\tilde{\boldsymbol{x}}, \hat{\boldsymbol{x}}) d\Gamma(\hat{\boldsymbol{x}}) - \mathbf{I}\right) \mathbf{u}(\tilde{\boldsymbol{x}}_n) = \int_{\Gamma} (\mathsf{U}(\tilde{\boldsymbol{x}}_n, \hat{\boldsymbol{x}})\mathbf{t}(\hat{\boldsymbol{x}}) - \mathsf{T}(\tilde{\boldsymbol{x}}_n, \hat{\boldsymbol{x}})\mathbf{u}(\hat{\boldsymbol{x}}))\, d\Gamma(\hat{\boldsymbol{x}}). \tag{6.10}$$

Therefore to cover finite and infinite domain problems Eq. (6.7) has to be changed to

$$\int_{\Gamma} \mathsf{U}(\tilde{\boldsymbol{x}}_n, \hat{\boldsymbol{x}})\, \mathbf{t}(\hat{\boldsymbol{x}})\, d\Gamma(\hat{\boldsymbol{x}}) - \int_{\Gamma} \mathsf{T}(\tilde{\boldsymbol{x}}_n, \hat{\boldsymbol{x}})(\mathbf{u}(\hat{\boldsymbol{x}}) - \mathbf{u}(\tilde{\boldsymbol{x}}_n))\, d\Gamma(\hat{\boldsymbol{x}}) - \mathbf{A}_n \mathbf{u}(\tilde{\boldsymbol{x}}_n) = 0 \tag{6.11}$$

where $\mathbf{A}_n = \mathbf{I}$ for infinite and $\mathbf{A}_n = \mathbf{0}$ for finite domain problems.

6.5 Discretisation of the Integral Equations

To be able to solve the integral equations numerically we need to discretise them. Discretisation involves two steps. First we express the integrals as a summation of integrals over NURBS patches (geometry discretisation) and then we substitute an approximation for the physical quantities (field approximation).

6.5.1 Geometry Discretisation

The discretised version of Eq. (6.11) can be written as

$$\begin{aligned} &\sum_{e=1}^{E} \int_{\Gamma_e} \mathsf{U}(\tilde{\boldsymbol{x}}_n, \hat{\boldsymbol{x}})\, \mathbf{t}^e(\hat{\boldsymbol{x}})\, d\Gamma_e(\hat{\boldsymbol{x}}) - \sum_{e=1}^{E} \int_{\Gamma_e} \mathsf{T}(\tilde{\boldsymbol{x}}_n, \hat{\boldsymbol{x}})\mathbf{u}^e(\hat{\boldsymbol{x}}) d\Gamma_e \\ &+ \left[\sum_{e=1}^{E} \left(\int_{\Gamma_e} \mathsf{T}(\tilde{\boldsymbol{x}}_n, \hat{\boldsymbol{x}}) d\Gamma_e\right) - \mathbf{A}_n\right] \mathbf{u}(\tilde{\boldsymbol{x}}_n) = 0 \end{aligned} \tag{6.12}$$

where e specifies the patch number and E is the total number of patches.

6.5.1.1 Plane Problems

For plane simulations the patches are curves. In Fig. 6.3 we show an example of a patch. The mapping from the local ξ to the global $\boldsymbol{x}$ coordinate system is given by (see Sect. 4.5)

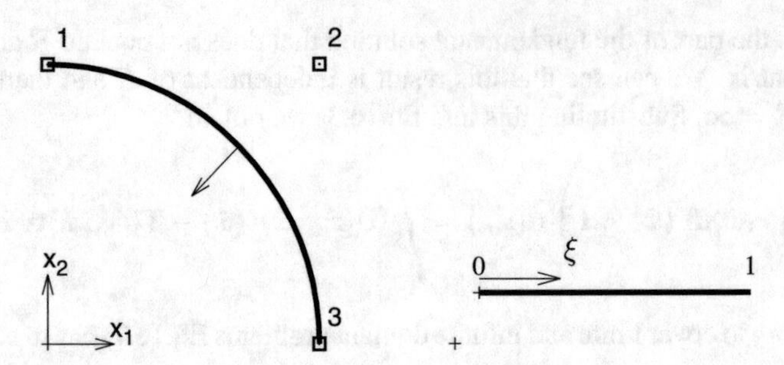

Fig. 6.3 Example of a patch for plane simulation with control points (numbered squares). Left: in the global, right: in the local coordinate system. Also shown is the "outward normal"

$$\boldsymbol{x}(\xi) = \sum_{i=1}^{I} R_i(\xi)\boldsymbol{x}_i \tag{6.13}$$

where $R_i(\xi)$ are NURBS basis functions and $\boldsymbol{x}_i$ are control point coordinates. The vector tangential to the curve is given by

$$\mathbf{v}_\xi = \frac{\partial \boldsymbol{x}}{\partial \xi} = \begin{pmatrix} \frac{\partial x_1}{\partial \xi} \\ \frac{\partial x_2}{\partial \xi} \end{pmatrix} = \sum_{i=1}^{I} \frac{\partial R_i(\xi)}{\partial \xi}\boldsymbol{x}_i \tag{6.14}$$

and the vector normal to the curve is

$$\mathbf{v}_\eta = \begin{pmatrix} \frac{\partial x_2}{\partial \xi} \\ -\frac{\partial x_1}{\partial \xi} \end{pmatrix}. \tag{6.15}$$

The Jacobian is

$$J = \sqrt{\mathrm{v}_{\eta 1}^2 + \mathrm{v}_{\eta 2}^2} \tag{6.16}$$

and the unit "outward normal" is given by

$$\mathbf{n} = \frac{1}{J}\mathbf{v}_\eta \tag{6.17}$$

The direction of the "outward normal" depends on how the control points are numbered (in the example clockwise).

6.5.1.2 3-D Problems

For 3-D simulations the patches are surfaces. In Fig. 6.4 we show an example of a patch. The mapping from the local $\boldsymbol{\xi}(\xi, \eta)$ to the global $\boldsymbol{x}$ coordinate system is given by

$$\boldsymbol{x}(\xi, \eta) = \sum_{i=1}^{I} R_i(\xi, \eta)\boldsymbol{x}_i. \tag{6.18}$$

Note that in the simplified notation the control points are numbered consecutively, first in the ξ- and then in the η-direction. The vectors tangential to the surface are given by

$$\mathbf{v}_\xi = \frac{\partial \boldsymbol{x}}{\partial \xi} = \begin{pmatrix} \frac{\partial x_1}{\partial \xi} \\ \frac{\partial x_2}{\partial \xi} \\ \frac{\partial x_3}{\partial \xi} \end{pmatrix} \quad \text{and} \quad \mathbf{v}_\eta = \frac{\partial \boldsymbol{x}}{\partial \eta} = \begin{pmatrix} \frac{\partial x_1}{\partial \eta} \\ \frac{\partial x_2}{\partial \eta} \\ \frac{\partial x_3}{\partial \eta} \end{pmatrix} \tag{6.19}$$

and the unit vector normal is

$$\mathbf{n} = \frac{\mathbf{v}_\xi \times \mathbf{v}_\eta}{J}. \tag{6.20}$$

The Jacobian is

$$J = |\mathbf{v}_\xi \times \mathbf{v}_\eta|. \tag{6.21}$$

The direction of the "outward normal" depends on how the control points are numbered.

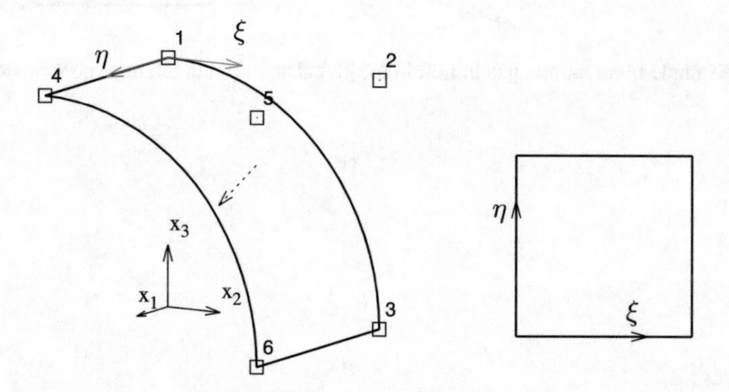

Fig. 6.4 Example of a patch for 3-D simulation with control points (numbered squares). Left: in the global, right: in the local coordinate system. Also shown is the "outward normal"

6.5.1.3 3-D Infinite Patches

Here we introduce a patch definition that is not standard in CAD but useful for the simulation in geomechanics where one sometimes has to simulate a surface that tends to infinity [10]. In this case we define an infinite patch as shown in Fig. 6.5. The mapping for a patch that extends to infinity in the η-direction is given by

$$\boldsymbol{x} = \sum_{j=1}^{2} \sum_{i=1}^{I} R_{ij}^{\infty}(\xi, \eta)\boldsymbol{x}_{ij} \tag{6.22}$$

where

$$R_{ij}^{\infty}(\xi, \eta) = R_i(\xi) M_j^{\infty}(\eta) \tag{6.23}$$

and the special infinite basis functions are

$$M_1^{\infty} = \frac{1 - 2\eta}{1 - \eta} \quad \text{and} \quad M_2^{\infty} = \frac{\eta}{1 - \eta}. \tag{6.24}$$

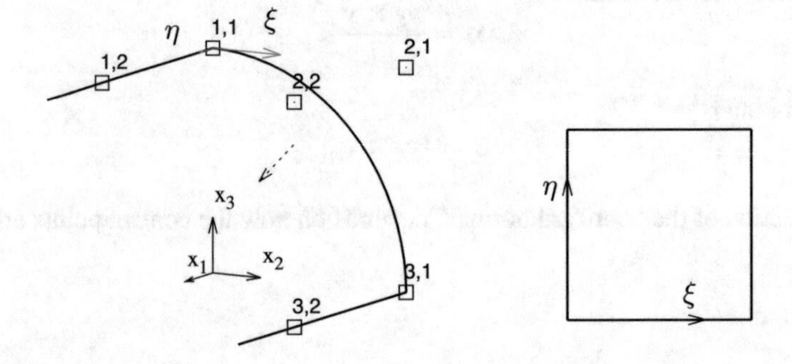

Fig. 6.5 Example of an infinite patch. Left in the global and right in the local coordinate system

The vectors in the tangential directions are given by

$$\mathbf{v}_\xi = \frac{\partial \boldsymbol{x}}{\partial \xi} = \sum_{j=1}^{2}\sum_{i=1}^{I} \frac{\partial R_i(\xi)}{\partial \xi} M_j^\infty(\eta)\boldsymbol{x}_{ij} \tag{6.25}$$

$$\mathbf{v}_\eta = \frac{\partial \boldsymbol{x}}{\partial \eta} = \sum_{j=1}^{2}\sum_{i=1}^{I} R_i(\xi)\frac{\partial M_j^\infty(\eta)}{\partial \eta}\boldsymbol{x}_{ij} \tag{6.26}$$

where

$$\frac{\partial M_1^\infty}{\partial \eta} = \frac{-1}{(1-\eta)^2} \quad \text{and} \quad \frac{\partial M_2^\infty}{\partial \eta} = \frac{1}{(1-\eta)^2}. \tag{6.27}$$

6.5.2 *Approximation of the Physical Values*

In the second step, we move $\mathbf{u}$ and $\mathbf{t}$ outside the integral by approximating them, in the case of unknown values, or defining them, in case of known values.

Remark It should be pointed out that there exist methods for solving the integral equations without the approximation of the unknowns. A unique feature of the Nyström method for example is, that the boundary integrals are evaluated directly by means of numerical quadrature without formulating an approximation of the unknown fields. In fact, the method is based on pointwise evaluations of the fundamental solutions on the boundary of the computational domain. Details of the implementation in an isogeometric framework can be found in [196].

We specify for each patch a boundary condition, which is either Dirichlet ($\mathbf{u}$ known, $\mathbf{t}$ unknown) or Neumann ($\mathbf{t}$ known, $\mathbf{u}$ unknown).

Unknown values can be approximated by

$$\begin{aligned}\hat{\mathbf{u}}^e(\boldsymbol{\xi}) &= \sum_{k=1}^{K} \hat{R}_k(\boldsymbol{\xi})\,\hat{\mathbf{u}}_k^e \\ \hat{\mathbf{t}}^e(\boldsymbol{\xi}) &= \sum_{k=1}^{K} \hat{R}_k(\boldsymbol{\xi})\,\hat{\mathbf{t}}_k^e.\end{aligned} \tag{6.28}$$

Known values can be defined by

$$\begin{aligned}\bar{\mathbf{u}}^e(\boldsymbol{\xi}) &= \sum_{k=1}^{\bar{K}} \bar{R}_k(\boldsymbol{\xi})\,\bar{\mathbf{u}}_k^e \\ \bar{\mathbf{t}}^e(\boldsymbol{\xi}) &= \sum_{k=1}^{\bar{K}} \bar{R}_k(\boldsymbol{\xi})\,\bar{\mathbf{t}}_k^e.\end{aligned} \tag{6.29}$$

$\hat{R}_k$ and $\bar{R}_k$ are basis functions which may be different for the unknown and known values and $\hat{\mathbf{u}}_k^e$, $\hat{\mathbf{t}}_k^e$ and $\bar{\mathbf{u}}_k^e$, $\bar{\mathbf{t}}_k^e$ are parameter values at control points.[2]

For infinite patches we have two options:

1. The physical value is constant in infinite direction, i.e. $\mathbf{u}(\xi, \eta) = \mathbf{u}(\xi, \eta = 0)$
2. The physical value decays in infinite direction, i.e. $\mathbf{u}(\xi, \eta) = (1 - \eta)\mathbf{u}(\xi, \eta = 0)$

The first corresponds to plane strain conditions in elasticity.

6.5.3 Refinement Philosophy (Geometry Independent Field Approximation)

In a simulation, the approximation of the unknowns usually has to be refined until a solution of acceptable accuracy is obtained. Here we discuss our basic philosophy of refinement.

In the majority of the simulations presented in this book we start with approximating the unknowns with the same basis functions as used for the definition of the geometry, i.e. $\hat{R}_k = R_k$. We then refine the basis functions $\hat{R}_k$ by changing the knot vector, using the options discussed in Chap. 3. In some practical examples we also use B-splines for the approximation and refine these.

Remark It should be noted here that in the isogeometric analysis as published in [87] a refinement of both the geometry and the approximation was envisaged (hence the term "iso" which is "equal" in Greek, meaning that the basis functions for describing the unknown and the geometry are the same). Our approach is different in that we leave the definition of the geometry, which – if taken from a CAD model – already exactly defines the design geometry, untouched and only change the approximation of the unknowns. Unnecessary geometric refinement decreases the efficiency of the simulation, especially when the order of the basis functions is increased. We will show examples of this later. In addition, with geometry independent field approximation the physical fields can be defined by B-splines, also for NURBS geometries. This can further improve the efficiency [113] and does usually not lead to a loss of accuracy (as demonstrated in [105, 112, 184]) because the approximation quality of B-spline and NURBS for general functions are similar as discussed in Sect. 4.4.

6.5.4 Location of Collocation Points

For the location of the collocation points we can use continuous or discontinuous collocation. In the context of the book we are only referring to the continuity at

[2]It should be noted here that for known values that have a simple variation we may even use lower order Lagrange polynomials for $\bar{R}_k$ in which case $\bar{\mathbf{u}}_k^e$ or $\bar{\mathbf{t}}_k^e$ are nodal values.

the edges of a patch, which has implications for the assembly, discussed next. In the continuous collocation we choose the Greville abscissae $\tilde{\xi}_k$ defined in Eq. (3.19) as the locations of the collocation points. In the discontinuous collocation we start with the Greville abscissae but then move the location of the discontinuous collocation points by a value of α inside the patch. One way to define this offset α is the following: first, the knots of a knot vector Ξ and the corresponding Greville abscissae are accumulated in a non-decreasing sequence $\hat{\Xi}$ with $\hat{\xi}_i = \xi_j \bigcup \tilde{\xi}_k$. For instance, $\Xi = (0, 0, 0, 1, 1, 1)$ with $\tilde{\xi}_i = \{0, 0.5, 1\}$ leads to $\hat{\Xi} = (0, 0, 0, 0, 0.5, 1, 1, 1, 1)$. For each $\tilde{\xi}_k$ that shall be shifted, we find its location counterpart $\hat{\xi}_{\hat{k}}$ in $\hat{\Xi}$ and determine the distances to its L neighboring $\hat{\xi}_i$ in both directions, where L is given by

$$L = \begin{cases} 1 & \text{if } p < 3, \\ 2 & \text{otherwise.} \end{cases} \tag{6.30}$$

The sum of these distances, divided by a constant, provides the offset

$$\alpha_k = \frac{\sum_{l=1}^{L} \left(\hat{\xi}_{\hat{k}-l} - \hat{\xi}_{\hat{k}} \right) + \sum_{l=1}^{L} \left(\hat{\xi}_{\hat{k}+l} - \hat{\xi}_{\hat{k}} \right)}{2L + 1}. \tag{6.31}$$

Note that α can have positive or negative values. For the example in Fig. 6.6 with $\tilde{\xi}_k = \{0, 0.5, 1\}$, the offsets are $\alpha_k = \{1/6, 0, 0\}$.

As we enrich the basis functions, the number of collocation points also increases. In Fig. 6.6 we show on the example of a 2-D patch the locations of the collocation points and the associated basis functions. Note that the collocation points are not evenly distributed, even for the continuous collocation.

6.5.5 System of Equations

Discretised equations. To establish the system of equations, we separate the unknown and known values in Eq. (6.12) and write for the n-th collocation point

$$\begin{aligned} &\sum_{e_D} \hat{\mathbf{U}}_n^e \hat{\mathbf{t}}^e - \sum_{e_N} \hat{\mathbf{T}}_n^e \hat{\mathbf{u}}^e + (\mathbf{T}_n - \mathbf{A}_n)\hat{\mathbf{u}}(\tilde{\boldsymbol{x}}_n) = \\ &-\sum_{e_N} \bar{\mathbf{U}}_n^e \bar{\mathbf{t}}^e + \sum_{e_D} \bar{\mathbf{T}}_n^e \bar{\mathbf{u}}^e - (\mathbf{T}_n - \mathbf{A}_n)\bar{\mathbf{u}}(\tilde{\boldsymbol{x}}_n) \end{aligned} \tag{6.32}$$

where e_D indicates that the sum is taken over patches where Dirichlet BCs are specified and e_N where Neumann BCs are specified.

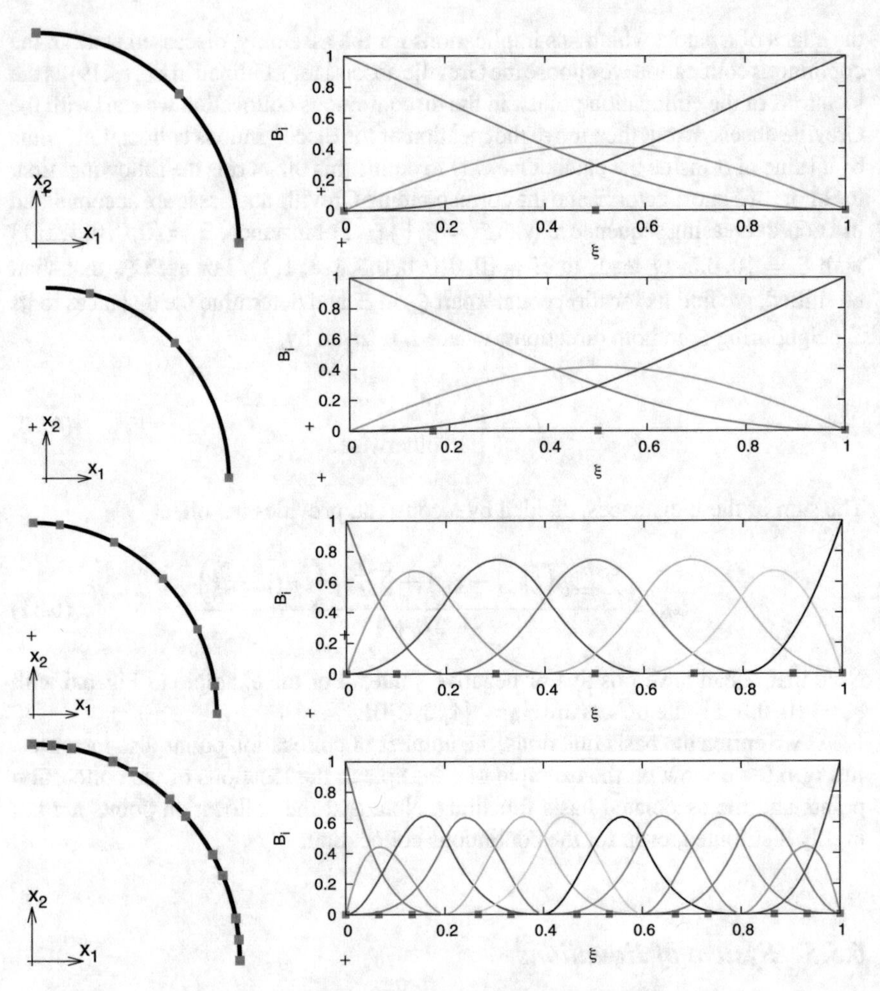

Fig. 6.6 Example of the proposed refinement methods on a patch: left: patch in the global coordinate system, right: basis functions for the approximation of the unknown. The location of anchors and collocation points are depicted as red squares. From top to bottom: original basis functions with continuous collocation, with discontinuous collocation at left edge, continuous after 4 knot insertions and after order elevation followed by 4 knot insertions

We define patch matrices as

$$\hat{\mathbf{U}}_n^e = \left[\hat{\mathbf{U}}_{n1}^e \cdots \hat{\mathbf{U}}_{nK}^e\right], \quad \hat{\mathbf{T}}_n^e = \left[\hat{\mathbf{T}}_{n1}^e \cdots \hat{\mathbf{U}}_{nK}^e\right] \quad \text{etc.} \tag{6.33}$$

where

$$\hat{\mathbf{U}}^e_{nk} = \int_{\Gamma_e} \mathsf{U}(\tilde{\boldsymbol{x}}_n, \hat{\boldsymbol{x}})\ \hat{R}_k d\Gamma_e(\hat{\boldsymbol{x}}), \quad \bar{\mathbf{U}}^e_{nk} = \int_{\Gamma_e} \mathsf{U}(\tilde{\boldsymbol{x}}_n, \hat{\boldsymbol{x}})\ \bar{R}_k d\Gamma_e(\hat{\boldsymbol{x}})$$
$$\hat{\mathbf{T}}^e_{nk} = \int_{\Gamma_e} \mathsf{T}(\tilde{\boldsymbol{x}}_n, \hat{\boldsymbol{x}})\ \hat{R}_k d\Gamma_e(\hat{\boldsymbol{x}}), \quad \bar{\mathbf{T}}^e_{nk} = \int_{\Gamma_e} \mathsf{T}(\tilde{\boldsymbol{x}}_n, \hat{\boldsymbol{x}})\ \bar{R}_k d\Gamma_e(\hat{\boldsymbol{x}}). \tag{6.34}$$

The patch vectors are defined as:

$$\hat{\mathbf{u}}^e = \begin{Bmatrix} \hat{\mathbf{u}}^e_1 \\ \vdots \\ \hat{\mathbf{u}}^e_K \end{Bmatrix}, \quad \hat{\mathbf{t}}^e = \begin{Bmatrix} \hat{\mathbf{t}}^e_1 \\ \vdots \\ \hat{\mathbf{t}}^e_K \end{Bmatrix} \quad \text{etc.} \tag{6.35}$$

The remaining matrix and vectors are defined as

$$\mathbf{T}_n = \sum_{e=1}^{E} \int_{\Gamma_e} \mathsf{T}(\tilde{\boldsymbol{x}}_n, \hat{\boldsymbol{x}})\ d\Gamma_e(\hat{\boldsymbol{x}}) \tag{6.36}$$

$$\hat{\mathbf{u}}(\tilde{\boldsymbol{x}}_n) = \sum_{k=1}^{K} \hat{R}_k(\boldsymbol{\xi}_n)\ \hat{\mathbf{u}}^{en}_k \tag{6.37}$$

$$\bar{\mathbf{u}}(\tilde{\boldsymbol{x}}_n) = \sum_{k=1}^{K} \bar{R}_k(\boldsymbol{\xi}_n)\ \bar{\mathbf{u}}^{en}_k \tag{6.38}$$

with en signifying the patch containing $\tilde{\boldsymbol{x}}_n$ with local coordinates $\boldsymbol{\xi}_n$.

Equation (6.32) can be re-written as

$$\sum_{\mathrm{e}_D} \mathbf{U}^e_n \hat{\mathbf{t}}^e - \sum_{\mathrm{e}_N} \mathbf{T}^e_n \hat{\mathbf{u}}^e + \mathbf{T}^{en}_{0n}\ \hat{\mathbf{u}}^{en}_n =$$
$$-\sum_{\mathrm{e}_N} \bar{\mathbf{U}}^e_n \bar{\mathbf{t}}^e + \sum_{\mathrm{e}_D} \bar{\mathbf{T}}^e_n \bar{\mathbf{u}}^e - \bar{\mathbf{T}}_{0n} \tag{6.39}$$

where

$$\mathbf{T}^{en}_{0n} = \big((\mathbf{T}_n - \mathbf{A}_n)\bar{R}_1(\boldsymbol{\xi}_n), (\mathbf{T}_n - \mathbf{A}_n)\bar{R}_2(\boldsymbol{\xi}_n), \cdots \big) \tag{6.40}$$
$$\bar{\mathbf{T}}_{0n} = (\mathbf{T}_n - \mathbf{A}_n)\bar{\mathbf{u}}(\tilde{\boldsymbol{x}}_n). \tag{6.41}$$

Assembly. Next we have to form a coefficient matrix that multiplies with the unknowns (left-hand side) and a vector that is related to known values (right-hand side). In the BEM it is possible to have a C^{-1} continuity (or discontinuity) at the locations where boundary elements are connected. Indeed, in the early days of the

BEM constant boundary elements were extensively used, which implied a C^{-1} continuity at element boundaries. Even with high-order basis functions discontinuous elements were used in [34] for 3-D, mainly because the assembly process is simplified. However, it was shown that the efficiency was poor, mainly because the number of degrees of freedom is increased considerably, since double nodes appear at inter-element boundaries. In addition discontinuous assembly introduces some error in the solution. For a discussion see for example [3, 62, 126].

The situation with the IGABEM is sightly different. Since patches are able to describe a much large portion of a smooth geometry, connection between patches are not as frequent as they occur between boundary elements with the conventional BEM. A discontinuous assembly makes sense where there are edges and corners. The main advantage of a discontinuous assembly is that there is no restriction on the choice of the basis functions for the approximation of the unknowns at the edges of a patch, which is particularly convenient when trimmed surfaces are considered.

Continuous assembly is not more complicated for 2-D problems as the points, where patches are connected, are the ends of patch curves and are interpolatory points. However, to ensure C^0-continuity in 3-D the following conditions have to exist on the boundaries, where two or more patches meet:

- There must be a unique global location of the collocation points where patches meet. Considering that collocation points are computed using local coordinates, this is not trivial. Note that this also implies that the parameter spaces of the patches must match.
- The basis functions used for the approximation of the unknown must be the same along the patch boundaries.

The user may choose different assembly strategies by simply specifying the method for the computation of the collocation points in patches. The most optimal way (in terms of efficiency and accuracy) would be to use continuous assembly for most patches where the transition between patches is smooth. Indeed, unless otherwise specified, continuous assembly is used in the examples presented in this book. The following explanation of the assembly process is applicable to both.

To start the process of assembling we assign a unique (global) numbering to the unknown parameters. We create a vector of *Element destinations* for each patch, which links the local (patch) to the global numbering of parameters. This allows the global and local numbering to be linked. For example, we can now express the global parameters in terms of local parameters by

$$\mathbf{u}_{n=Ldest(e,i)} = \mathbf{u}_i^e \tag{6.42}$$

$$\mathbf{t}_{n=Ldest(i,e)} = \mathbf{t}_i^e \tag{6.43}$$

where $Ldest(e,i)$ are coefficients of matrix $\mathbf{Ldest}$ that links ith parameter of the patch e to the global parameter n. For the assembly we need also information if the parameter is known or unknown. For this we use an *Element assembly* matrix $\mathbf{Lass}$ with coefficients $Lass(e,i)$, which are assigned a one if the ith unknown parameter of patch e is $\mathbf{t}$ or greater than one if $\mathbf{u}$ is unknown. For the rigid body trick, we also need information which patch en contains the global parameter n (array $Len(n)$).

Assembly with discontinuous BCs. In some cases the boundary condition changes between patches. For example, we may have a Dirichlet BC on one side of a point and a Neumann BC on the other side. If we use discontinuous assembly this problem is solved. If continuous assembly is used, we change the assembly procedure for the point, where discontinuous BCs occur. Since we enforce C^0-continuity of the primary variable $\mathbf{u}$ we assemble the coefficients associated with U into the left-hand side only for the patch that was assigned Dirichlet BC. For the patch that has been assigned Neumann BC (=2) and is connected to the one assigned Dirichlet BC (=1), there is no assembly in the left-hand side, since on this side $\mathbf{t}$ is known. For this we need a global vector $Ncode(n)$ that specifies the global BC for each degree of freedom.

The algorithm for the assembly is shown in Algorithm 1. Finally one ends up with a system of equations

$$\mathbf{L}\mathbf{x} = \mathbf{r} \tag{6.44}$$

where $\mathbf{x}$ contains a mixture of $\mathbf{u}$ and $\mathbf{t}$ parameters.

Algorithm 1 Algorithm for assembly of left and right-hand side

Require: Patch matrices $\mathbf{U}^e, \mathbf{T}^e, \bar{\mathbf{U}}^e, \bar{\mathbf{T}}^e, \mathbf{T}_0, \bar{\mathbf{T}}_0, Ldest, Lass, Ncode, Len$
1: initialise matrix $\mathbf{L}$ and vector $\mathbf{r}$
2: **for** e **to** Number of patches **do**
3: **for** n **to** Number of global parameters **do**
4: **for** i **to** Number patch parameters **do**
5: **if** $Lass(e,i) = 1$ **then**
6: $\mathbf{r}(n) = \mathbf{r}(n) - \bar{\mathbf{T}}^e(n,i)$
7: $\mathbf{L}(n, Ldest(e,i)) = \mathbf{L}(n, Ldest(e,i)) + \mathbf{U}^e(n,i)$
8: **else**
9: $\mathbf{r}(n) = \mathbf{r}(n) + \bar{\mathbf{U}}^e(n,i)$
10: **if** $Ncode(n) \neq 1$ **then**
11: $\mathbf{L}(n, Ldest(e,i)) = \mathbf{L}(n, Ldest(e,i)) - \mathbf{T}^e(n,i)$
12: **end if**
13: **end if**
14: **end for**
15: **end for**
16: **end for**
17: Apply rigid body mode:
18: **for** n **to** Number of global parameters **do**
19: $\mathbf{r}(n) = \mathbf{r}(n) - \bar{\mathbf{T}}_0(n)$
20: $en = Len(n)$
21: **for** i **to** Number of patch parameters **do**
22: $\mathbf{L}(n, Ldest(en,i)) = \mathbf{L}(n, Ldest(en,i)) + \mathbf{T}_0(n,i,en)$
23: **end for**
24: **end for**
25: **return** $\mathbf{L}, \mathbf{r}$

6.6 Solution of System of Equations

The resulting system of equations will be fully populated and not symmetric. This is in contrast to the FEM where the system is sparsely populated and symmetric, leading to less solution effort and this has been claimed as one of the main drawbacks of the BEM even though the number of unknowns is lower by an order of magnitude as compared with the FEM. However, for large 3-D simulations this is still an issue.

The storage requirement for the BEM is N^2 and the solution effort using standard Gauss elimination is approximately $\frac{1}{3}N^3$ floating point operations. One can see that the storage requirement and solution effort increases rapidly with problem size. In most practical problems one is not interested in an "exact" solution (that is exact really only up to round-off errors introduced during Gauss elimination) but a solution with a specified accuracy. In addition there are other errors that are introduced during the solution, the most important being the errors made during numerical integration. Therefore it makes sense to solve the system of equations only to a specified error. To reduce the computational effort and storage requirement several methods have been introduced that provide solutions that are not "exact" but contain a negligible error, specified by the user. We will discuss one of these here, namely the concept of Hierarchical matrices ($\mathcal{H}$-matrices). First we introduce methods that reduce the storage requirement and secondly methods which lead to the reduction in computation time.

6.6.1 Properties of BEM Matrices

Although the matrices are not symmetric, they have a particular characteristic that can be exploited. The Kernels, that are used to establish the system matrices, are asymptotically smooth and decay rapidly with the distance between the source point and the field point. To explain the exploitation of this, we first discuss a method that can be used to approximate a matrix, namely cross-approximation.

6.6.2 Cross Approximation

Suppose an $m \times m$ matrix $\mathbf{L}$ has a rank r. Then it can be exactly recovered as

$$\mathbf{L} = \mathbf{A}\mathbf{B}^{\mathrm{T}} \tag{6.45}$$

where $\mathbf{A}$ is an $r \times m$ matrix that contains r columns of vector $\mathbf{a}$ and $\mathbf{B}^{\mathrm{T}}$ is an $m \times r$ matrix containing r rows of vector $\mathbf{b}$.

Remark In linear algebra, the rank of a matrix corresponds to the number of linearly independent rows or columns.

The steps in cross approximation are equivalent to Gauss elimination. For a selected pivot element $\delta = L_{i^*j^*}$, we perform the following operation

$$L_{ij} := L_{ij} - a_i b_j \tag{6.46}$$

where $a_i = L_{ij^*}/\delta$ and $b_j = L_{i^*j}$.

6.6.3 *Adaptive Cross-Approximation (ACA)*

The cross approximation has the following disadvantages, especially for large matrices:

- All elements of **L** have to be known.
- Matrix **L** has to be updated.
- The rank of **L** has to be known in advance.

The idea exploited next is that the whole matrix **L** is never established and stored. Instead only certain rows and columns of the matrix are used to get a low rank approximation. Therefore, the algorithm has to be changed in that the first row-index is determined by the user and the error in the approximation is checked after each iteration. The modified algorithm is shown in Algorithm 2.

Algorithm 2 Algorithm for adaptive cross approximation

Require: Matrix **L**, Allowable error: ϵ_{ACA}
1: Select initial row index i^1
2: **for** k= 1 to K **do**
3: $b(1,:)^k = L(i^k,:)$
4: **for** n = 1 to $k-1$ **do**
5: $b(1,:)^k := b(1,:)^k - a(i^k,1)^n * b(1,:)^n$
6: **end for**
7: j^k := Maximum coefficient of $\mathbf{b}^k$
8: $\delta = b(i^k, j^k)^k$
9: $a(:,1)^k = L(:,j^k)/\delta$
10: **for** n = 1 to $k-1$ **do**
11: $a(:,1)^k := b(:,1)^k - a(:,1)^n * b(1,j^k)^n/\delta$
12: **end for**
13: i^k := Maximum coefficient of $\mathbf{a}^k$
14: $\tilde{\mathbf{L}} = \sum_{n=1}^{k} \mathbf{a}^n \mathbf{b}^n$
15: Compute Error Norm: $\epsilon = \frac{\|\mathbf{a}^k\| \cdot \|\mathbf{b}^k\|}{\|\tilde{\mathbf{L}}\|}$
16: **if** $\epsilon < \epsilon_{\text{ACA}}$ **then**
17: exit
18: **end if**
19: **end for**

Numerical example. Here we show the ACA on a numerical example. The matrix **L** to be approximated is given by

$$\mathbf{L} = \begin{bmatrix} 6.00 & 4.00 & 5.00 & 4.00 & 3.00 & 5.00 & 3.00 & 4.00 \\ 8.00 & 6.00 & 7.00 & 5.00 & 4.00 & 7.00 & 4.00 & 5.00 \\ 10.00 & 7.00 & 8.00 & 7.00 & 5.00 & 9.00 & 5.00 & 6.00 \\ 6.00 & 4.00 & 5.00 & 4.00 & 3.00 & 5.00 & 3.00 & 4.00 \\ 8.00 & 5.00 & 6.00 & 6.00 & 4.00 & 7.00 & 4.00 & 5.00 \\ 10.00 & 7.00 & 9.00 & 6.00 & 5.00 & 8.00 & 5.00 & 7.00 \\ 6.00 & 4.00 & 5.00 & 4.00 & 3.00 & 5.00 & 3.00 & 4.00 \\ 10.00 & 7.00 & 8.00 & 7.00 & 5.00 & 9.00 & 5.00 & 6.00 \end{bmatrix}. \tag{6.47}$$

For $k = 1$ we obtain

$$\mathbf{A} = \begin{bmatrix} 1.00 \\ 1.33 \\ 1.67 \\ 1.00 \\ 1.33 \\ 1.67 \\ 1.00 \\ 1.67 \end{bmatrix} \tag{6.48}$$

$$\mathbf{B}^{\mathrm{T}} = \begin{bmatrix} 6.00 & 4.00 & 5.00 & 4.00 & 3.00 & 5.00 & 3.00 & 4.00 \end{bmatrix}. \tag{6.49}$$

which gives the first approximation to $\mathbf{L}$

$$\tilde{\mathbf{L}} = \begin{bmatrix} 6.00 & 4.00 & 5.00 & 4.00 & 3.00 & 5.00 & 3.00 & 4.00 \\ 8.00 & 5.33 & 6.67 & 5.33 & 4.00 & 6.67 & 4.00 & 5.33 \\ 10.00 & 6.67 & 8.33 & 6.67 & 5.00 & 8.33 & 5.00 & 6.67 \\ 6.00 & 4.00 & 5.00 & 4.00 & 3.00 & 5.00 & 3.00 & 4.00 \\ 8.00 & 5.33 & 6.67 & 5.33 & 4.00 & 6.67 & 4.00 & 5.33 \\ 10.00 & 6.67 & 8.33 & 6.67 & 5.00 & 8.33 & 5.00 & 6.67 \\ 6.00 & 4.00 & 5.00 & 4.00 & 3.00 & 5.00 & 3.00 & 4.00 \\ 10.00 & 6.67 & 8.33 & 6.67 & 5.00 & 8.33 & 5.00 & 6.67 \end{bmatrix}. \tag{6.50}$$

For $k = 2$ we obtain

$$\mathbf{A} = \begin{bmatrix} 1.00 & 0.00 \\ 1.33 & 0.50 \\ 1.67 & 1.00 \\ 1.00 & 0.00 \\ 1.33 & 0.50 \\ 1.67 & -0.50 \\ 1.00 & 0.00 \\ 1.67 & 1.00 \end{bmatrix} \tag{6.51}$$

$$\mathbf{B}^{\mathrm{T}} = \begin{bmatrix} 6.00 & 4.00 & 5.00 & 4.00 & 3.00 & 5.00 & 3.00 & 4.00 \\ 0.00 & 0.33 & -0.33 & 0.33 & 0.00 & 0.67 & 0.00 & -0.67 \end{bmatrix} \tag{6.52}$$

which results in the second approximation

$$\tilde{\mathbf{L}} = \begin{bmatrix} 6.00 & 4.00 & 5.00 & 4.00 & 3.00 & 5.00 & 3.00 & 4.00 \\ 8.00 & 5.50 & 6.50 & 5.50 & 4.00 & 7.00 & 4.00 & 5.00 \\ 10.00 & 7.00 & 8.00 & 7.00 & 5.00 & 9.00 & 5.00 & 6.00 \\ 6.00 & 4.00 & 5.00 & 4.00 & 3.00 & 5.00 & 3.00 & 4.00 \\ 8.00 & 5.50 & 6.50 & 5.50 & 4.00 & 7.00 & 4.00 & 5.00 \\ 10.00 & 6.50 & 8.50 & 6.50 & 5.00 & 8.00 & 5.00 & 7.00 \\ 6.00 & 4.00 & 5.00 & 4.00 & 3.00 & 5.00 & 3.00 & 4.00 \\ 10.00 & 7.00 & 8.00 & 7.00 & 5.00 & 9.00 & 5.00 & 6.00 \end{bmatrix} . \tag{6.53}$$

The error of this approximation is $\epsilon = 0.038538$.

6.6.4 Hierarchical Matrices

ACA by itself would be not useful and would probably lead to a full rank approximation for $\mathbf{L}$ unless we exploit the fact that the fundamental solutions, with which the coefficients of $\mathbf{L}$ are computed, decay rapidly with the distance between the source point ($\tilde{x}$) and the field point ($\hat{x}$). Using the context of Hierarchical Matrices ($\mathcal{H}$-matrices) introduced by Hackbusch [77], we link the matrix entries to geometry and separate them into *near field* (when the distances between $\tilde{x}$ and $\hat{x}$ are small) and *far field* clusters where the distances are large. A clustering example is shown in Fig. 6.7 where the domain is systematically subdivided to determine collocation points that are close to each other. It proceeds in hierarchical manner until a specified minimum *leaf size* (=minimum number of collocation points in one box) is reached. Next we establish a criterion that determines if a cluster is admissible for treatment with ACA:

$$\min(\mathrm{diam}(S), \mathrm{diam}(F)) < \eta\, \mathrm{dist}(S, F) \tag{6.54}$$

where $\mathrm{diam}(S)$ is the diameter of the box containing source points and $\mathrm{diam}(F)$ the one containing field points. $\mathrm{dist}(S, F)$ is the smallest distance between them and η is a admissibility parameter (see Fig. 6.8). A possible partition of a matrix resulting from the clustering in Fig. 6.7 is shown in Fig. 6.9 where admissible blocks are coloured in green. The idea is to apply ACA to the admissible blocks and obtain a low rank approximation for them. In order to make sure collocation points lie within a bounding box we exploit the *convex hull* property of NURBS and use the control points instead of the collocation points to determine the limits of a bounding box. However, for patches with maximal smoothness the bounding box based on the control points would be rather large (see also Fig. 4.3). Hence, we define the bounding boxes based in the patch's Bézier elements. This is shown in Fig. 6.10 for a 2-D patch.

In Fig. 6.11, we show some examples of $\mathcal{H}$-matrices for a practical problem with increasing size. The numbers in the boxes indicate the rank of the approximation.

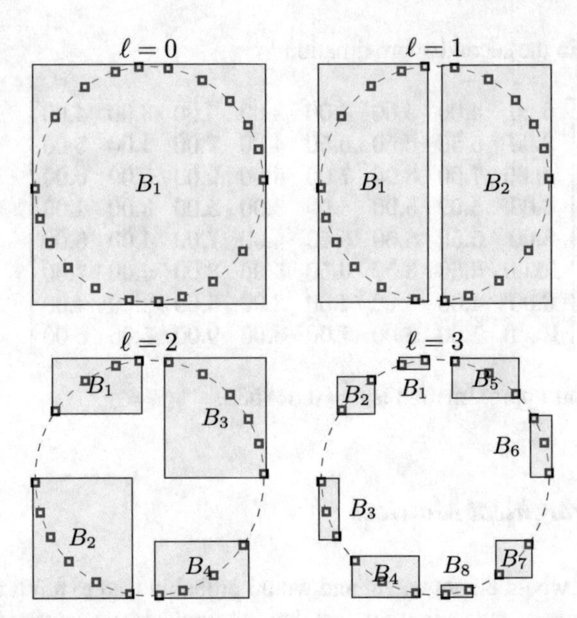

Fig. 6.7 Example of clustering with collocation points located on a circle and with a minimum leaf size of 2

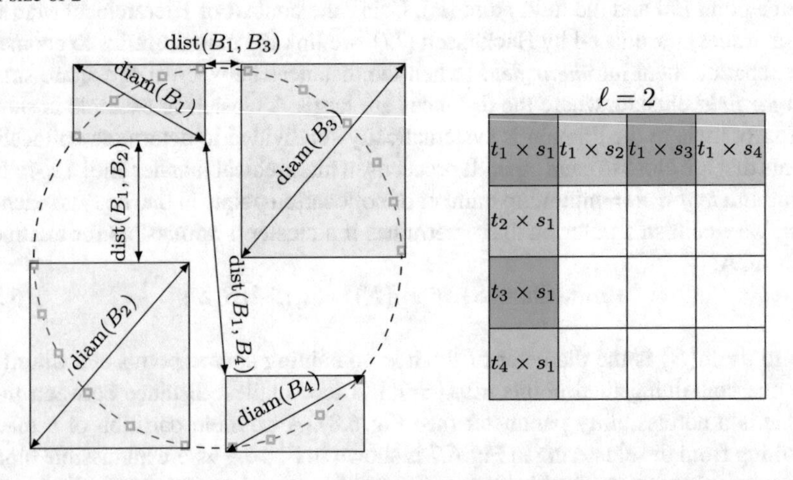

Fig. 6.8 Geometrical definitions for the check for admissibility between the box B_1 and the boxes B_i, $i = \{2, 3, 4\}$ of level $\ell = 2$. Blocks admissible for a low-rank approximation are coloured in green, whereas non-admissible blocks are coloured in red. White blocks remain to be checked

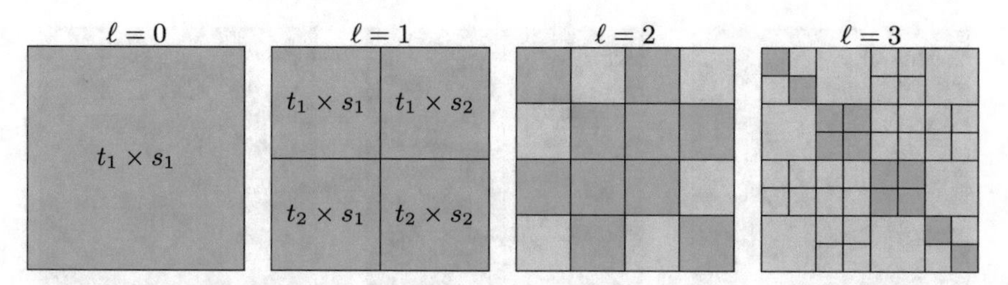

Fig. 6.9 Matrix partition into blocks up to level $\ell = 3$. Blocks admissible for a low-rank approximation are coloured in green

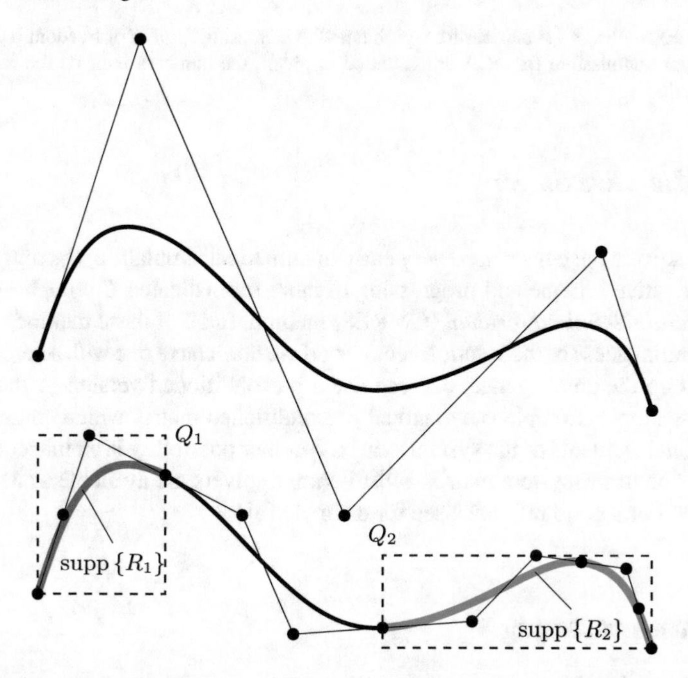

Fig. 6.10 Cubic NURBS curve with its control grid (top) and Bézier segments (bottom) determined by the knot vector for the convex hull. The dashed boxes Q_1 and Q_2 denote the bounding box for the support of selected NURBS basis functions

One can see that the number of admissible blocks increases rapidly with size. Due to the low rank approximations the size of the original matrix is reduced significantly. Compression rates that can be achieved are discussed in Chap. 9. However, the advantage of $\mathcal{H}$-matrices is not only in the reduction of storage but also that operations, such as matrix vector multiplications, can be carried out very fast, with almost linear complexity, i.e., $\mathcal{O}(n \log n)$. For details see [77]. We exploit this in the next discussion on solvers which involve matrix-vector products.

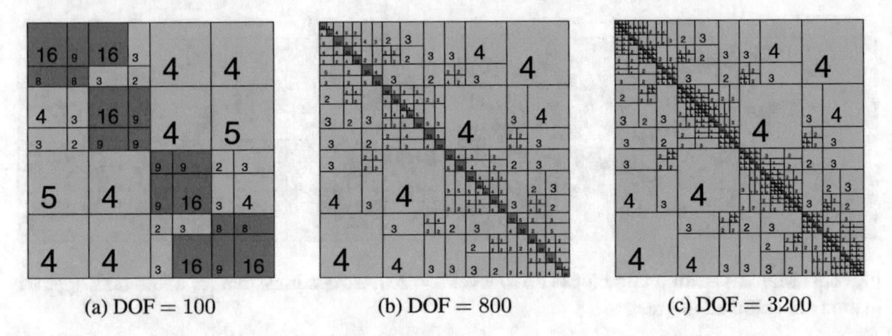

(a) DOF = 100 (b) DOF = 800 (c) DOF = 3200

Fig. 6.11 Examples of $\mathcal{H}$-matrices for a problem with increasing degrees of freedom (DOF) from [195]. Blocks admissible for ACA are coloured in green, the numbers indicate the rank of the approximation

6.6.5 *Iterative Solvers*

Iterative solvers have been used very early in numerical simulation, starting with the *Jacobi* iterative scheme and progressing to more sophisticated *Conjugate gradient* and *General minimised residual* (GMRES) method. In all of these methods we start with an initial guess of the solution vector (or if we don't have one with a zero vector). To speed up the convergence, one can use a preconditioned version of the system matrix. A simple example is a diagonal preconditioned matrix which contains only the diagonal elements of the system matrix. Another possibility is an inaccurate LU-factorisation of the system matrix. Most iterative solvers are available as MATLAB functions. For a good reference see for example [6].

6.7 Post-processing

After the solution, the parameter values at the collocation points are known. To get the physical values at any boundary point we first have to retrieve the patch parameters.

Algorithm 3 Algorithm for retrieval of patch parameters

Require: Patch matrices $\mathbf{Ldest}$, $\mathbf{Lass}$, solution vector $\mathbf{x}$
1: **for** e **to** Number of patches **do**
2: **for** i **to** Number patch parameters **do**
3: **if** $Lass(e, i) = 1$ **then**
4: $t(i, e) = \mathrm{x}(Ldest(e, i))$
5: $u(i, e) = \bar{u}(i, e)$
6: **else**
7: $u(i, e) = \mathrm{x}(Ldest(e, i))$
8: $t(i, e) = \bar{t}(i, e)$
9: **end if**
10: **end for**
11: **end for**
12: **return** $\mathbf{u}^e, \mathbf{t}^e$

The patch parameters are retrieved by

$$\mathbf{u}_i^e = \mathbf{x}_{Ldest(e,i)} \quad \text{if Neumann BC} \tag{6.55}$$

$$\mathbf{t}_i^e = \mathbf{x}_{Ldest(e,i)} \quad \text{if Dirichlet BC}\,. \tag{6.56}$$

The retrieval of patch parameters is shown in Algorithm 3. We recall that patch parameters are not physical values. Depending if the values are defined as BC or the solution, the physical values at a point $\boldsymbol{\xi}$ on patch e are obtained by

$$\mathbf{u}^e(\boldsymbol{\xi}) = \sum_{k=1}^{K} \hat{R}_k(\boldsymbol{\xi})\, \hat{\mathbf{u}}_k^e, \quad \mathbf{t}^e(\boldsymbol{\xi}) = \sum_{k=1}^{K} \hat{R}_k(\boldsymbol{\xi})\, \hat{\mathbf{t}}_k^e \quad \text{for the solution and} \tag{6.57}$$

$$\mathbf{u}^e(\boldsymbol{\xi}) = \sum_{k=1}^{K} \bar{R}_k(\boldsymbol{\xi})\, \bar{\mathbf{u}}_k^e, \quad \mathbf{t}^e(\boldsymbol{\xi}) = \sum_{k=1}^{K} \bar{R}_k(\boldsymbol{\xi})\, \bar{\mathbf{t}}_k^e \quad \text{for defined BCs.} \tag{6.58}$$

The value of $\mathbf{u}$ at any point $\boldsymbol{x}$ inside the domain is computed by

$$\mathbf{u}(\boldsymbol{x}) = \int_\Gamma \left[\mathsf{U}(\boldsymbol{x}, \hat{\boldsymbol{x}})\, \mathbf{t}(\hat{\boldsymbol{x}}) - \mathsf{T}(\boldsymbol{x}, \hat{\boldsymbol{x}})\, \mathbf{u}(\hat{\boldsymbol{x}})\right] d\Gamma(\hat{\boldsymbol{x}}). \tag{6.59}$$

Equation (6.59) can be written in matrix form:

$$\mathbf{u}(\boldsymbol{x}) = \mathbf{E}\mathrm{x} + \bar{\mathbf{u}}(\boldsymbol{x}) \tag{6.60}$$

where $\mathbf{E}$ is a matrix that is assembled in a similar way as shown in Algorithm 1 with the source point coordinates replaced by the coordinates of the internal point and $\mathbf{x}$ is the solution vector. $\bar{\mathbf{u}}(\boldsymbol{x})$ is the value of $\mathbf{u}$ at point $\boldsymbol{x}$ due to known values at the boundary. The flow vector $\mathbf{q}$ at a point $\boldsymbol{x}$ can be computed by

$$\mathbf{q}(\boldsymbol{x}) = \int_{\Gamma} [\mathsf{S}(\boldsymbol{x}, \hat{\boldsymbol{x}})\, \mathbf{t}(\hat{\boldsymbol{x}}) - \mathsf{R}(\boldsymbol{x}, \hat{\boldsymbol{x}})\, \mathbf{u}(\hat{\boldsymbol{x}})]\, d\Gamma(\hat{\boldsymbol{x}}) \tag{6.61}$$

which can be written in matrix notation

$$\mathbf{q}(\boldsymbol{x}) = \mathbf{S}\mathbf{x} + \bar{\mathbf{q}}(\boldsymbol{x}) \tag{6.62}$$

where $\mathbf{S}$ is an assembled matrix and $\bar{\mathbf{q}}(\boldsymbol{x})$ is the value of $\mathbf{q}$ at point $\boldsymbol{x}$ due to known values at the boundary.

The strains (in Voigt notation) are computed by

$$\{\varepsilon\}(\boldsymbol{x}) = \int_{\Gamma} \left[\hat{\mathsf{S}}(\boldsymbol{x}, \hat{\boldsymbol{x}})\, \mathbf{t}(\hat{\boldsymbol{x}}) - \hat{\mathsf{R}}(\boldsymbol{x}, \hat{\boldsymbol{x}})\, \mathbf{u}(\hat{\boldsymbol{x}})\right] d\Gamma(\hat{\boldsymbol{x}}) \tag{6.63}$$

and the stresses are given by:

$$\{\boldsymbol{\sigma}\}(\boldsymbol{x}) = \int_{\Gamma} [\mathsf{S}(\boldsymbol{x}, \hat{\boldsymbol{x}})\, \mathbf{t}(\hat{\boldsymbol{x}}) - \mathsf{R}(\boldsymbol{x}, \hat{\boldsymbol{x}})\, \mathbf{u}(\hat{\boldsymbol{x}})]\, d\Gamma(\hat{\boldsymbol{x}}). \tag{6.64}$$

S, R are derived fundamental solutions for flow or stress and $\hat{\mathsf{S}}$ and $\hat{\mathsf{R}}$ are derived solutions for strain. They are presented in the Appendix A. Equations (6.63) and (6.64) can also be written in matrix form. For the strain we can write

$$\{\varepsilon\} = \hat{\mathbf{E}}\mathbf{x} + \{\bar{\varepsilon}\} \tag{6.65}$$

where $\hat{\mathbf{E}}$ is an assembled matrix and $\{\bar{\varepsilon}\}$ specifies the value of $\{\varepsilon\}$ due to known values at the boundary. For the stress we have

$$\{\boldsymbol{\sigma}\} = \mathbf{S}\mathbf{x} + \{\bar{\boldsymbol{\sigma}}\} \tag{6.66}$$

To obtain accurate solutions for points near the boundary with a minimum of effort, *Regularisation* is introduced in the chapters dealing with specific applications.

6.8 Summary and Conclusions

In this chapter, we have discussed how the integral equations can be solved numerically. This involves several steps. First the integral equations have to be discretised. We choose the point collocation method whereby the integral equations are only satisfied at discrete points. It should be noted here that in contrast to the conventional BEM the collocation points are not nodal points but their locations are computed in the local patch coordinate system. Next, the fact has to be addressed that one of the integrals only exists as a Cauchy principal value and this is solved with an engineering approach.

The geometry of the boundary is defined using patches and NURBS basis functions. The patches differ from the conventional boundary elements as they are able to describe much more complex geometry with few parameters. Therefore patches tend to be significantly larger than boundary elements.

The final step in the discretisation is the approximation of the unknown parameters with basis functions. Here we introduced a geometry independent field approximation, which means that the approximation of the unknown parameters is independent of the definition of the geometry. The reason for this approach is that in most cases the definition of the geometry either represents the design geometry in CAD or – in the case of the test examples presented in this book – is already exactly represented by NURBS. In most cases we start with the basis functions, that have been used to define the geometry and then refine them until a satisfactory solution has been obtained.

The assembly of the system equations was explained. Since the left-hand side matrix is fully populated and not symmetric the solution effort increases very steeply with increasing number of unknowns if Gauss elimination is used. This has given rise to the development of fast solution methods, that introduce sparsity, using the fact that coefficients decay rapidly with the distance from the diagonal.

Finally, the recovery of physical results from the parameter values at the boundary has been discussed. Values inside the domain can be computed using the original and derived integral equations, but one has to be careful when the internal point is very near to the boundary because of the singularity of the Kernel. Regularisation, introduced next, helps here to reduce the integration effort and improve the quality of the results near the boundary.

It should be noted that it is a feature of the BEM that this chapter is valid for all physical problems discussed in this book.

Chapter 7
Numerical Integration

The integrals in the discretised integral equations involve Kernel basis function products and can only be evaluated analytically for low-order basis functions and for simple geometries. For practical problems, they have to be evaluated numerically. Numerical integration has been used in the FEM extensively. In contrast to the FEM, where integrands are well behaved, integration in the BEM is more problematic as the integrands tend to infinity with various degrees. Gauss Quadrature has been found to work well for the FEM and practical experience indicates that good results can also be obtained with the BEM. Gauss Quadrature was first used by Lachat and Watson [103] when they introduced higher-order basis functions for the description of the unknowns. It was emphasised by Watson [2] some time ago that it is crucially important to keep the error in the integration low, especially when one is near the singularity, which means that one has to increase the number of Gauss points or subdivide the region of integration as the singularity is approached.

In this chapter, we show how the integrals can be evaluated efficiently and accurately. Therefore we discuss the integration over patches for the assembly of the system matrices as discussed in the previous chapter. The integrals to be evaluated over a patch e are

$$\begin{aligned}
\mathbf{U}^e_{nk} &= \int_{\Gamma_e} \mathsf{U}\left(\tilde{\boldsymbol{x}}_n, \hat{\boldsymbol{x}}\right) \; R_k \; d\Gamma_e(\hat{\boldsymbol{x}}) \\
\mathbf{T}^e_{nk} &= \int_{\Gamma_e} \mathsf{T}\left(\tilde{\boldsymbol{x}}_n, \hat{\boldsymbol{x}}\right) \; R_k \; d\Gamma_e(\hat{\boldsymbol{x}}) \\
\mathbf{T}^e_{n} &= \int_{\Gamma_e} \mathsf{T}\left(\tilde{\boldsymbol{x}}_n, \hat{\boldsymbol{x}}\right) \; d\Gamma_e(\hat{\boldsymbol{x}})
\end{aligned} \tag{7.1}$$

where depending on the BC specified for the patch R_k is replaced by $\hat{R}_k$ or $\bar{R}_k$.

G. Beer et al., *The Isogeometric Boundary Element Method*, Lecture Notes in Applied and Computational Mechanics 90,
https://doi.org/10.1007/978-3-030-23339-6_7

7.1 Numerical Integration Methods

The two main integration methods used in this book are the classical Gauss Quadrature and the Gauss–Laguerre integration, where the latter is used for integrating a logarithmic singularity.

7.1.1 Gauss Quadrature

Gauss Quadrature is a quadrature rule constructed to yield an exact result for polynomials of order $2g - 1$ or less, by a suitable choice of g points (Gauss points) at which the function is evaluated.

We can write for the integral of a function $f(\bar{\xi})$

$$\int_{-1}^{1} f(\bar{\xi})d\bar{\xi} = \sum_{i=1}^{g} W_i f(\bar{\xi}_i) \tag{7.2}$$

where $\bar{\xi} \in [-1, 1]$ and g is the number of Gauss points. Their locations $\bar{\xi}_i$ and the weights W_i are published in [174].

Remark Note that we distinguish between the local coordinate system $\bar{\xi}$ used for the Gauss integration and the patch coordinate system ξ.

Equation (7.2) is only exact for polynomials up to $2g - 1$. For NURBS, which are rational functions, it is always an approximation. To determine the appropriateness of Gauss Quadrature for the integration of NURBS basis functions we test the accuracy of the integration of NURBS with varying order over a straight line, i.e. we compute

$$Int = \int_{-1}^{1} R_2(\xi)d\bar{\xi} \tag{7.3}$$

where R_2 is the second basis function that has been assigned a weight of 0.707 to make it a rational function.

Remark For this and the following tests the error of the integration is computed by

$$\text{Error} = \frac{|Int - Exact|}{|Exact|}. \tag{7.4}$$

We compute the "exact" value of the integral ($Exact$) numerically by increasing the number of Gauss points and subdividing the integration region until the difference in the result is less than 10^{-9}.

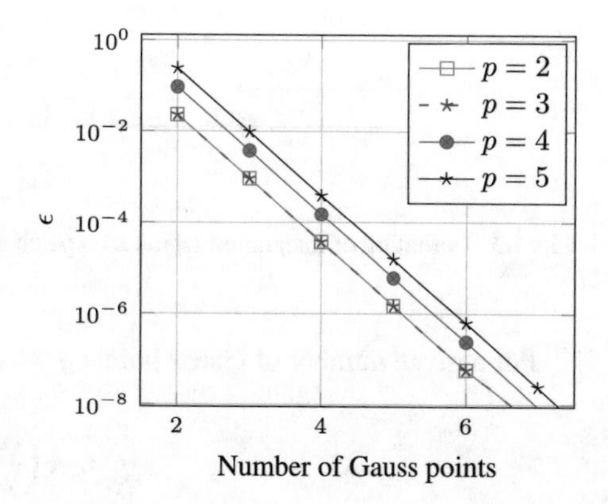

Fig. 7.1 Reduction of error ϵ with increasing number of Gauss points for NURBS basis functions of increasing order integrated over a straight line. The second basis function was assigned a weight of 0.707

In Fig. 7.1 we plot the error in the integration with increasing order of R. It is interesting that the same results are obtained for orders 2 and 3.

Stroud and Secrest [174] published the following error bound for estimating the error in integrating the function $f(\xi)$ with Gauss Quadrature

$$\epsilon = |\int_{-1}^{1} f(\bar{\xi})d\bar{\xi} - \sum_{i=1}^{g} W_i f(\bar{\xi}_i)| \leqslant \frac{4}{2^{2g}(2g)!} H_g \tag{7.5}$$

where H_g depends on the $2g$th derivative of f

$$H_g \leqslant \frac{\partial^{2g} f(\bar{\xi})}{\partial \bar{\xi}^{2g}}. \tag{7.6}$$

The function that we need to integrate contains the Kernel, the basis function and the Jacobian and it would be impossible to get an error estimate. The value of the function that changes most rapidly is the singular part of the Kernel which varies $(\frac{1}{r^n})$. We can get a very rough estimate of the error by assuming the patch to be a straight line of length L and by placing the source of the Kernel at a distance R on the left as shown in Fig. 7.2.

Substituting $f(\bar{\xi}_i) = \frac{1}{r^n}$ into Eq. (7.6) and after some manipulation Watson [2] presented formulae for determining the number of Gauss points required to ensure a uniform precision of integration depending on the proximity of the source point to the integration region. The idea is that we adjust the number of Gauss points depending on the ratio of L to R.

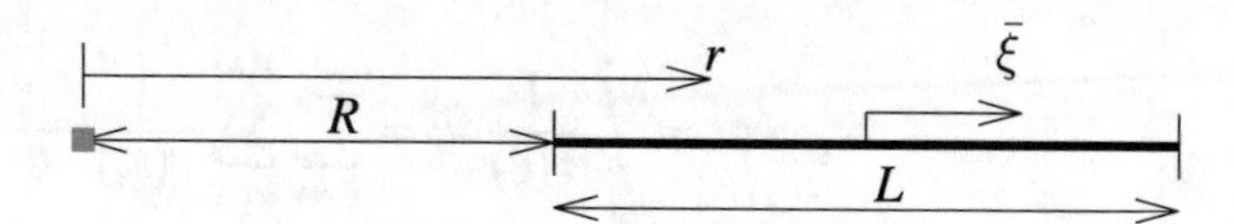

Fig. 7.2 Simplified setup to obtain a rough estimate of the integration error

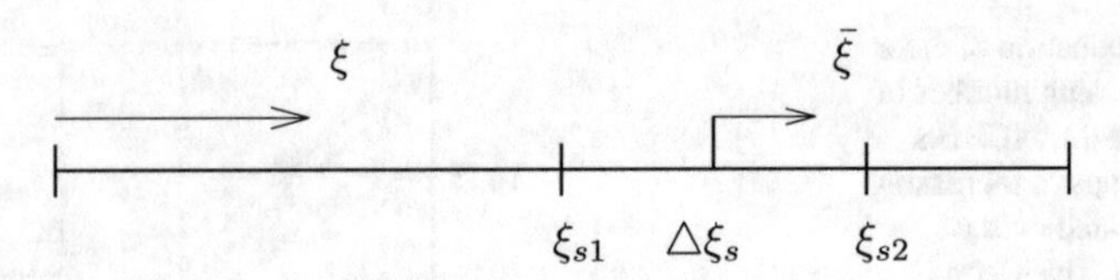

Fig. 7.3 Definition of integration region s on patch e

For a given number of Gauss points g we can write for $f = \frac{1}{r}$

$$\frac{L}{R} \leqslant 4\left(\frac{\alpha}{2}\right)^{\frac{1}{2g}} \tag{7.7}$$

where α is a constant that determines the precision of integration (a value of 0.01 is suggested).

If we limit the maximum number of Gauss points to g_{max} then we have to subdivide the region of integration. The number of subdivisions S required is given by

$$S \geqslant Integer\left(\frac{L}{4R}\left(\frac{2}{\alpha}\right)^{\frac{1}{g_{max}}}\right). \tag{7.8}$$

The transformation between the NURBS coordinates ξ and Gauss coordinates $\bar{\xi}$ is for subregion s (see Fig. 7.3)

$$\xi = \frac{\triangle\xi_s}{2}(1+\bar{\xi}) + \xi_{s1} \tag{7.9}$$

where $\triangle\xi_s$ is the size of the subregion and ξ_{s1} is the starting coordinate. The Jacobian of this transformation is $\frac{\triangle\xi_s}{2}$.

This is a very simplified determination of the required number of Gauss points as it only considers the singular part of the Kernel and that the source of the Kernel is aligned to the patch, which is the most favourable case. To check the relevance of the proposed equations to predict the number of Gauss points for a given accuracy we perform a numerical experiment.

Assume the set-up as shown in Fig. 7.4 with $L = 1$ but now we allow the source point to be at different locations within the left lower quadrant. We compute the "exact" value of the integral numerically as outlined above and then compare with the value computed with Gauss quadrature using Eq. (7.7) with $\alpha = 0.01$ to determine the number of Gauss points $g(s)$ for each subregion s and Eq. (7.8) to determine the number of subregions S. The numerical result is obtained by

$$Int = \int_{-1}^{1} \frac{1}{r(\xi)} d\xi = \sum_{s=1}^{S}\sum_{i=1}^{g(s)} \frac{1}{r(\xi_i)} \frac{\triangle\xi_s}{2} W_i. \tag{7.10}$$

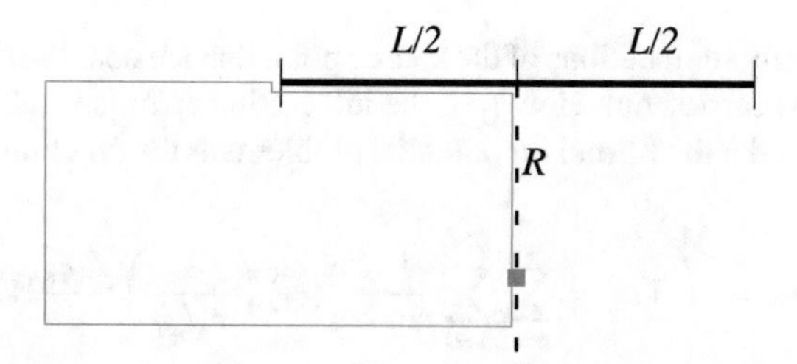

Fig. 7.4 Test setup for predicting the integration error showing the area where the source point can be located in red. For determining the table of admissible R/L values we locate the source point along the dashed line

Figure 7.5(a) shows contours of the error, where the value of the error is displayed at the location of the source point. It can be seen that, as expected, the error is not

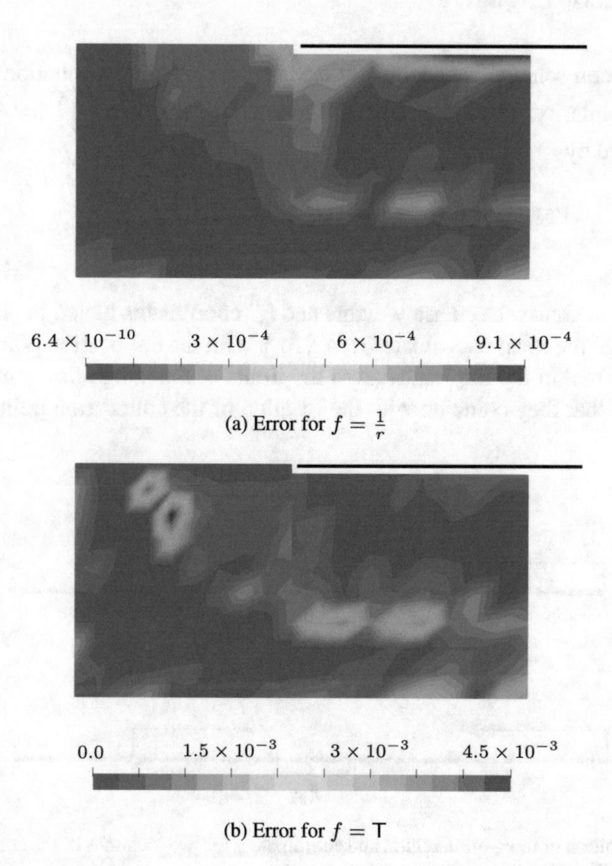

Fig. 7.5 Integration error for $f = \frac{1}{r}$ and $f = \mathsf{T}$ at different locations of the source point according to the integration scheme by Watson [2]

uniform, especially for locations of the source point that are near the line along which the integration is carried out. However, the integration error is small (below 10^{-3}).

Next, we consider the Kernel for potential problems as the function to be integrated

$$Int = \int_{-1}^{1} \mathsf{T} d\xi = \sum_{s=1}^{S} \sum_{i=1}^{g(s)} \frac{1}{2\pi} \frac{\partial r}{\partial n}(\xi_i) \left(\frac{1}{r(\xi_i)} \right) \frac{\Delta \xi_s}{2} W_i. \tag{7.11}$$

The result is shown in Fig. 7.5(b). Note that in this case the integral along the extension of the line, over which we integrate, is zero as $\frac{\partial r}{\partial n}$ is zero. The maximum error is increased and for this case the method does not seem to be very good in predicting a uniform accuracy.

7.1.2 Gauss–Laguerre

This integration scheme is specifically designed to integrate a function with a logarithmic singularity. The integral of a function $f(\xi)$ times $\ln\left(\frac{1}{\xi}\right)$ for $\xi \in [0, 1]$ is approximated by

$$\int_{0}^{1} f(\xi) \ln\left(\frac{1}{\xi}\right) d\xi \approx \sum_{i=1}^{g} W_i^L f(\xi_i^L). \tag{7.12}$$

where W_i^L are Gauss–Laguerre weights and ξ_i^L coordinates tabled in [174].

We change the local coordinate from ξ to γ with the collocation point located at $\gamma = 0$. Note that in the implementation the limits of the integration regions will be chosen such that they coincide with the location of the collocation points as shown in Fig. 7.6.

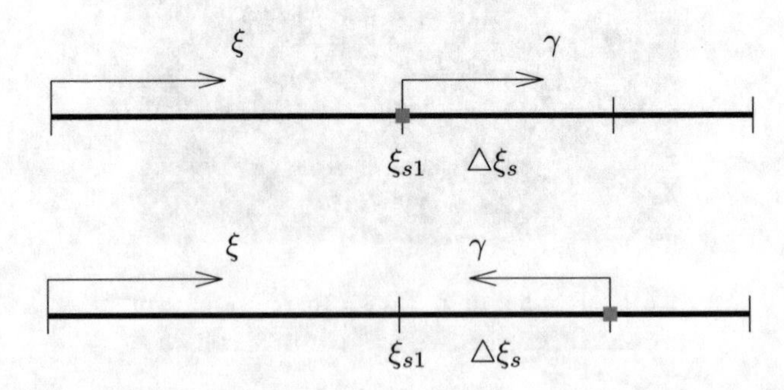

Fig. 7.6 Definition of integration region and coordinate γ for two locations of the collocation point depicted by a filled square

The transformation between ξ- and γ-coordinates depends on the location of the collocation point and is given by

$$\begin{aligned} \xi &= \Delta\xi_s\gamma + \xi_{s1} && \text{if the collocation point is on the left} \\ \xi &= \Delta\xi_s(1-\gamma) + \xi_{s1} && \text{if it is on the right.} \end{aligned} \quad (7.13)$$

7.2 Computation of the Shortest Distance of a Point to a Patch

In the following we will require to compute the shortest distance of the point $\boldsymbol{x}_P$ to the patch.

We explain this here for a 3-D patch, the application to a 2-D patch can be easily deduced. For this we use a Newton–Raphson Algorithm 4 with reference to Fig. 7.7.

For starting the iteration, we choose a point $\boldsymbol{x}_{S0}$ on the surface of the patch with the local coordinates ξ_0, η_0. We obtain an approximation of the local coordinate of the desired point on the surface of the patch, by computing the vector dot product of the vector $(\boldsymbol{x}_P - \boldsymbol{x}_S)$ with the unit vectors in ξ- and η-directions, dividing by the length of the patch in the relevant direction. We then iteratively refine until the minimum distance R does not change within a specified tolerance. To make sure that the iteration does not diverge for extreme geometries, we may apply a damping factor.

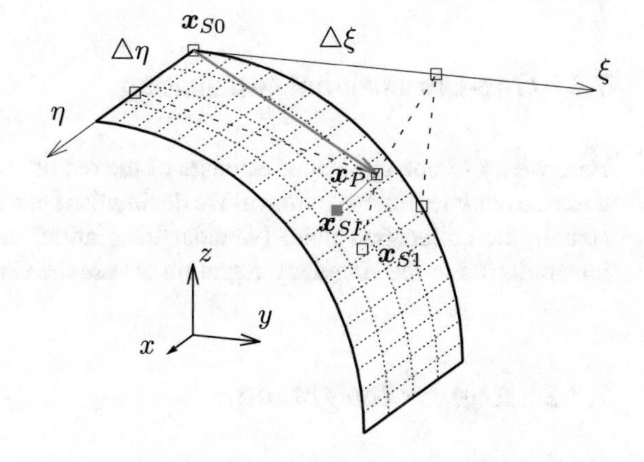

Fig. 7.7 Explanation of the procedure to compute the minimum distance between $\boldsymbol{x}_P$ and the patch surface, showing the location after the first iteration $\boldsymbol{x}_{S1}$ and final location $\boldsymbol{x}_{SI}$ of the point on the surface. The selected starting point coordinates are $\xi_0 = 0, \eta_0 = 0$

Algorithm 4 Algorithm for computing minimum distance to patch surface

Require: Point coordinates $\boldsymbol{x}_P$, Geometry information of patch, Max. number of iterations I , damping factor (damp) and Tolerance.

1: Select starting point coordinates ξ_0, η_0 and compute $\boldsymbol{x}_{S0}$
2: Compute $R_0 = \|\boldsymbol{x}_P - \boldsymbol{x}_{S0}\|$
3: **for** $i = 0$ **to** Max. number of iterations I **do**
4: Compute vectors $\mathbf{v}_1$ and $\mathbf{v}_2$ in ξ- and η-directions.
5: Compute patch lengths in ξ and η directions, L_ξ, L_η
6: $\triangle\xi = \mathbf{v}_1 \cdot (\boldsymbol{x}_P - \boldsymbol{x}_{Si})/L_\xi$
7: $\triangle\eta = \mathbf{v}_2 \cdot (\boldsymbol{x}_P - \boldsymbol{x}_{Si})/L_\eta$
8: $\xi_{i+1} = \xi_i + \triangle\xi$*damp
9: **if** $\xi_{i+1} > 1$ **then**
10: $\xi_{i+1} = 1$
11: **else if** $\xi_{i+1} < 0$ **then**
12: $\xi_{i+1} = 0$
13: **end if**
14: $\eta_{i+1} = \eta_i + \triangle\eta$*damp
15: **if** $\eta_{i+1} > 1$ **then**
16: $\eta_{i+1} = 1$
17: **else if** $\eta_{i+1} < 0$ **then**
18: $\eta_{i+1} = 0$
19: **end if**
20: Compute new point $\boldsymbol{x}_{Si+1}(\xi_{i+1}, \eta_{i+1})$ on surface.
21: Compute distance $R_{i+1} = \|\boldsymbol{x}_P - \boldsymbol{x}_{Si+1}\|$
22: **if** $R_{i+1} - R_i$ <Tolerance **then**
23: Exit loop
24: **end if**
25: **end for**
26: **return** Minimum distance R and local coordinates of point on surface ξ, η.

7.3 One-Dimensional Integration

Here we try to obtain a better estimate of the required number of Gauss points and describe our integration approach. We distinguish between integration regions which contain the collocation point (singular integration) and ones that do not (regular integration). For the singular integration we use the Gauss–Laguerre scheme.

7.3.1 Regular Integration

The Kernel T has the higher singularity and we therefore concentrate on this Kernel. We can improve the integration scheme proposed by Watson using a different strategy. Instead of trying to estimate the integration error using a formula, we determine it numerically. Specifically, we calculate the integration error as a function of the number of Gauss points and the distance R of points, which we now locate at midpoint and perpendicular to L instead along it (see Fig. 7.4). In this way, we consider the worst case instead of the most favourable case with the Watson formula.

The results for integration of the Kernel T (for potential problems) over a straight line with length L is shown in Fig. 7.8. It can be seen that using a log scale the error is represented by straight lines. We can now choose limiting values of R/L for a particular number of Gauss points depending on the maximum integration error that we want to accept. In other words, we use the diagram to determine the number of Gauss points necessary for a given R/L ratio by intersecting horizontal lines for the required precision of integration with the error lines for particular values of R/L.

The results are summarised in Table 7.1. To interpret the table we note for example that for a target accuracy of 10^{-2} and 3 Gauss points the minimum distance of the source point to the integration region (R) cannot be smaller than $0.5L$ to the integration region (i.e. $R/L > 0.5$).

The regular integration Eq. (7.1) can be now rewritten as

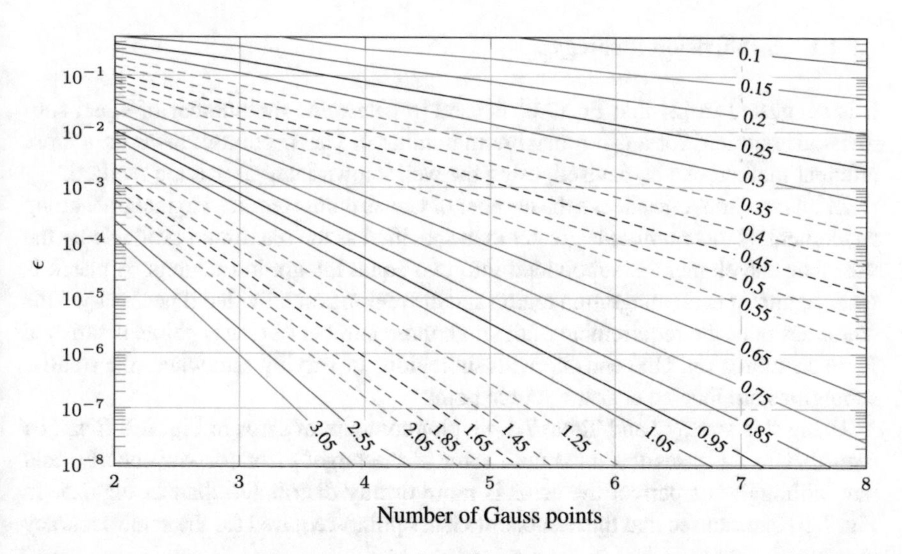

Fig. 7.8 Plot of error ϵ in the integration as a function of number of Gauss points and the proximity of the point R/L which is indicated by the labels of the graphs

Table 7.1 Limits on R/L for different number of Gauss points and acceptable maximum error (Error)

Gauss points	2	3	4	5	6	7	8
Error $= 10^{-2}$	0.85	0.50	0.33	0.30	0.25	0.25	0.28
Error $= 10^{-3}$	1.45	0.85	0.55	0.45	0.35	0.30	0.30
Error $= 10^{-4}$	2.55	1.25	0.85	0.65	0.50	0.40	0.35

$$\mathbf{U}^e_{nk} = \sum_{s=1}^{S}\sum_{i=1}^{g(s)} \mathsf{U}\left(\tilde{\boldsymbol{x}}_n, \hat{\boldsymbol{x}}(\bar{\xi}_i)\right) R_k(\xi(\bar{\xi}_i))\frac{\Delta\xi_s}{2}J(\bar{\xi}_i)W_i$$
$$\mathbf{T}^e_{nk} = \sum_{s=1}^{S}\sum_{i=1}^{g(s)} \mathsf{T}\left(\tilde{\boldsymbol{x}}_n, \hat{\boldsymbol{x}}(\bar{\xi}_i)\right) R_k(\xi(\bar{\xi}_i))\frac{\Delta\xi_s}{2}J(\bar{\xi}_i)W_i \tag{7.14}$$
$$\mathbf{T}^e_{n} = \sum_{s=1}^{S}\sum_{i=1}^{g(s)} \mathsf{T}\left(\tilde{\boldsymbol{x}}_n, \hat{\boldsymbol{x}}(\bar{\xi}_i)\right) \frac{\Delta\xi_s}{2}J(\bar{\xi}_i)W_i$$

where S is the number of integration regions, $g(s)$ is the number of Gauss points for region s and J is the Jacobian of the transformation from global coordinates x, y to the local ξ coordinate. The strategy to set up the integration regions is discussed next.

7.3.1.1 Subdivision Strategy

It is suggested in [2] that Eq. (7.8) is used to determine the number of equal subdivisions required for a given maximum number of Gauss points. However, a more efficient method can be devised, using the well known interval halving method.

In this method, one checks the number of Gauss points required to integrate along an element. If this number is greater than specified as the maximum available by the user then the element is subdivided into two equal integration regions. A check is made again for each integration region and the region is subdivided if necessary. This continues until the requirement of the maximum number of Gauss points is satisfied for each subregion. One ends up with subregions of varying size where the smaller subregions are located near the source point.

Using this strategy and Table 7.1 we plot contours of error in Fig. 7.9. It can be seen that in all cases the maximum error is the target error (or very nearly) and that, although not perfect the error is more evenly distributed than in Fig. 7.5. In Fig. 7.10 one can see that the number of Gauss points required for the same accuracy is reduced.

7.3.2 Singular Integration

For the case where the collocation point is inside an integration region the Kernel U tends to infinity with $\ln\left(\frac{1}{r}\right)$ and a Gauss–Laguerre scheme has to be employed.

For this we isolate the logarithmic term and the fundamental solution can be written as

$$\mathsf{U} = \mathsf{U}_L \ln\left(\frac{1}{r}\right) + \mathsf{U}_R \tag{7.15}$$

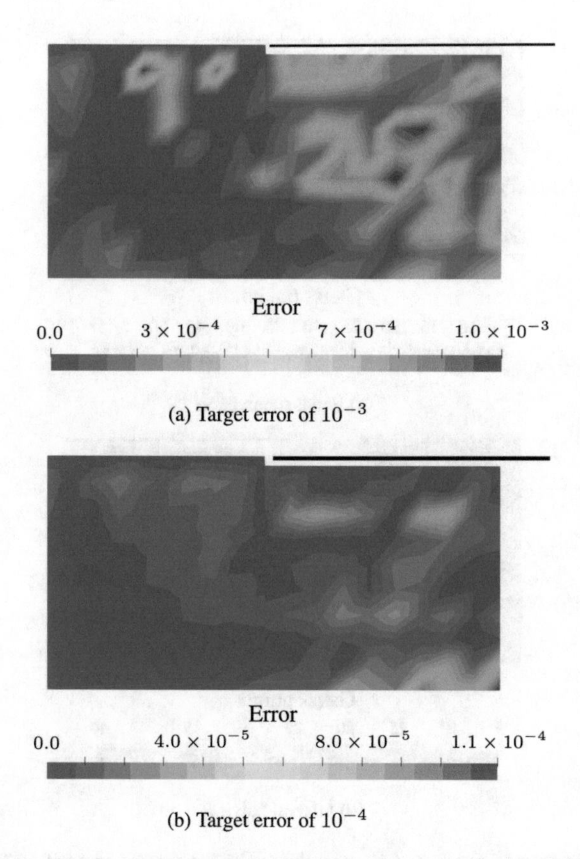

(a) Target error of 10^{-3}

(b) Target error of 10^{-4}

Fig. 7.9 Error in the integration of Kernel T for a target error of 10^{-3} and 10^{-4}

where U_L is the part of the fundamental solution that is multiplied with ln and U_R is the remainder. We can re-write the integral in Eq. (7.1) for the integration region s

$$\mathbf{U}^e_{nk} = \int_{-1}^{1} \left[\mathsf{U}\left(\tilde{\boldsymbol{x}}_n, \hat{\boldsymbol{x}}(\bar{\xi})\right) - \mathsf{U}_L \ln \frac{1}{\gamma} \right] R_k(\xi(\bar{\xi})) \frac{\Delta \xi_s}{2} J d\bar{\xi}$$
$$+ \int_{0}^{1} [\mathsf{U}_L\ R_k(\xi(\gamma)) \Delta \xi_s J] \ln \frac{1}{\gamma} d\gamma. \tag{7.16}$$

The singularity is cancelled out in the first integral, which can now be integrated using standard Gauss, whereas for the second integral the Gauss–Laguerre scheme can be applied leading to

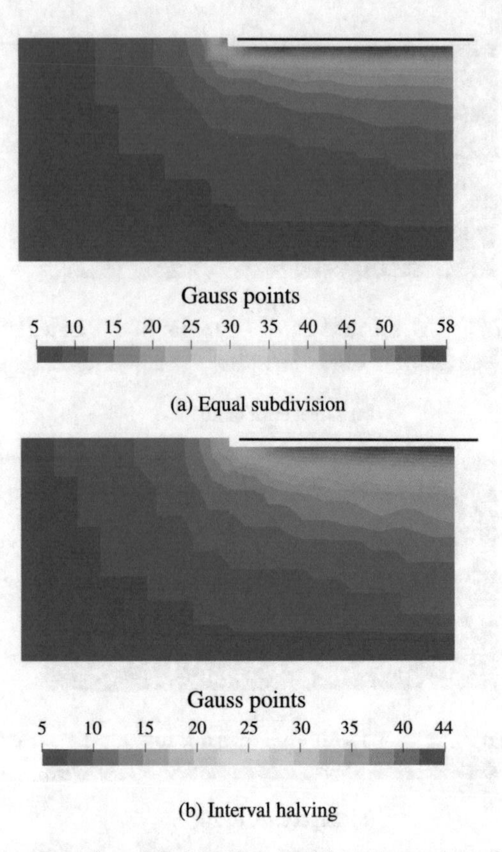

(a) Equal subdivision

(b) Interval halving

Fig. 7.10 Comparison of number of Gauss points used for a target error of 10^{-4} and (a) equal subdivision and (b) interval halving method

$$\mathbf{U}_{nk}^{e} = \sum_{i=1}^{g} \left[\mathsf{U}\left(\tilde{\boldsymbol{x}}_n, \hat{\boldsymbol{x}}(\bar{\xi}_i)\right) - \mathsf{U}_L \ln \frac{1}{\gamma(\bar{\xi}_i)} \right] R_k(\xi(\bar{\xi}_i)) \frac{\Delta\xi_s}{2} J(\bar{\xi}_i) W_i$$
$$+ \sum_{i=1}^{g^L} \left[\mathsf{U}_L \, R_k(\xi(\gamma_i)) \Delta\xi_s J(\xi(\gamma_i)) \right] W_i^L . \tag{7.17}$$

To find out the number of Gauss points required we integrate the Kernel U multiplied with a NURBS with increasing order over a straight line of length 1, where the second basis function has been applied a weight of 0.707. When calculating the error we assume that the "exact" value is obtained with 8 Gauss points. The result is shown in Fig. 7.11. It can be seen that for most cases 4 Gauss points should suffice to provide an accuracy of 10^{-4}.

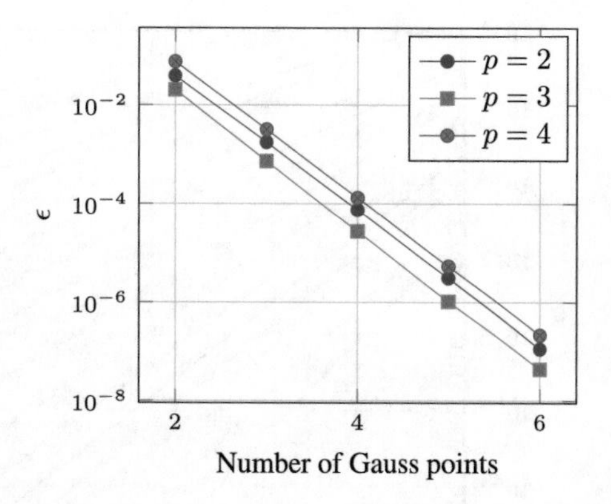

Fig. 7.11 Error ϵ for the singular integration of U times a NURBS over a straight line

7.4 Two-Dimensional Integration

This applies to all 3-D problems discussed in the book. We distinguish again between integration elements which contain the collocation point (singular integration) and ones that do not (regular integration).

7.4.1 Regular Integration

For the regular integration Kernel T has the higher singularity ($\mathcal{O}(\frac{1}{r^2})$) and we therefore concentrate on this Kernel. We re-calculate the integration error over a patch with dimensions 1×1 as a function of the number of Gauss points and the distance R of points, where the points are located on a line perpendicular to the patch in the middle of one side (see Fig. 7.12).

The results for integration of the Kernel T for potential problems is shown in Fig. 7.13.

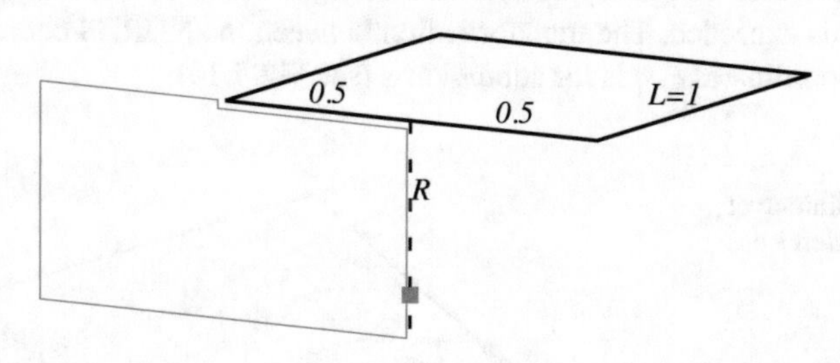

Fig. 7.12 Test setup for calculating the integration error depending on $\frac{R}{L}$. The location of the source point is along the dashed line. The red area indicates the possible location of source points for the error plots

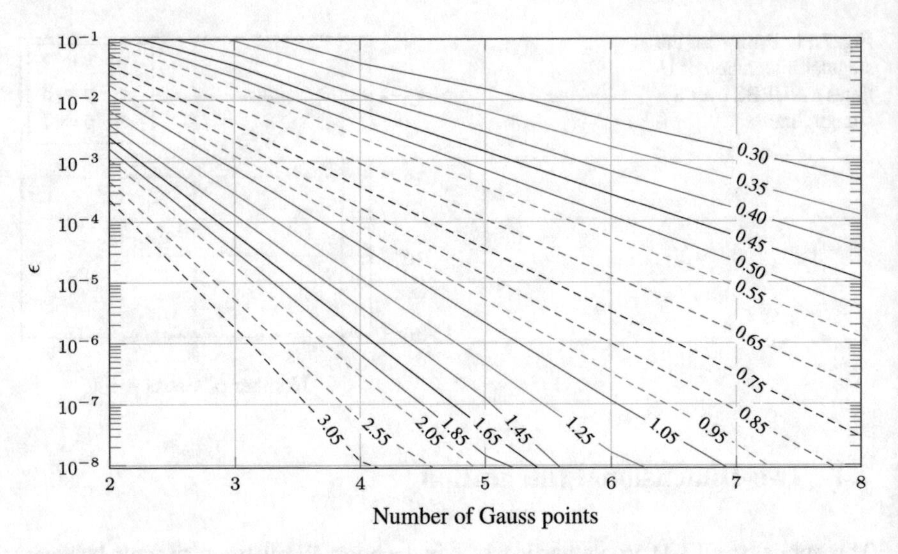

Fig. 7.13 Plot of error ϵ in the integration of T as a function of number of Gauss points and the proximity of the point R/L which is indicated by the labels of the graphs

Table 7.2 Limits on R/L for different number of Gauss points and acceptable maximum error

Gauss points	2	3	4	5	6	7	8
Error $= 10^{-2}$	0.95	0.65	0.45	0.35	0.30	0.30	0.30
Error $= 10^{-3}$	1.85	0.95	0.65	0.50	0.45	0.35	0.35
Error $= 10^{-4}$	3.05	1.45	0.95	0.75	0.55	0.50	0.40

Table 7.2 provides limits of R/L for the number of Gauss points and a given accuracy. We assume that this table is also valid for elasticity problems. We use the table with $L = L_\xi$ in the ξ-direction to obtain the required number of Gauss points g_ξ in that direction and $L = L_\eta$ in the η-direction to obtain the number of Gauss points g_η. As before we specify a maximum number of Gauss points and subdivide if this number is exceeded. The transformation between the NURBS coordinates ξ, η and Gauss coordinates $\bar{\xi}, \bar{\eta}$ is for subregion s (see Fig. 7.14)

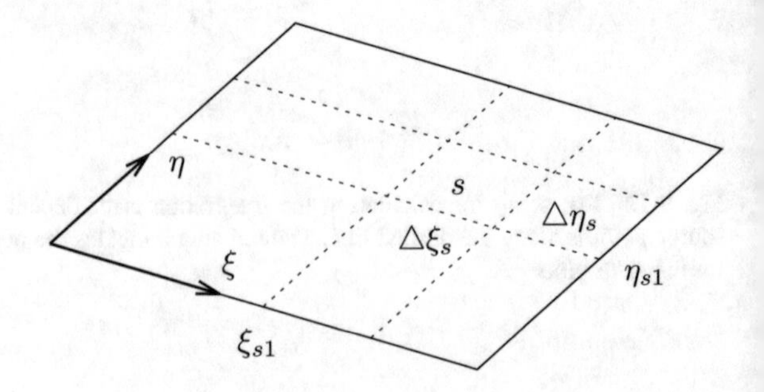

Fig. 7.14 Definition of integration region s on patch e

$$\xi = \frac{\triangle\xi_s}{2}(1+\bar{\xi}) + \xi_{s1} \tag{7.18}$$

$$\eta = \frac{\triangle\eta_s}{2}(1+\bar{\eta}) + \eta_{s1} \tag{7.19}$$

where $\triangle\xi_s \times \triangle\eta_s$ is the size of the subregion and ξ_{s1}, η_{s1} are the starting coordinates. The Jacobian of this transformation is $\frac{\triangle\xi_s}{2} \times \frac{\triangle\eta_s}{2}$.

For the regular integration Eq. (7.1) can be now rewritten as

$$\begin{aligned}
\mathbf{U}^e_{nk} &= \sum_{s=1}^{S}\sum_{i=1}^{g_\xi(s)}\sum_{j=1}^{g_\eta(s)} \mathsf{U}\left(\tilde{\boldsymbol{x}}_n, \hat{\boldsymbol{x}}(\bar{\xi}_i, \bar{\eta}_j)\right)\ R_k(\xi(\bar{\xi}_i), \eta(\bar{\eta}_j))\ J_{sij} \\
\mathbf{T}^e_{nk} &= \sum_{s=1}^{S}\sum_{i=1}^{g_\xi(s)}\sum_{j=1}^{g_\eta(s)} \mathsf{T}\left(\tilde{\boldsymbol{x}}_n, \hat{\boldsymbol{x}}(\bar{\xi}_i, \bar{\eta}_j)\right)\ R_k(\xi(\bar{\xi}_i), \eta(\bar{\eta}_j))\ J_{sij} \\
\mathbf{T}^e_{n} &= \sum_{s=1}^{S}\sum_{i=1}^{g_\xi(s)}\sum_{j=1}^{g_\eta(s)} \mathsf{T}\left(\tilde{\boldsymbol{x}}_n, \hat{\boldsymbol{x}}(\bar{\xi}_i, \bar{\eta}_j)\right)\ J_{sij}
\end{aligned} \tag{7.20}$$

where

$$J_{sij} = \frac{\triangle\xi_s}{2}\frac{\triangle\eta_s}{2} J(\bar{\xi}_i, \bar{\eta}_j) W_i W_j, \tag{7.21}$$

S is the number of integration regions and J is the Jacobian of the transformation from global coordinates $\boldsymbol{x}$ to local $\boldsymbol{\xi}$ coordinates.

7.4.1.1 Subdivision Strategy

Similar to the interval halving method, we introduce a Quadtree method to efficiently evaluate the integral (see Algorithm 5).

In this method, one determines the number of Gauss points required to integrate in each direction. If this number is greater than specified by the user, then the integration region in this direction is subdivided into 2 equal parts, otherwise no action is taken in this direction. One ends up with 4 or 2 or 0 subregions.

A check is made again for each subregion and if necessary this region is again subdivided. This continues until the requirement of the maximum number of Gauss points in each direction is satisfied for each subregion. One ends up with subregions of varying size where the smaller subregions are located near the source point. In the algorithm, the **Function** *Quadtree* calls itself until all requirements are satisfied.

In Fig. 7.15 we show the subdivisions and Gauss point locations for the case when the source point is near the integration region. In Fig. 7.16 we see the distribution of error for a target error of 10^{-3} and the number of Gauss points required. This is compared with the normal method where subdivision is made into subregions of equal size in Fig. 7.17. We see that a much larger number of Gauss points is required if the Quadtree method is not applied.

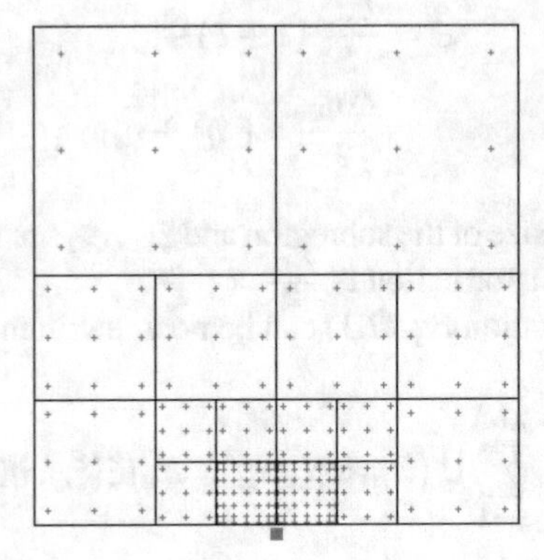

Fig. 7.15 Example of the Quadtree subdivision method: subdivisions and Gauss point locations for the source point depicted by a filled square

Algorithm 5 Algorithm for the Quadtree method

Require: Local coordinates defining the integration region, Proximity of source point (R/L_ξ, R/L_η), Integration table, Maximum number of Gauss points (MaxGP)

1: Check the number of Gauss points g_ξ, g_η required for integration to the specified accuracy
2: **if** (g_ξ and g_η<= MaxGP) **then**
3: Compute Gauss point coordinates
4: **return**
5: **end if**
6: **Function** *Quadtree*
7: **if** g_ξ > MaxGP **then**
8: Divide the integration region in ξ direction into 2 equal parts.
9: **end if**
10: **if** g_η > MaxGP **then**
11: Divide the integration region in η direction into 2 equal parts.
12: **end if**
13: **for** n=1 to total number of subregions **do**
14: Check the number of Gauss points g_ξ, g_η required for integration to the specified accuracy
15: **if** g_ξ and g_η <= MaxGP **then**
16: Compute Gauss point coordinates and store for subregion
17: **else**
18: Call *Quadtree*
19: **end if**
20: **end for**
21: **End function** *Quadtree*
22: **return** Gauss point locations for all subregions.

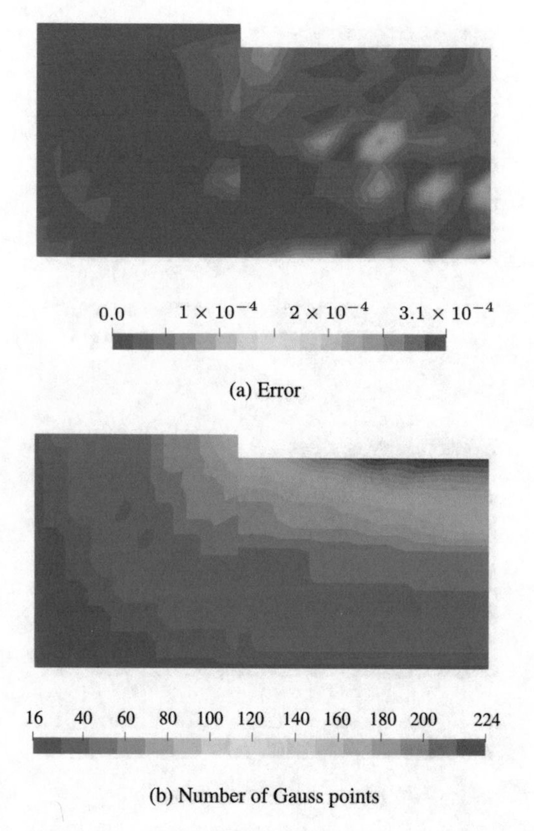

Fig. 7.16 Distribution of integration error for a target error of 10^{-3} and number of Gauss points required for the Quadtree method

7.4.2 Singular Integration

If the collocation point is inside the element, the integrand involving the Kernel U is weakly singular. To evaluate this integral we subdivide the integration region into triangular subregions where the Jacobian tends to zero as the collocation point is approached. In Fig. 7.18 we show the subdivision of an integration region into 4 subregions. This is the case when the collocation point is inside the region. If the collocation point is located on one of the sides, only 3 subregions are necessary and if it is located at one of the corners only 2 subregions are required.

The mapping from the Gauss coordinate system $\bar{\xi}, \bar{\eta}$ to the patch coordinate system ξ, η can be achieved by using linear B-spline basis functions B_i with knot vectors

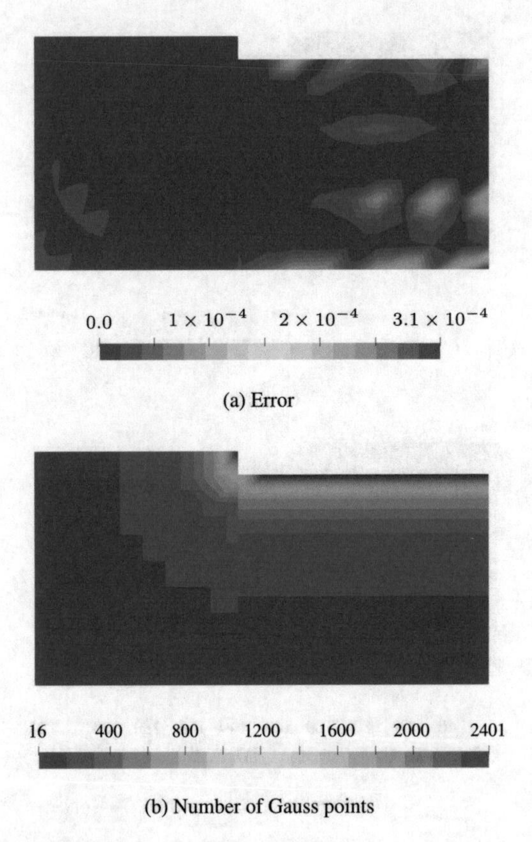

Fig. 7.17 Distribution of integration error for a target error of 10^{-3} and number of Gauss points required for the normal subdivision method

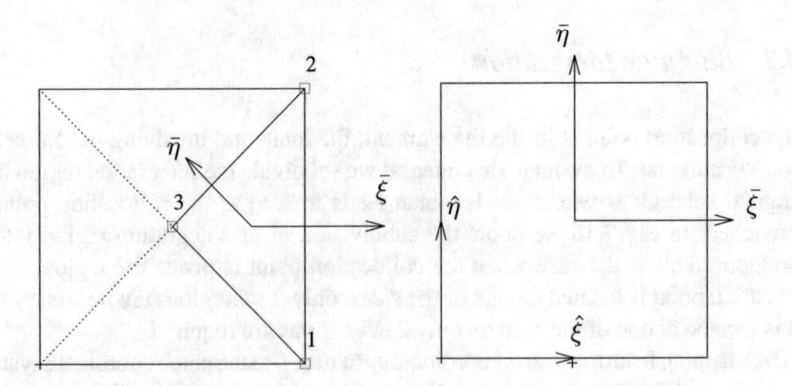

Fig. 7.18 Singular integration: subdivision of integration region into 4 triangular subregions. The mapping to the local coordinates $\bar{\xi}, \bar{\eta}$ of the highlighted sub-region is shown

$\Xi_1 = (0,0,1,1)$ and $\Xi_2 = (0,0,1,1)$. There are two mappings involved. One from the ξ, η to $\hat{\xi}, \hat{\eta}$ coordinate system

$$\xi = \sum_{i=1}^{4} B_i(\hat{\xi}, \hat{\eta})\, \xi_i \qquad \text{and} \qquad \eta = \sum_{i=1}^{4} B_i(\hat{\xi}, \hat{\eta})\, \eta_i \tag{7.22}$$

where ξ_i, η_i are the local coordinates of the apexes of the triangle, and we specify the third and fourth point to be coincidental with the collocation point. The second transformation is from the $\hat{\xi}, \hat{\eta}$ to $\bar{\xi}, \bar{\eta}$ coordinate system

$$\hat{\xi} = \frac{1}{2}(\bar{\xi} + 1) \qquad \text{and} \qquad \hat{\eta} = \frac{1}{2}(\bar{\eta} + 1) \tag{7.23}$$

The Jacobi matrix of the mapping (7.22) is for triangle n_t

$$\mathbf{J}_{\xi,n_t} = \begin{pmatrix} \partial\xi/\partial\hat{\xi} & \partial\eta/\partial\hat{\xi} \\ \partial\xi/\partial\hat{\eta} & \partial\eta/\partial\hat{\eta} \end{pmatrix} \tag{7.24}$$

where

$$\begin{aligned} \frac{\partial\xi}{\partial\hat{\xi}} &= \sum_{i=1}^{4} \frac{\partial B_i(\hat{\xi},\hat{\eta})}{\partial\hat{\xi}}\, \xi_i & \frac{\partial\eta}{\partial\hat{\xi}} &= \sum_{i=1}^{4} \frac{\partial B_i(\hat{\xi},\hat{\eta})}{\partial\hat{\xi}}\, \eta_i \\ \frac{\partial\xi}{\partial\hat{\eta}} &= \sum_{i=1}^{4} \frac{\partial B_i(\hat{\xi},\hat{\eta})}{\partial\hat{\eta}}\, \xi_i & \frac{\partial\eta}{\partial\hat{\eta}} &= \sum_{i=1}^{4} \frac{\partial B_i(\hat{\xi},\hat{\eta})}{\partial\hat{\eta}}\, \eta_i . \end{aligned} \tag{7.25}$$

The Jacobian is $J_{\xi,n_t} = \det|\mathbf{J}_\xi|$ and because points 3 and 4 coincide with the collocation point the Jacobian tends to zero as it is approached. The Jacobian of the transformation (7.23) is $J_{\hat{\xi}} = 0.25$. The singular integration can now be written as

$$\Delta\mathbf{U}^e_{nk} = \sum_{n_t=1}^{N_t} \sum_{i=1}^{g_\xi} \sum_{j=1}^{g_\eta} \mathsf{U}\left(\tilde{\boldsymbol{x}}_n, \hat{\boldsymbol{x}}(\bar{\xi}_i, \bar{\eta}_j)\right) R_k(\xi(\bar{\xi}_i), \eta(\bar{\eta}_j))\, J_{n_t ij} \tag{7.26}$$

where N_t is the number of triangles and

$$J_{n_t ij} = 0.25\, J_{\xi,n_t}(\bar{\xi}_i, \bar{\eta}_j) J(\bar{\xi}_i, \bar{\eta}_j) W_i W_j . \tag{7.27}$$

The integration error is now dependent on the aspect ratio $\frac{L_\eta}{L_\xi}$. Two examples of a triangular subdivision with different aspect ratios are shown in Fig. 7.19. Figure 7.20 shows the integration error depending on the aspect ratio. It can be seen that the aspect ratio has a significant influence, which means that integration regions with bad aspect ratio have to be subdivided.

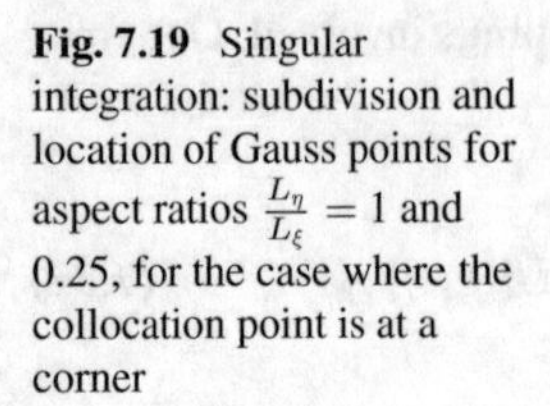

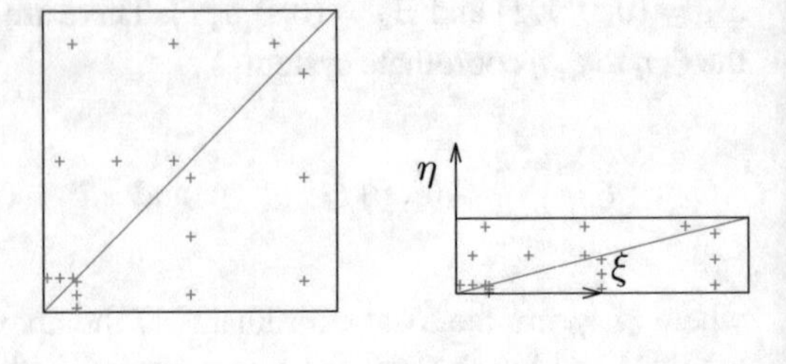

Fig. 7.19 Singular integration: subdivision and location of Gauss points for aspect ratios $\frac{L_\eta}{L_\xi} = 1$ and 0.25, for the case where the collocation point is at a corner

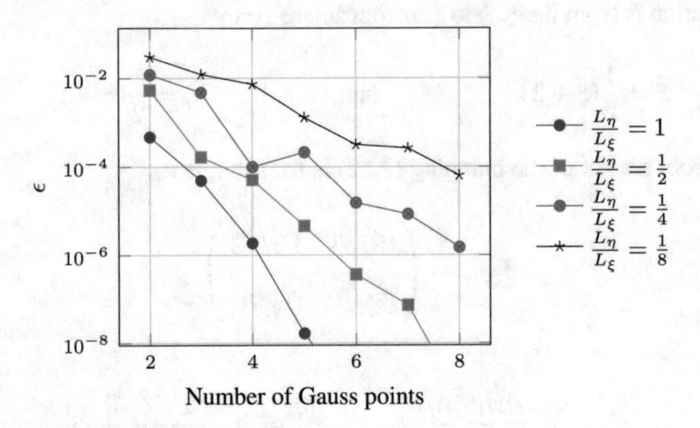

Fig. 7.20 Singular integration: effect of aspect ratio $\frac{L_\eta}{L_\xi}$ on the integration error

7.5 Summary and Conclusions

While numerical integration in the FEM is trivial and "less may be more" (i.e. reduced integration), this is a critical aspect of the BEM and requires careful consideration. If this is not done properly spectacularly wrong results can be obtained. It was Watson that first introduced numerical integration into the BEM and carefully looked at ensuring that the required precision of integration is maintained. The reason for this is that the Kernels inside the integrals tend to infinity with various degrees and NURBS are not polynomials but rational functions. Because the integrands are not polynomials, Gauss Quadrature is not exact. Lacking a better alternative, however, it is still being used, albeit with a careful selection of the number of Gauss points.

We discussed regular integration first, where the source point is outside the integration region, so the integrand increases rapidly towards the source point but is not infinite. We approach this by subdividing the integration region and by determining the number of Gauss points necessary to achieve a given precision of integration. There are various ways for determining the number of Gauss points required and an error estimator is available for polynomials. However, the estimator is not really able to predict the actual error, considering that the integrands are complex expressions of the Kernel times the basis function times the Jacobian. Therefore we introduced a numerical approach for determining the number of Gauss points required. This is not perfect but seems to work well.

Numerical integration is a substantial part of the total computational effort, so clever strategies can lead to savings. Here we introduced the interval halving and Quadtree method that result in an intelligent subdivision into integration regions so that the total number of Gauss points can be reduced. We showed an example in 3-D with the source point very near to the patch, where the number of Gauss points was reduced by an order of magnitude.

Next, we discussed singular integration, i.e. the case where the source point is inside an integration region. This involves integrands that are weakly singular and the integral exists but regular Gauss cannot be applied. For 2-D problems, we can use Gauss–Laguerre instead, that integrates a logarithmic singularity. For 3-D problems, we geometrically eliminate the singularity by subdividing the integration region in such a way that the Jacobian tends to zero.

This is probably the most important chapter of the book, as maintaining a high precision of the integration is crucial to getting good quality results.

Chapter 8
Steady State Potential Problems

8.1 Introduction

Steady state potential problems are problems where the primary variable is a scalar and the results are independent of time. The integral equations have been derived in Chap. 2 and their numerical solution discussed in Chap. 6. Here we discuss problem specifics and the application to practical examples. Problems that can be solved include hydraulic flow and the flow of heat and electricity, but we will concentrate on hydraulic flow, in particular seepage, which is governed by the Laplace equation

$$k_{ij}\, u_{,ij} = 0 \tag{8.1}$$

where k_{ij} is the conductivity tensor. For the isotropic case, k_{ij} reverts to the scalar k.

The system of integral equations to be solved is

$$c(\tilde{\boldsymbol{x}})\, u(\tilde{\boldsymbol{x}}) = \int_{\Gamma} \mathsf{U}(\tilde{\boldsymbol{x}}, \hat{\boldsymbol{x}})\, t(\hat{\boldsymbol{x}})\, d\Gamma(\hat{\boldsymbol{x}}) - \int_{\Gamma} \mathsf{U}(\tilde{\boldsymbol{x}}, \hat{\boldsymbol{x}})\, u(\hat{\boldsymbol{x}})\, d\Gamma(\hat{\boldsymbol{x}}) \tag{8.2}$$

where u is the potential and t is the flow normal to the boundary.

The following boundary conditions may be specified:

- Dirichlet BC: Temperature or potential is prescribed
- Neumann BC: Flow is prescribed

We will also discuss free surface flow where both Dirichlet and Neumann conditions are prescribed.

G. Beer et al., *The Isogeometric Boundary Element Method*, Lecture Notes in Applied and Computational Mechanics 90,
https://doi.org/10.1007/978-3-030-23339-6_8

8.2 Post-processing

After the discretised equation (8.2) has been solved for the unknowns on the boundary as explained in Chap. 6 we compute the values inside the domain by post-processing. The potential inside the domain can be computed by

$$u(\boldsymbol{x}) = \int_{\Gamma} [\mathsf{U}(\boldsymbol{x}, \hat{\boldsymbol{x}})t(\hat{\boldsymbol{x}}) - \mathsf{T}(\boldsymbol{x}, \hat{\boldsymbol{x}})\, u(\hat{\boldsymbol{x}})]\, d\Gamma(\hat{\boldsymbol{x}}). \tag{8.3}$$

The integral in Eq. (8.3) tends to infinity as the boundary is approached, so u cannot be computed exactly on the boundary. Furthermore, to ensure an adequate precision of the integration, the number of Gauss points has to be increased as the internal point moves closer to the boundary, resulting in a high number of Gauss points required for internal points that are close to the boundary.

This problem can be alleviated by a *Regularisation*. We consider that the distribution of u can be split up into a part that is constant throughout the domain and one that depends on the location. If the potential is constant we observe that the associated flow has to be zero.

For a finite problem, if u is the same everywhere on the boundary, then this corresponds to a constant potential inside the domain. Considering that the flow t is zero everywhere on the boundary, Eq. (8.3) can be used to solve the problem

$$u(\boldsymbol{x}) = -\int_{\Gamma} \mathsf{T}(\boldsymbol{x}, \hat{\boldsymbol{x}})\, u(\hat{\boldsymbol{x}}) d\Gamma(\hat{\boldsymbol{x}}). \tag{8.4}$$

Substituting $u(\boldsymbol{x}) = u(\hat{\boldsymbol{x}}) = u(\boldsymbol{x}_p)$, where $\boldsymbol{x}_p$ is a point on the boundary, and rearranging we obtain

$$0 = u(\boldsymbol{x}_p) + \int_{\Gamma} \mathsf{T}(\boldsymbol{x}, \hat{\boldsymbol{x}})\, u(\boldsymbol{x}_p) d\Gamma(\hat{\boldsymbol{x}}). \tag{8.5}$$

Adding Eqs. (8.3) and (8.5), we obtain

$$u(\boldsymbol{x}) = u(\boldsymbol{x}_p) + \int_{\Gamma} [\mathsf{U}(\boldsymbol{x}, \hat{\boldsymbol{x}})t(\hat{\boldsymbol{x}}) - \mathsf{T}(\boldsymbol{x}, \hat{\boldsymbol{x}})\, (u(\hat{\boldsymbol{x}}) - u(\boldsymbol{x}_p))]\, d\Gamma(\hat{\boldsymbol{x}}). \tag{8.6}$$

We see that the Kernel T is multiplied with a term that approaches zero as the point on the boundary is approached and therefore can be evaluated numerically with Gauss Quadrature.

To get a constant potential field for an infinite domain problem, we surround the problem with an artificial boundary Γ_∞, assign a constant value of $u(\boldsymbol{x}) = u(\boldsymbol{x}_p)$ there and obtain

$$0 = u(\boldsymbol{x}_p) + \int_\Gamma \mathsf{T}(\boldsymbol{x}, \hat{\boldsymbol{x}})\, u(\boldsymbol{x}_p) d\Gamma(\hat{\boldsymbol{x}}) + u(\boldsymbol{x}_p) A \tag{8.7}$$

where A is the azimuthal integral that has been already introduced in Chap. 6

$$A = \int_{\Gamma_\infty} \mathsf{T}(\boldsymbol{x}, \hat{\boldsymbol{x}}) d\Gamma_\infty(\hat{\boldsymbol{x}}) = -1. \tag{8.8}$$

Adding Eqs. (8.3) and (8.7) yields

$$u(\boldsymbol{x}) = u(\boldsymbol{x}_p) + \int_\Gamma \left[\mathsf{U}(\boldsymbol{x}, \hat{\boldsymbol{x}}) t(\hat{\boldsymbol{x}}) - \mathsf{T}(\boldsymbol{x}, \hat{\boldsymbol{x}}) \left(u(\hat{\boldsymbol{x}}) - u(\boldsymbol{x}_p)\right)\right] d\Gamma(\hat{\boldsymbol{x}}) + u(\boldsymbol{x}_p) A. \tag{8.9}$$

Although any boundary value could be used for $u(\boldsymbol{x}_p)$, we have good experience with using the value at the boundary point nearest to the internal point.

The flow vector $\mathbf{q}$ is related to the potential or temperature by

$$q_i = -k_{ij}\, u_{,j}. \tag{8.10}$$

At a point $\boldsymbol{x}$ inside the domain, it can be computed by

$$\mathbf{q}(\boldsymbol{x}) = \int_\Gamma \mathsf{S}(\boldsymbol{x}, \hat{\boldsymbol{x}}) t(\hat{\boldsymbol{x}}) d\Gamma(\hat{\boldsymbol{x}}) - \int_\Gamma \mathsf{R}(\boldsymbol{x}, \hat{\boldsymbol{x}})\, u(\hat{\boldsymbol{x}}) d\Gamma(\hat{\boldsymbol{x}}). \tag{8.11}$$

where the derived fundamental solutions S, R are presented in Appendix A. We note that the derived fundamental solutions are strongly singular which presents some problems if the internal point is close to the boundary. If the point is very close to the boundary, the number of Gauss points required may increase significantly or the computed values may be in error.

Again we solve this by *Regularisation*. We note that for a constant flow field the temperature variation must be linear. For a constant flow $\mathbf{q}(\boldsymbol{x}) = \mathbf{q}(\hat{\boldsymbol{x}}) = \mathbf{q}(\boldsymbol{x}_p)$, we must have the following conditions on the boundary

$$t^*(\hat{\boldsymbol{x}}) = \mathbf{q}(\boldsymbol{x}_p) \cdot \mathbf{n}(\hat{\boldsymbol{x}}) \tag{8.12}$$

$$u^*(\hat{\boldsymbol{x}}) = u(\boldsymbol{x}_p) + \frac{\partial u(\boldsymbol{x}_p)}{\partial \boldsymbol{x}} \cdot (\hat{\boldsymbol{x}} - \boldsymbol{x}_p). \tag{8.13}$$

Substitution into Eq. (8.11) gives for finite problems

$$\mathbf{q}(\boldsymbol{x}) = \int_\Gamma \mathsf{S}(\boldsymbol{x}, \hat{\boldsymbol{x}}) t^*(\hat{\boldsymbol{x}}) d\Gamma(\hat{\boldsymbol{x}}) - \int_\Gamma \mathsf{R}(\boldsymbol{x}, \hat{\boldsymbol{x}})\, u^*(\hat{\boldsymbol{x}}) d\Gamma(\hat{\boldsymbol{x}}). \tag{8.14}$$

and for infinite problems

$$\mathbf{q}(\boldsymbol{x}) = \int_{\Gamma} \mathsf{S}(\boldsymbol{x}, \hat{\boldsymbol{x}}) t^*(\hat{\boldsymbol{x}}) d\Gamma(\hat{\boldsymbol{x}}) - \int_{\Gamma} \mathsf{R}(\boldsymbol{x}, \hat{\boldsymbol{x}})\, u^*(\hat{\boldsymbol{x}}) d\Gamma(\hat{\boldsymbol{x}}) - \mathbf{A}_1 + \mathbf{A}_2. \quad (8.15)$$

Where the azimuthal integrals are

$$\mathbf{A}_1 = \int_{\Gamma_0} \mathsf{S}(\boldsymbol{x}, \hat{\boldsymbol{x}}) t^*(\hat{\boldsymbol{x}}) d\Gamma_0 = 0.5\, \mathbf{q}(\boldsymbol{x}_p) \quad (8.16)$$

$$\mathbf{A}_2 = \int_{\Gamma_0} \mathsf{R}(\boldsymbol{x}, \hat{\boldsymbol{x}}) u^*(\hat{\boldsymbol{x}}) d\Gamma_0 = 0.5 \begin{pmatrix} \partial u(\boldsymbol{x}_p)/\partial x \\ \partial u(\boldsymbol{x}_p)/\partial y \\ \partial u(\boldsymbol{x}_p)/\partial z \end{pmatrix}. \quad (8.17)$$

Adding Eqs. (8.11) and (8.14) we obtain for finite problems

$$\begin{aligned} \mathbf{q}(\boldsymbol{x}) = \mathbf{q}(\boldsymbol{x}_p) &+ \int_{\Gamma} \mathsf{S}(\boldsymbol{x}, \hat{\boldsymbol{x}}) \left[t(\hat{\boldsymbol{x}}) - t^*(\hat{\boldsymbol{x}}) \right] d\Gamma(\hat{\boldsymbol{x}}) \\ &- \int_{\Gamma} \mathsf{R}(\boldsymbol{x}, \hat{\boldsymbol{x}}) \left[u(\hat{\boldsymbol{x}}) - u^*(\hat{\boldsymbol{x}}) \right] d\Gamma(\hat{\boldsymbol{x}}). \end{aligned} \quad (8.18)$$

8.3 Computation of Boundary Values

For the regularisation we require the values of u, $\frac{\partial u}{\partial x}$ and $\mathbf{q}$ on the boundary. The computation of u on the boundary has already been discussed in Chap. 6. The derivative of u can be obtained by first calculating local derivatives (normal and tangential to the boundary) and then transferring them to global derivatives. We discuss this separately for plane and 3-D problems.

8.3.1 Plane Problems

We calculate first the derivative tangential to the boundary (in $\bar{x}$-direction) as:

$$u_{,\bar{x}}(\boldsymbol{x}_P) = \frac{\partial u}{\partial \xi} \frac{1}{J} = \sum_{k=1}^{K^u} \frac{\partial R_k^u(\xi_P)}{\partial \xi}\, u_k^{ep} \frac{1}{J} \quad (8.19)$$

where J is the Jacobian of the transformation between local ξ and global $\boldsymbol{x}$-coordinate systems and the superscript ep means the patch that contains $\boldsymbol{x}_P$. R_k^u are basis functions for the approximation of u. The derivative normal to the boundary ($\bar{y}$) is obtained for isotropic material by solving:

$$q_{\bar{y}} = -ku_{,\bar{y}}(\boldsymbol{x}_P) = t \tag{8.20}$$

for $u_{,\bar{y}}$, resulting in:

$$u_{,\bar{y}}(\boldsymbol{x}_P) = -\frac{t(\xi_P)}{k} \tag{8.21}$$

For anisotropic problems we have:

$$\begin{pmatrix} q_{\bar{x}} \\ q_{\bar{y}} \end{pmatrix} = \begin{pmatrix} k_{\bar{x}\bar{x}} & k_{\bar{x}\bar{y}} \\ k_{\bar{x}\bar{y}} & k_{\bar{y}\bar{y}} \end{pmatrix} \begin{pmatrix} u_{,\bar{x}} \\ u_{,\bar{y}} \end{pmatrix} \tag{8.22}$$

where $k_{\bar{y}\bar{y}}$, $k_{\bar{x}\bar{x}}$ and $k_{\bar{x}\bar{y}}$ are conductivities in the local directions.

Substituting $q_{\bar{y}} = t(\xi_P)$ we can solve for $u_{,\bar{y}}$:

$$u_{,\bar{y}}(\boldsymbol{x}_P) = -\frac{1}{k_{\bar{y}\bar{y}}} \left(t(\xi_P) + k_{\bar{x}\bar{y}} \frac{u_{,\xi}}{J} \right) \tag{8.23}$$

The derivatives in global directions can be obtained by:

$$\begin{pmatrix} u_{,x} \\ u_{,y} \end{pmatrix} = \mathbf{J}^{-1} \begin{pmatrix} u_{,\bar{x}} \\ u_{,\bar{y}} \end{pmatrix} \tag{8.24}$$

where the Jacobi matrix is given by:

$$\mathbf{J} = \begin{pmatrix} \mathbf{v}_{\bar{x}}^{\mathrm{T}} \\ \\ \mathbf{n}^{\mathrm{T}} \end{pmatrix} \tag{8.25}$$

with

$$\mathbf{v}_{\bar{x}} = \sum_{i=1}^{I} \frac{\partial R_i}{\partial \xi} \boldsymbol{x}_i \frac{1}{J} \tag{8.26}$$

8.3.2 3-D Problems

For 3-D problems, we first compute the flow in local directions tangential and normal to the boundary. The orthogonal axes $(\bar{x}, \bar{y}, \bar{z})$ are determined as follows (see Fig. 8.1):

- The direction of the ξ-axis (defined by unit vector $\mathbf{v}_\xi$) is taken as the direction of the $\bar{x}$ axis ($\mathbf{v}_{\bar{x}}$).
- The direction of the $\bar{y}$-axis ($\mathbf{v}_{\bar{y}}$) is determined by taking the cross-product between the vector in direction of the $\bar{x}$-axis and the vector normal to the boundary surface i.e. $\mathbf{v}_{\bar{y}} = \mathbf{v}_{\bar{x}} \times \mathbf{n}$.
- The direction $\bar{z}$ is taken in the direction normal to the boundary.

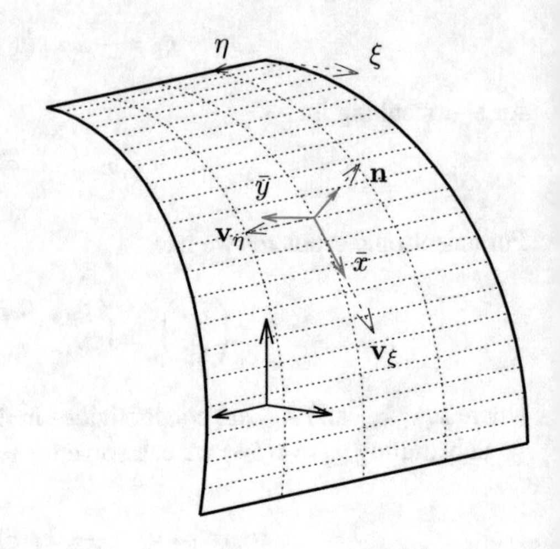

Fig. 8.1 Determination of local axes for computing boundary values

The derivative in $\bar{x}$-direction is computed by:

$$u_{,\bar{x}}(\boldsymbol{x}_P) = \frac{\partial u}{\partial \xi}\frac{1}{J} = \sum_{k=1}^{K^u} \frac{\partial R_k^u(\xi_P)}{\partial \xi}\, u_k^{ep}\frac{1}{J} \tag{8.27}$$

and the one in the $\bar{y}$-direction is

$$\begin{aligned} u_{,\bar{y}}(\boldsymbol{x}_P) &= \frac{\partial u}{\partial \xi}\frac{\partial \xi}{\partial \bar{y}} + \frac{\partial u}{\partial \eta}\frac{\partial \eta}{\partial \bar{y}} \\ &= \left[\sum_{k=1}^{K^u} \frac{\partial R_k^u(\xi_P,\eta_P)}{\partial \xi} u_k^{ep}\right]\frac{\partial \xi}{\partial \bar{y}} + \left[\sum_{k=1}^{K^u} \frac{\partial R_k^u(\xi_P,\eta_P)}{\partial \eta} u_k^{ep}\right]\frac{\partial \eta}{\partial \bar{y}}. \end{aligned} \tag{8.28}$$

For isotropic material the derivative in $\bar{z}$-direction is given by

$$u_{,\bar{z}}(\boldsymbol{x}_P) = \frac{-t(\boldsymbol{x}_P)}{k}. \tag{8.29}$$

The geometric derivatives are given by:

$$\frac{\partial \xi}{\partial \bar{x}} = \frac{1}{J_\xi} \tag{8.30}$$

$$\frac{\partial \xi}{\partial \bar{y}} = \frac{\cos\theta}{J_\xi\, \sin\theta} \tag{8.31}$$

$$\frac{\partial \eta}{\partial \bar{y}} = \frac{1}{J_\eta\, \sin\theta} \tag{8.32}$$

where

$$J_\xi = |\sum_{i=1}^{I} \frac{\partial R_i}{\partial \xi} \boldsymbol{x}_i^e| \tag{8.33}$$

$$J_\eta = |\sum_{i=1}^{I} \frac{\partial R_i}{\partial \eta} \boldsymbol{x}_i^e| \tag{8.34}$$

and

$$\cos\theta = \mathbf{v}_\xi \cdot \mathbf{v}_\eta \tag{8.35}$$

$$\sin\theta = \mathbf{v}_\eta \cdot \mathbf{v}_{\bar{y}} \tag{8.36}$$

where $\mathbf{v}_\eta$ is a unit vector in the η-direction.

The derivatives in global directions can be obtained by:

$$\begin{pmatrix} u_{,x} \\ u_{,y} \\ u_{,z} \end{pmatrix} = \mathbf{J}^{-1} \begin{pmatrix} u_{,\bar{x}} \\ u_{,\bar{y}} \\ u_{,\bar{z}} \end{pmatrix} \tag{8.37}$$

where

$$\mathbf{J} = \begin{pmatrix} \mathbf{v}_{\bar{x}}^{\mathrm{T}} \\ \mathbf{v}_{\bar{y}}^{\mathrm{T}} \\ \mathbf{n}^{\mathrm{T}} \end{pmatrix} \tag{8.38}$$

with

$$\mathbf{v}_{\bar{x}} = \sum_{i=1}^{I} \frac{\partial R_i}{\partial \xi} \boldsymbol{x}_i \frac{1}{J_\xi} \tag{8.39}$$

$$\mathbf{v}_{\bar{y}} = \sum_{i=1}^{I} \frac{\partial R_i}{\partial \eta} \boldsymbol{x}_i \frac{1}{J_\eta}. \tag{8.40}$$

8.4 Confined Flow Problems

We define confined problems where the boundary conditions are either Dirichlet or Neumann, in contrast to unconfined problems where both boundary conditions may be specified.

8.4.1 Example 1: Flow Around Obstacle

The first example relates to the flow past a circular obstacle. Figure 8.2 shows the problem definition: a uniform flow field $\mathbf{q}_0 = [0, 1]$ is specified and we want to determine on how the flow field is changed when a circular obstacle is placed in the flow. Since no flow can penetrate the obstacle the flow normal to the boundary of the obstacle $t_0 = q_{0i} \cdot n_i$ must be zero. The solution therefore consists of two parts: the original flow plus the change in flow due to the obstacle

$$\mathbf{q} = \mathbf{q}_0 + \triangle\mathbf{q}. \tag{8.41}$$

The change in flow $\triangle\mathbf{q}$ is computed by applying the Neumann boundary condition, $-t_0 = -q_0 \cdot n_i$, so that the component of the final flow in Eq. (8.41) normal to the boundary is zero. We assume a radius of the obstacle of $R = 1$ and isotropic material properties with $k = 1$.

An analytical solution to this problem can be easily derived using cylindrical coordinates. The analytical solutions for the potential u and the flow in radial and tangential direction (q_r, q_t) at a point with the cylindrical coordinates r, θ are given by

$$\begin{aligned} u &= \quad -q_0/k\,(r + R^2/r)\,\sin\theta, \\ q_r &= \quad q_0/k(1 - R^2/r^2)\sin\theta, \\ q_t &= \quad q_0/k(1 - R^2/r^2)\cos\theta. \end{aligned} \tag{8.42}$$

The discretisation of the obstacle is shown in Fig. 8.3 and consists of 4 patches defined by basis functions of order 2. This *exactly* defines a circle.

A Neumann boundary condition with $t = -q_{02} \cdot n_2$ was applied for all patches and the same basis functions as for the description of the geometry were used for

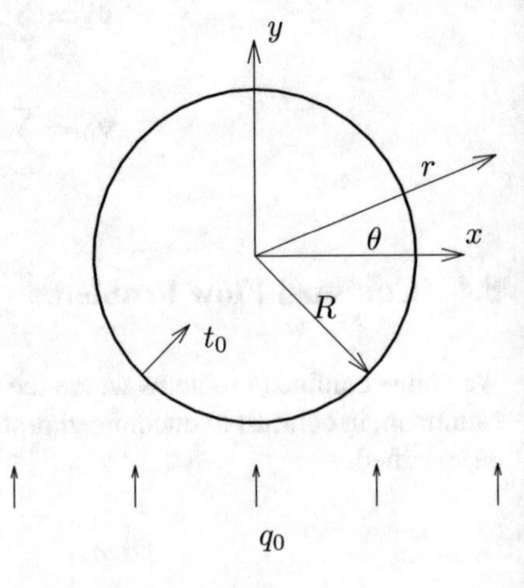

Fig. 8.2 Example 1: flow past a circular obstacle, problem statement and definition of cylindrical coordinates

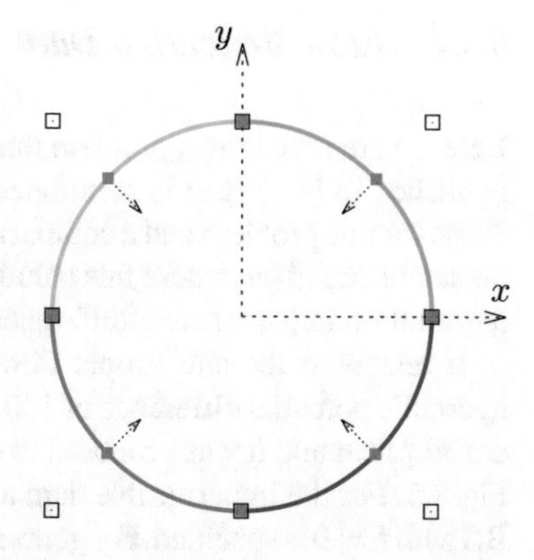

Fig. 8.3 Example 1: discretisation of circular obstacle with 4 NURBS patches. Control points are depicted by hollow squares and Collocation points by filled squares

the approximation of the unknown u. The resulting collocation point locations are depicted in Fig. 8.3. The simulation has only 8 degrees of freedom. Since the geometry and therefore also the normal to the boundary is exactly represented, the boundary conditions are also exact. Furthermore, we note that the analytical solution for u can also be exactly represented by the basis functions. Indeed, the error of the obtained solution is 10^{-7} which can be attributed solely to the precision of the input data and specified accuracy of the numerical integration. The flow vectors and contours of the potential are shown in Fig. 8.4.

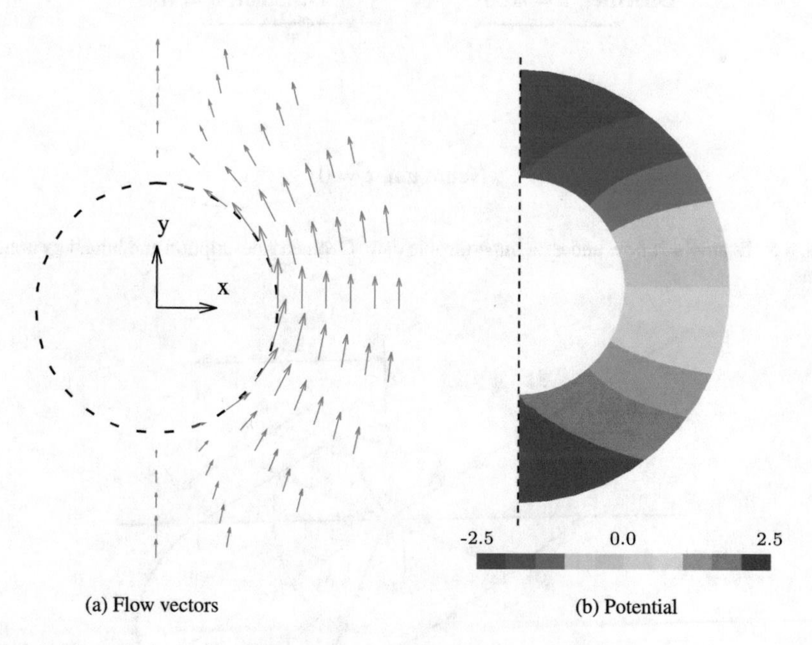

(a) Flow vectors (b) Potential

Fig. 8.4 Example 1: resulting flow vectors and contours of potential, displayed on the right of the symmetry plane

8.4.2 Flow Beneath a Dam

Here we are revisiting a problem that has been solved by the conventional BEM and published in [4, 39]. It is mentioned in these references that an analytical solution exists for this problem and a comparison is made with this solution, but unfortunately no details are given where this solution was published. We compare our results with the solution in [4] where a difference of 1% to the analytical solution is claimed.

It relates to the anisotropic flow under an impermeable dam across there is a hydraulic potential difference of 100 units. The streamlines for the analytical solution are elliptical and it was proposed to define the geometry of the problem as shown in Fig. 8.5. For the impermeable dam as well as for the elliptical boundary a Neumann BC with $t = 0$ is specified. For generating the input data the geometry transformations in Fig. 8.6, taken from [4], are used. The elliptical boundary can be modelled with one trimmed NURBS curve of order 2. For the trimming we determine the local coordinates of the intersection with the horizontal boundary and then trim away the top part as shown in Fig. 8.7. The geometry of the problem was defined by 8

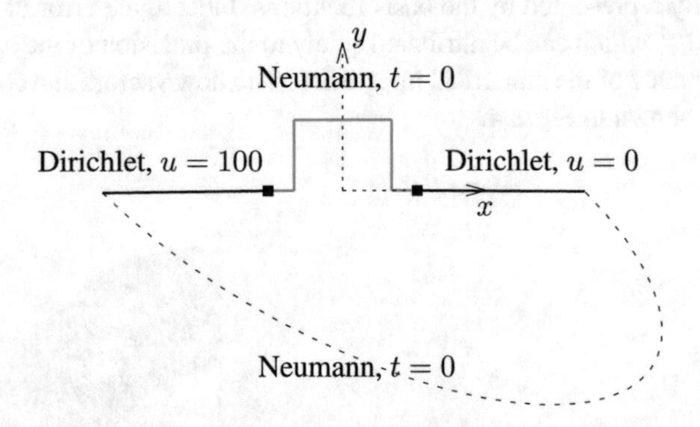

Fig. 8.5 Example 2: flow under an impermeable dam. Geometry description and boundary conditions

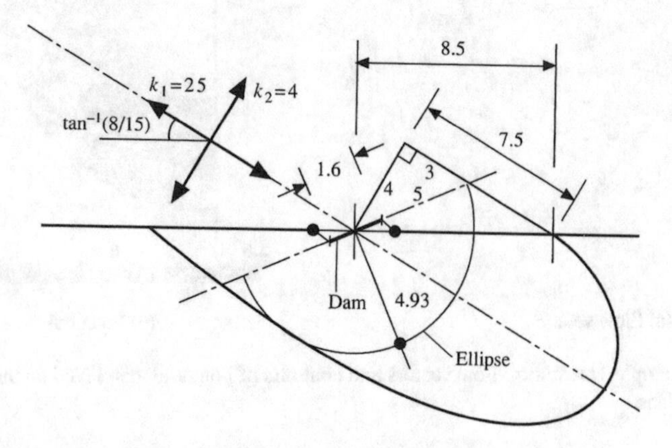

Fig. 8.6 Example 2: geometry definitions and material properties, taken from [4]

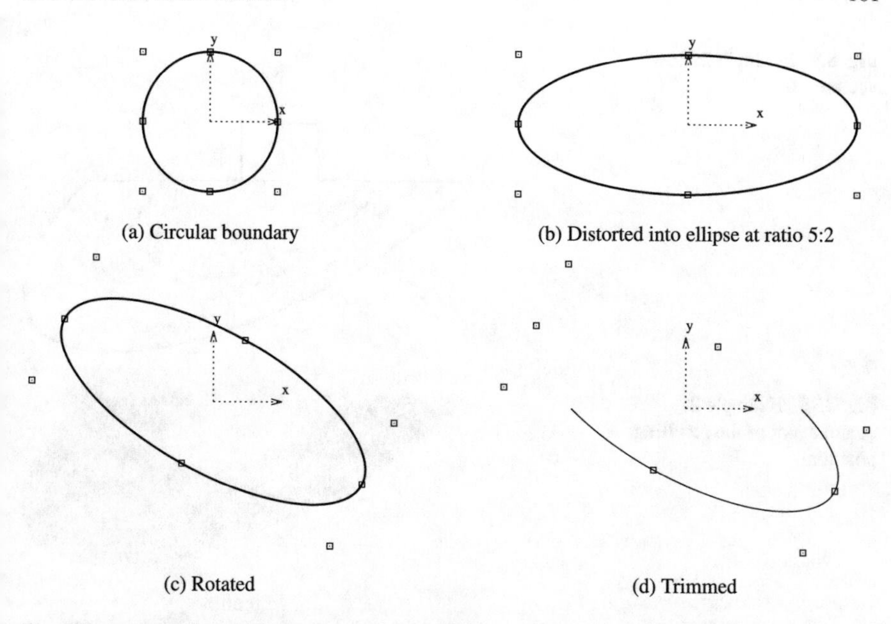

Fig. 8.7 Example 2: steps in defining the NURBS curve for the elliptical boundary: from (a) a circle to (d) the final trimmed ellipse

NURBS patches. For the approximation of the unknowns the basis functions used for describing the geometry were refined using knot insertion, which resulted in the collocation points as shown in Fig. 8.8. The simulation has only 21 unknowns.

The results of the simulation are shown in Fig. 8.9 as flow vectors and in Fig. 8.10 as contours of potential. The results agree well with the solution in [4] which had 74 degrees of freedom.

Fig. 8.8 Example 2: discretisation used showing control points as hollow squares and collocation points as filled squares

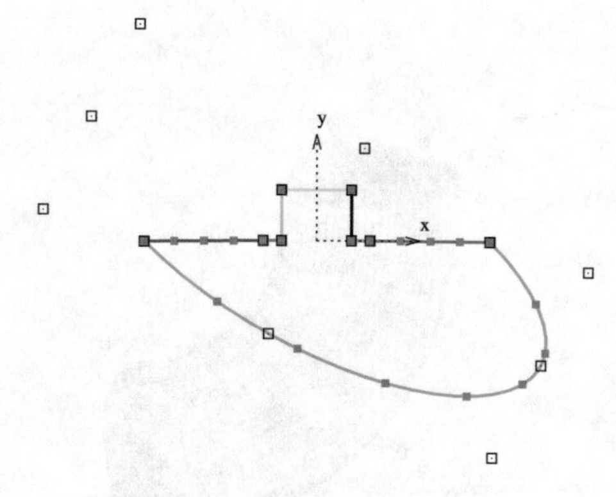

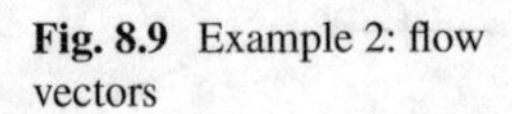
Fig. 8.9 Example 2: flow vectors

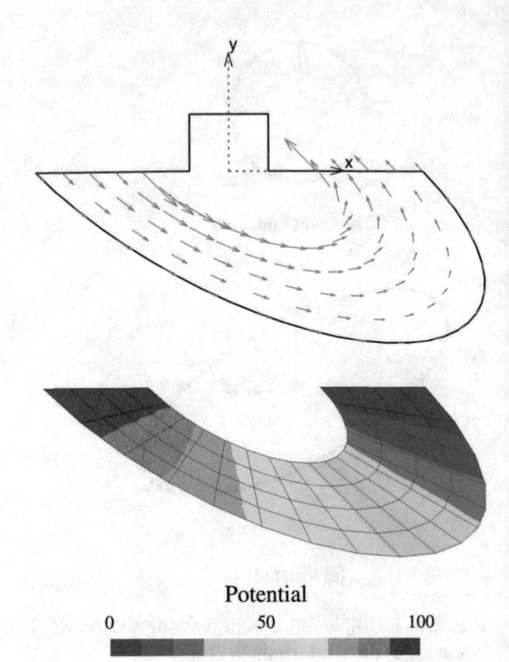

Fig. 8.10 Example 2: contour plot of the resulting potential

8.4.3 Flow Through Pipe

The next example is intended to demonstrate how 3-D objects can be described with very few parameters and a solution obtained with few unknowns. It relates the potential flow through a pipe. The geometry of the pipe is a quarter torus and this can be exactly described with 48 control points as shown in Fig. 8.11 on the left. It should be noted that if trimming is used the number of control points can be

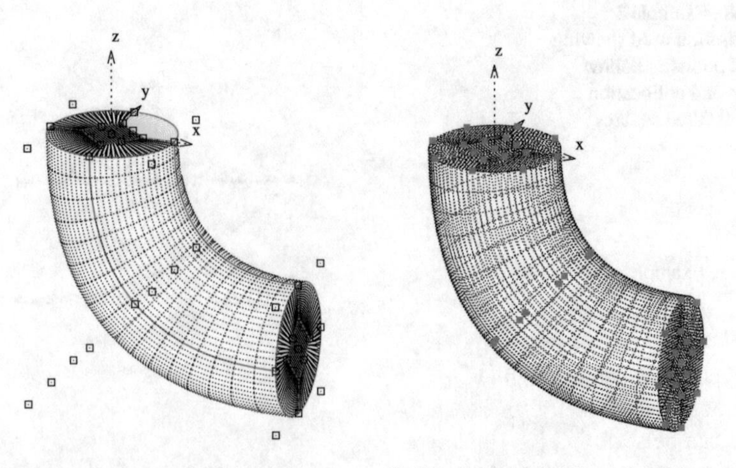

Fig. 8.11 Example 4: description of the geometry (left) and (right) location of collocation points without refinement

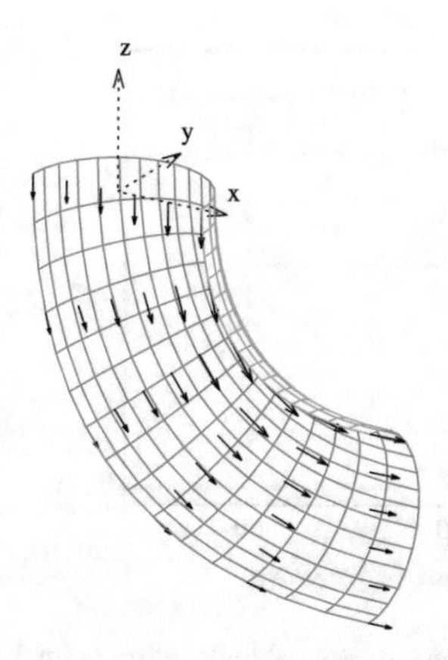

Fig. 8.12 Example 4: resulting flow vectors

decreased significantly. An analysis with no refinement (i.e. taking the basis functions used for describing the geometry as the ones used for describing the variation of the unknowns) results in the collocation points as shown in the figure on the right. The model has 42 degrees of freedom. The results of the simulation are shown as flow vectors in Fig. 8.12. Next we refine the solution by order elevating the basis functions for approximating the unknowns along the torus by one order (i.e. from quadratic to cubic). The resulting collocation points as well as the results are shown in Fig. 8.13.

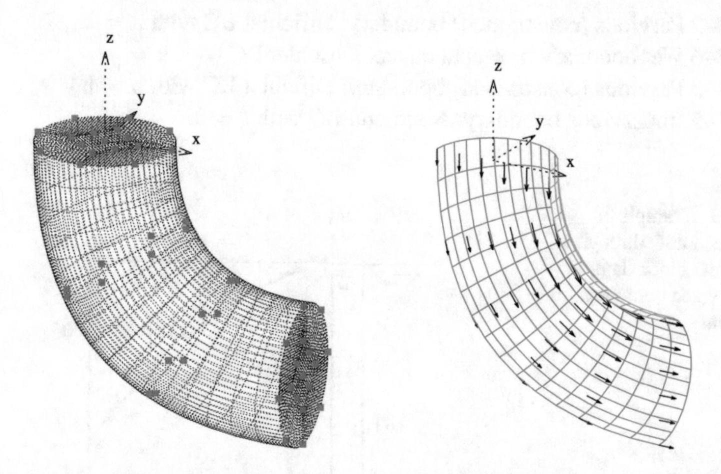

Fig. 8.13 Example 4: location of collocation points after refinement (left) and (right) resulting flow vectors

The convergence of the flow at collocation point 1 is shown in Fig. 8.14 for the conventional isogeometric approach and the geometry independent field approximation. It can be seen that the results are identical and convergence is achieved after only a few refinements.

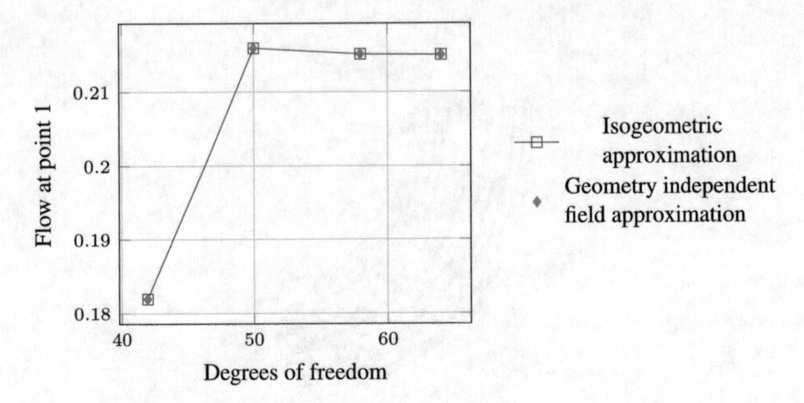

Fig. 8.14 Example 4: convergence of flow at collocation point 1

8.5 Unconfined Flow Problems

8.5.1 Seepage Through Dam

Here we present an example in free surface potential flow, which occurs for example in seepage through a dam as shown in Fig. 8.15.

The following boundary conditions are applicable:

- On 1–2 Pervious (constrained) boundary: Dirichlet BC with $u = \mathrm{h}_1$
- On 3–4 Wet boundary or seepage face: Dirichlet BC with $u = y$
- On 4–5 Pervious (constrained) boundary: Dirichlet BC with $u = \mathrm{h}_2$
- On 1–5 Impervious boundary: Neumann BC with $t = 0$.

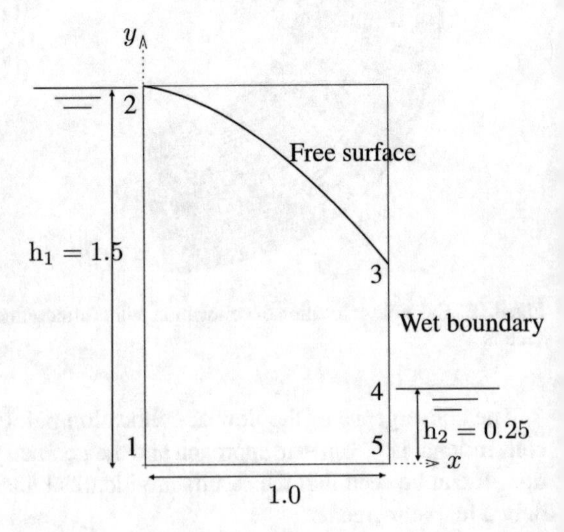

Fig. 8.15 Example 3: unconfined flow through a rectangular block dam, geometry and boundary conditions

The free surface (2–3) is defined by two boundary conditions:

- The flow normal to the boundary is zero (i.e. Neumann BC with $t = 0$).
- The potential has to be the same as the elevation (i.e. Dirichlet BC with $u = y$).

Furthermore, the point where the free surface meets the wet boundary the condition at the point 3 is that the flow normal to line 3–4 has to be zero. It is clear that these boundary conditions cannot be solved explicitly. Instead an iterative algorithm is employed whereby the Neumann BC is specified on the free surface and the enforcement of the Dirichlet BC as well as the condition at point 3 is achieved after a number of iterations. We employ a modified Newton–Raphson, i.e. we keep the left-hand side constant during the iterations.

The following solution strategy is employed:

- We start with a location of the free surface and define this with a spline curve of order 1.
- We enrich the basis functions of the curve by knot insertion or order elevation, to allow for greater flexibility in describing the geometry of the free surface.
- After the first solution we calculate the potential at collocation points and compare with their elevation (y-coordinate).
- If the elevation is higher than the potential we move the associated control point down, if it is lower we move it up.
- We check the flow normal to the line 3–4.
- If the flow is negative we move point 3 down if it is positive we move it up.
- We solve again with the updated geometry.
- We iterate until the differences between elevation and potential is less than 1% and the condition at point 3 is satisfied.

We test this algorithm on an example where a FEM solution is reported in [43]. It relates to the seepage through a rectangular block dam as shown in Fig. 8.15. The boundary conditions are as indicated above and an isotropic material with $k = 1$ is assumed. We discretise the problem with 5 linear patches. The basis functions for the approximation the unknown are chosen to be the same as for the definition of the geometry for all patches except for patch 4 (line 3–4 in Fig. 8.15), where we refine them by inserting 3 knots. We start the iteration with the location of the free surface at the top of the block.

Three different strategies were employed:

1. We enrich the basis functions of the curve defining the free surface by inserting 7 equal-spaced knots. The problem has 15 degrees of freedom.
2. We enrich the basis functions by order elevation, i.e. elevating the order 1 time from order 1 to 2. The problem has now only 9 degrees of freedom.
3. We enrich the basis functions by elevating the order 3 times from order 1 to 4. The problem has now 11 degrees of freedom.

The initial location of the control points and the collocation points as well as the change in the geometry of the free surface and the final location of the control points are shown in the Figs. 8.16, 8.17 and 8.18. In Fig. 8.19, we show the flow vectors and contours of the potential for the most accurate solution. We can see that the condition $\frac{\partial q}{\partial n} = 0$ at point 3 is well observed.

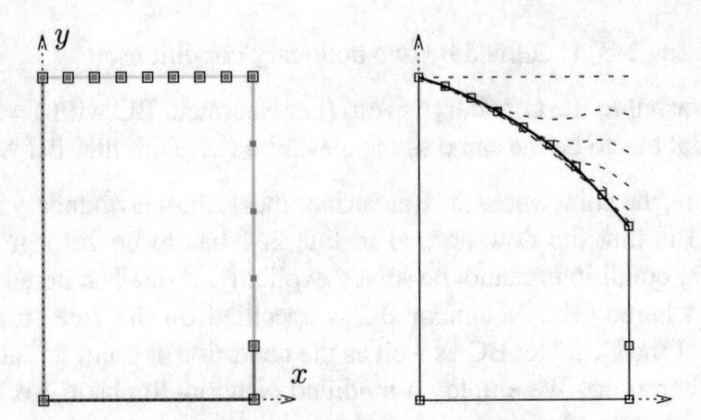

Fig. 8.16 Example 3 left: discretisation with enrichment by knot insertion showing control points (hollow squares) and collocation points (filled squares), right: change in geometry of free surface and final location of control points

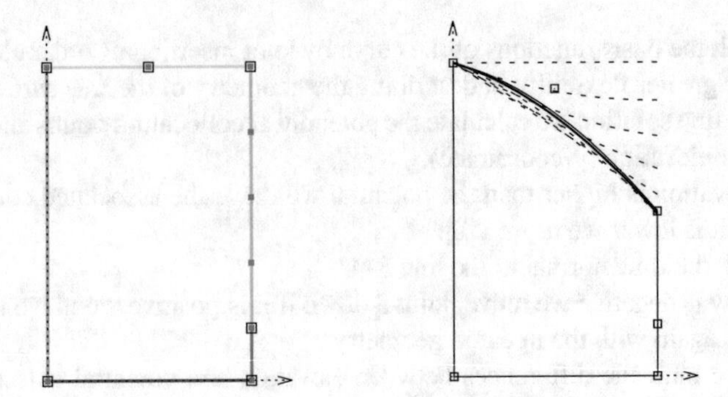

Fig. 8.17 Example 3 left: discretisation with enrichment by order elevation to order 2 showing control points (hollow squares) and collocation points (filled squares), right: change in geometry of free surface and final location of control point

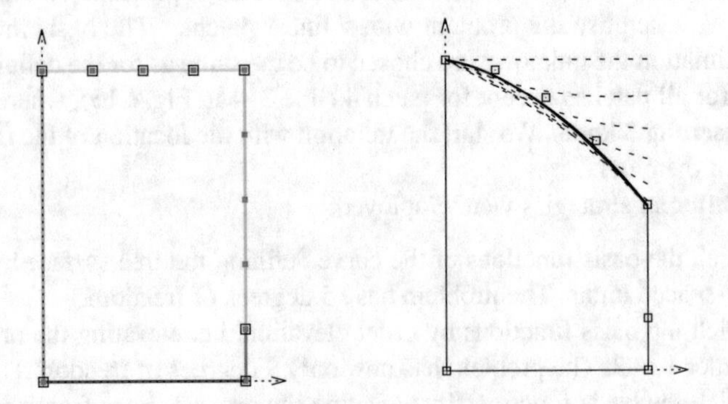

Fig. 8.18 Example 3 left: discretisation with enrichment by order elevation to order 4 showing control points (hollow squares) and collocation points (filled squares), right: change in geometry of free surface and final location of control points

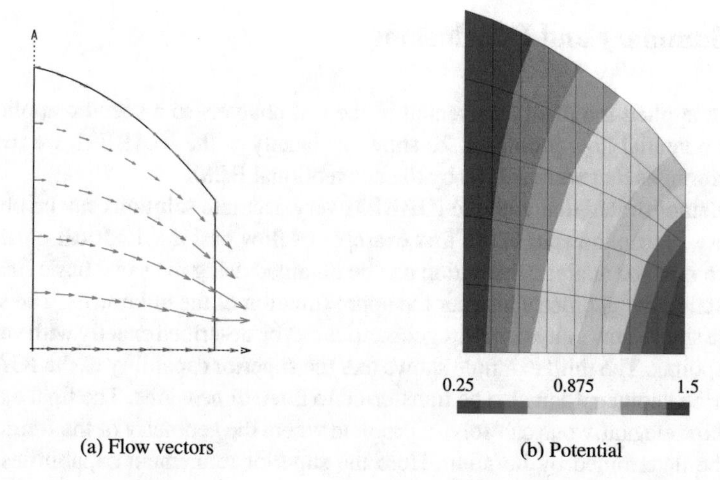

(a) Flow vectors (b) Potential

Fig. 8.19 Example 3: results

Finally we compare the calculated free surface geometry to the one published in [43] (reference solution) in Fig. 8.20. It can be seen that good agreement is obtained by all enrichments, with the last one being closest.

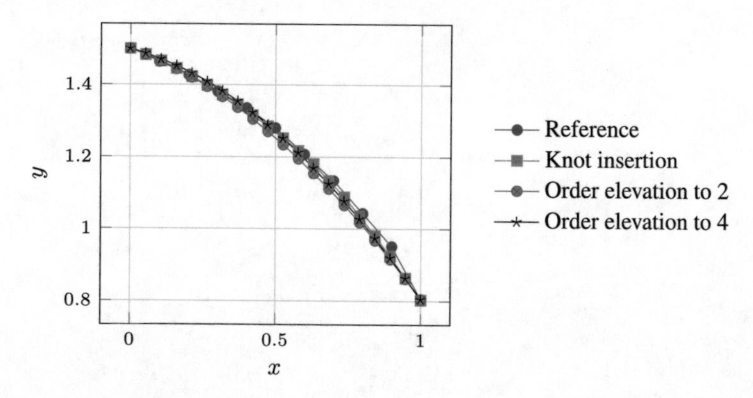

Fig. 8.20 Example 3: comparison of the results obtained by knot insertion and order elevation with the reference solution

8.6 Summary and Conclusions

Here we applied the theory presented in the last chapters to a specific application, namely potential flow problems. To show the beauty of the IGABEM, we revisited some examples that were solved by the conventional BEM.

We demonstrated that with the IGABEM very accurate solutions can be obtained with very few unknowns. In the first example of flow past a cylindrical obstacle, it is shown how the analytical solution can be obtained using the same basis functions that describe the geometry also for the approximation of the unknowns. The second example shows how more complex geometries can be described exactly with very few control points. The third example shows that the superior capability of the IGABEM to describe geometry can also be transferred to three dimensions. The final example shows how elegantly one can solve a problem where the geometry of the free surface has to be determined by iteration. Here the superior refinement capabilities of B-splines and NURBS come to the fore.

Chapter 9
Static Linear Solid Mechanics

9.1 Introduction

For linear elasticity, we have to solve the following integral equations, derived in Chap. 2

$$\int_{\Gamma} \mathsf{U}(\tilde{\boldsymbol{x}}_n, \hat{\boldsymbol{x}})\mathbf{t}(\hat{\boldsymbol{x}})d\Gamma(\hat{\boldsymbol{x}}) - \int_{\Gamma} \mathsf{T}(\tilde{\boldsymbol{x}}_n, \hat{\boldsymbol{x}})(\mathbf{u}(\hat{\boldsymbol{x}}) - \mathbf{u}(\tilde{\boldsymbol{x}}_n))d\Gamma(\hat{\boldsymbol{x}}) = \mathbf{A}_n\mathbf{u}(\tilde{\boldsymbol{x}}_n) \quad (9.1)$$

where $\mathbf{A}_n = \mathbf{I}$ for infinite and $\mathbf{A}_n = \mathbf{0}$ for finite domain problems, $\mathbf{u}$ is the displacement vector and $\mathbf{t}$ is the traction vector, which is related to the stress tensor by

$$t_i = \sigma_{ij}\, n_j. \quad (9.2)$$

We define the strain tensor as

$$\boldsymbol{\epsilon} = \epsilon_{ij} = \frac{1}{2}(u_{i,j} + u_{j,i}). \quad (9.3)$$

The stress tensor is related to the strain tensor by

$$\boldsymbol{\sigma} = \sigma_{ij} = C_{ijkl}\epsilon_{kl} \quad (9.4)$$

where C_{ijkl} is the elastic constitutive tensor.

For convenience, we introduce the Voigt notation to express the stress tensor as a pseudo-vector

$$\{\sigma\} = \begin{Bmatrix} \sigma_x \\ \sigma_y \\ \sigma_z \\ \tau_{xy} \\ \tau_{zy} \\ \tau_{xz} \end{Bmatrix}. \quad (9.5)$$

G. Beer et al., *The Isogeometric Boundary Element Method*, Lecture Notes in Applied and Computational Mechanics 90,
https://doi.org/10.1007/978-3-030-23339-6_9

For the strain pseudo-vector we have

$$\{\boldsymbol{\epsilon}\} = \begin{Bmatrix} \epsilon_x \\ \epsilon_y \\ \epsilon_z \\ \gamma_{xy} \\ \gamma_{zy} \\ \gamma_{xz} \end{Bmatrix}. \tag{9.6}$$

With this new notation the constitutive equation can be written as

$$\{\boldsymbol{\sigma}\} = \mathbf{D}\{\boldsymbol{\epsilon}\} \tag{9.7}$$

where $\mathbf{D}$ is the elastic constitutive matrix

$$\mathbf{D} = c_1 \begin{bmatrix} 1 & c_2 & c_2 & 0 & 0 & 0 \\ c_2 & 1 & c_2 & 0 & 0 & 0 \\ c_2 & c_2 & 1 & 0 & 0 & 0 \\ 0 & 0 & 0 & G/c_1 & 0 & 0 \\ 0 & 0 & 0 & 0 & G/c_1 & 0 \\ 0 & 0 & 0 & 0 & 0 & G/c_1 \end{bmatrix} \tag{9.8}$$

with $c_1 = \frac{E(1-\nu)}{(1+\nu)(1-2\nu)}$, $c_2 = \frac{\nu}{(1-\nu)}$ and $G = \frac{E}{2(1+\nu)}$. ν is the Poisson's ratio and E the modulus of elasticity.

In the case of *plain strain*, the pseudo-vectors reduce to

$$\{\boldsymbol{\sigma}\} = \begin{Bmatrix} \sigma_x \\ \sigma_y \\ \sigma_z \\ \tau_{xy} \end{Bmatrix} \quad \text{and} \quad \{\boldsymbol{\epsilon}\} = \begin{Bmatrix} \epsilon_x \\ \epsilon_y \\ 0 \\ \gamma_{xy} \end{Bmatrix}. \tag{9.9}$$

The elastic constitutive matrix is reduced to

$$\mathbf{D} = c_1 \begin{bmatrix} 1 & c_2 & c_2 & 0 \\ c_2 & 1 & c_2 & 0 \\ c_2 & c_2 & 1 & 0 \\ 0 & 0 & 0 & G/c_1 \end{bmatrix} \tag{9.10}$$

In the case of *plain stress* the pseudo-vectors reduce to

$$\{\boldsymbol{\sigma}\} = \begin{Bmatrix} \sigma_x \\ \sigma_y \\ \tau_{xy} \end{Bmatrix} \quad \text{and} \quad \{\boldsymbol{\epsilon}\} = \begin{Bmatrix} \epsilon_x \\ \epsilon_y \\ \gamma_{xy} \end{Bmatrix}. \tag{9.11}$$

In this case the constants c_1, c_2 for matrix **D** are determined by substituting $\nu/(1+\nu)$ for the Poisson's ratio.

The methods used for the numerical solution of Eq. (9.1) have been shown in Chap. 6. Here the post-processing is discussed in more detail.

9.2 Post-processing

The values of displacement at a point $\boldsymbol{x}$ inside the domain can be computed by

$$\mathbf{u}(\boldsymbol{x}) = \int_{\Gamma} [\mathsf{U}(\boldsymbol{x}, \hat{\boldsymbol{x}})\mathbf{t}(\hat{\boldsymbol{x}}) - \mathsf{T}(\boldsymbol{x}, \hat{\boldsymbol{x}})\,\mathbf{u}(\hat{\boldsymbol{x}})]\, d\Gamma(\hat{\boldsymbol{x}}). \tag{9.12}$$

To avoid a large number of Gauss points and errors for points that are near to the boundary, we use *Regularisation*.

For a finite problem, if $\mathbf{u}$ is the same everywhere on the boundary, then this corresponds to rigid body motion. Considering that for this case the traction $\mathbf{t}$ is zero everywhere on the boundary, Eq. (9.12) can be written as

$$\mathbf{u}(\boldsymbol{x}) = -\int_{\Gamma} \mathsf{T}(\boldsymbol{x}, \hat{\boldsymbol{x}})\,\mathbf{u}(\hat{\boldsymbol{x}}) d\Gamma(\hat{\boldsymbol{x}}). \tag{9.13}$$

Substituting $\mathbf{u}(\boldsymbol{x}) = \mathbf{u}(\hat{\boldsymbol{x}}) = \mathbf{u}(\boldsymbol{x}_p)$, where $\boldsymbol{x}_p$ is a point on the boundary, and rearranging we obtain

$$0 = \mathbf{u}(\boldsymbol{x}_p) + \int_{\Gamma} \mathsf{T}(\boldsymbol{x}, \hat{\boldsymbol{x}})\,\mathbf{u}(\boldsymbol{x}_p) d\Gamma(\hat{\boldsymbol{x}}). \tag{9.14}$$

Adding Eqs. (9.12) and (9.14), we obtain

$$\mathbf{u}(\boldsymbol{x}) = \mathbf{u}(\boldsymbol{x}_p) + \int_{\Gamma} [\mathsf{U}(\boldsymbol{x}, \hat{\boldsymbol{x}})\mathbf{t}(\hat{\boldsymbol{x}}) - \mathsf{T}(\boldsymbol{x}, \hat{\boldsymbol{x}})\,(\mathbf{u}(\hat{\boldsymbol{x}}) - \mathbf{u}(\boldsymbol{x}_p))]\, d\Gamma(\hat{\boldsymbol{x}}). \tag{9.15}$$

We see that as $\hat{\boldsymbol{x}}$ approaches $\boldsymbol{x}_p$ the term multiplying with T tends to zero.

To get a rigid body motion for an infinite domain problem, we surround the problem with an artificial boundary Γ_0 at infinity, assign a constant value of $\mathbf{u}(\boldsymbol{x}) = \mathbf{u}(\boldsymbol{x}_p)$ there and obtain:

$$0 = \mathbf{u}(\boldsymbol{x}_p) + \int_{\Gamma} \mathsf{T}(\boldsymbol{x}, \hat{\boldsymbol{x}})\,\mathbf{u}(\boldsymbol{x}_p) d\Gamma(\hat{\boldsymbol{x}}) + \mathbf{u}(\boldsymbol{x}_p)\mathbf{A} \tag{9.16}$$

where $\mathbf{A}$ is the azimuthal integral that has been already introduced in Chap. 6

$$\mathbf{A} = \int_{\Gamma_\infty} \mathsf{T}(\boldsymbol{x}, \hat{\boldsymbol{x}}) d\Gamma_0(\hat{\boldsymbol{x}}) = -\mathbf{I}. \tag{9.17}$$

Adding Eqs. (9.12) and (9.16) we obtain

$$\mathbf{u}(\boldsymbol{x}) = \mathbf{u}(\boldsymbol{x}_p) + \int_\Gamma [\mathsf{U}(\boldsymbol{x}, \hat{\boldsymbol{x}})\mathbf{t}(\hat{\boldsymbol{x}}) - \mathsf{T}(\boldsymbol{x}, \hat{\boldsymbol{x}})\left(\mathbf{u}(\hat{\boldsymbol{x}}) - \mathbf{u}(\boldsymbol{x}_p)\right)] \, d\Gamma(\hat{\boldsymbol{x}}) + \mathbf{u}(\boldsymbol{x}_p)\mathbf{A}. \tag{9.18}$$

The stress inside the domain can be computed by

$$\{\boldsymbol{\sigma}(\boldsymbol{x})\} = \int_\Gamma \mathsf{S}(\boldsymbol{x}, \hat{\boldsymbol{x}})\mathbf{t}(\hat{\boldsymbol{x}}) d\Gamma(\hat{\boldsymbol{x}}) - \int_\Gamma \mathsf{R}(\boldsymbol{x}, \hat{\boldsymbol{x}})\, \mathbf{u}(\hat{\boldsymbol{x}}) d\Gamma(\hat{\boldsymbol{x}}) \tag{9.19}$$

where the derived fundamental solutions S, R are presented in the Appendix A. To introduce *Regularisation* we note that for a constant stress field the variation of the displacement field must be linear. For a constant stress $\boldsymbol{\sigma}(\boldsymbol{x}) = \boldsymbol{\sigma}(\hat{\boldsymbol{x}}) = \boldsymbol{\sigma}(\boldsymbol{x}_p)$, we must have the following conditions on the boundary

$$t_i^*(\hat{\boldsymbol{x}}) = \sigma_{ij}(\boldsymbol{x}_p) \cdot n_j(\hat{\boldsymbol{x}}) \tag{9.20}$$

$$\mathbf{u}^*(\hat{\boldsymbol{x}}) = \mathbf{u}(\boldsymbol{x}_P) + \frac{\partial \mathbf{u}(\boldsymbol{x}_p)}{\partial \boldsymbol{x}} \cdot (\hat{\boldsymbol{x}} - \boldsymbol{x}_p). \tag{9.21}$$

Substitution into Eq. (9.19) gives for finite problems

$$\{\boldsymbol{\sigma}(\boldsymbol{x})\} = \int_\Gamma \mathsf{S}(\boldsymbol{x}, \hat{\boldsymbol{x}})\mathbf{t}^*(\hat{\boldsymbol{x}}) d\Gamma(\hat{\boldsymbol{x}}) - \int_\Gamma \mathsf{R}(\boldsymbol{x}, \hat{\boldsymbol{x}})\, \mathbf{u}^*(\hat{\boldsymbol{x}}) d\Gamma(\hat{\boldsymbol{x}}). \tag{9.22}$$

and for infinite problems

$$\{\boldsymbol{\sigma}(\boldsymbol{x})\} = \int_\Gamma \mathsf{S}(\boldsymbol{x}, \hat{\boldsymbol{x}})\mathbf{t}^*(\hat{\boldsymbol{x}}) d\Gamma(\hat{\boldsymbol{x}}) - \int_\Gamma \mathsf{R}(\boldsymbol{x}, \hat{\boldsymbol{x}})\mathbf{u}^*(\hat{\boldsymbol{x}}) d\Gamma(\hat{\boldsymbol{x}}) - \mathbf{A}_1 + \mathbf{A}_2. \tag{9.23}$$

Where the azimuthal integrals for plane problems are

$$\mathbf{A}_1 = \frac{1}{8(\nu - 1)} \begin{bmatrix} 4\nu - 5 & 1 - 4\nu & 0 \\ 1 - 4\nu & 4\nu - 5 & 0 \\ 0 & 0 & 8\nu - 6 \end{bmatrix} \{\boldsymbol{\sigma}(\boldsymbol{x}_P)\} \tag{9.24}$$

and

$$\mathbf{A}_2 = \frac{E}{8(\nu^2 - 1)} \begin{bmatrix} 3 & 1 & 0 \\ 1 & 3 & 0 \\ 0 & 0 & 1 \end{bmatrix} \{\epsilon(\boldsymbol{x}_P)\}. \tag{9.25}$$

For 3-D problems we have

$$\mathbf{A}_1 = \frac{\{\boldsymbol{\sigma}(\boldsymbol{x}_P)\}}{15(\nu-1)} \begin{bmatrix} 5\nu-7 & 1-5\nu & 1-5\nu & 0 & 0 & 0 \\ 1-5\nu & 5\nu-7 & 1-5\nu & 0 & 0 & 0 \\ 1-5\nu & 1-5\nu & 5\nu-7 & 0 & 0 & 0 \\ 0 & 0 & 0 & 10\nu-8 & 0 & 0 \\ 0 & 0 & 0 & 0 & 10\nu-8 & 0 \\ 0 & 0 & 0 & 0 & 0 & 10\nu-8 \end{bmatrix} \tag{9.26}$$

and

$$\mathbf{A}_2 = \frac{\{\boldsymbol{\epsilon}(\boldsymbol{x}_P)\}E}{30(\nu^2-1)} \begin{bmatrix} 16 & 2+10\nu & 2+10\nu & 0 & 0 & 0 \\ 2+10\nu & 16 & 2+10\nu & 0 & 0 & 0 \\ 2+10\nu & 2+10\nu & 16 & 0 & 0 & 0 \\ 0 & 0 & 0 & 7-5\nu & 0 & 0 \\ 0 & 0 & 0 & 0 & 7-5\nu & 0 \\ 0 & 0 & 0 & 0 & 0 & 7-5\nu \end{bmatrix} \tag{9.27}$$

Adding (9.19) and (9.22), we obtain for finite problems

$$\begin{aligned} \{\boldsymbol{\sigma}(\boldsymbol{x})\} = & \int_\Gamma \mathsf{S}(\boldsymbol{x},\hat{\boldsymbol{x}})\,[\mathbf{t}(\hat{\boldsymbol{x}}) - \mathbf{t}^*(\hat{\boldsymbol{x}})]\,d\Gamma(\hat{\boldsymbol{x}}) \\ & - \int_\Gamma \mathsf{R}(\boldsymbol{x},\hat{\boldsymbol{x}})\,[\mathbf{u}(\hat{\boldsymbol{x}}) - \mathbf{u}^*(\hat{\boldsymbol{x}})]\,d\Gamma(\hat{\boldsymbol{x}}). \end{aligned} \tag{9.28}$$

Adding (9.19) and (9.23), we obtain for infinite problems

$$\begin{aligned} \{\boldsymbol{\sigma}(\boldsymbol{x})\} = & \int_\Gamma \mathsf{S}(\boldsymbol{x},\hat{\boldsymbol{x}})\,[\mathbf{t}(\hat{\boldsymbol{x}}) - \mathbf{t}^*(\hat{\boldsymbol{x}})]\,d\Gamma(\hat{\boldsymbol{x}}) \\ & - \int_\Gamma \mathsf{R}(\boldsymbol{x},\hat{\boldsymbol{x}})\;[\mathbf{u}(\hat{\boldsymbol{x}}) - \mathbf{u}^*(\hat{\boldsymbol{x}})]\,d\Gamma(\hat{\boldsymbol{x}}) - \mathbf{A}_1 + \mathbf{A}_2. \end{aligned} \tag{9.29}$$

The computation of $\mathbf{u}(\boldsymbol{x}_P)$, $\frac{\partial \mathbf{u}(\boldsymbol{x}_P)}{\partial \boldsymbol{x}}$, $\boldsymbol{\sigma}(\boldsymbol{x}_P), \boldsymbol{\epsilon}(\boldsymbol{x}_P)$ on a point on the boundary is discussed next.

9.3 Computation of Boundary Values

The calculation of $\mathbf{u}(\boldsymbol{x}_P)$ on the boundary was discussed in Chap. 6. Here we discuss the calculation of its derivatives and stress/strain, separately for plane and 3-D problems.

9.3.1 Plane Problems

To obtain the global derivatives, we calculate first the strains in a local coordinate system $(\bar{x}, \bar{y})$ which is specified by unit vectors $\mathbf{v}_{\bar{x}}$ (in the ξ-direction) and $\mathbf{n}$. The strain in tangential direction is given by

$$\epsilon_{\bar{x}\bar{x}}(\boldsymbol{x}_P) = \frac{\partial u_{\bar{x}}}{\partial \bar{x}} = \left[\frac{\partial \mathbf{u}}{\partial \xi} \cdot \mathbf{v}_{\bar{x}}\right] \frac{\partial \xi}{\partial \bar{x}} = \left[\sum_{k=1}^{K^u} \frac{\partial R_k^u(\xi_P)}{\partial \xi} \mathbf{u}_k^{ep}\right] \cdot \mathbf{v}_{\bar{x}} \frac{1}{J}. \tag{9.30}$$

where the superscript u specifies the basis function R is used for the approximation of the displacements, J is the Jacobian and the superscript *ep* refers to the patch that contains $\boldsymbol{x}_P$.

The stress in the $\bar{x}$-direction is given by

$$\sigma_{\bar{x}\bar{x}} = c_1(\epsilon_{\bar{x}\bar{x}} + c_2\epsilon_{\bar{y}\bar{y}}) \tag{9.31}$$

with c_1 and c_2 the material constants having been defined in Sect. 9.1. The stress in $\bar{y}$-direction is

$$\sigma_{\bar{y}\bar{y}} = t_{\bar{y}} = \mathbf{t} \cdot \mathbf{n}. \tag{9.32}$$

To obtain the strain in $\bar{y}$-direction we use the following equation

$$\sigma_{\bar{y}\bar{y}} = c_1(\epsilon_{\bar{y}\bar{y}} + c_2\epsilon_{\bar{x}\bar{x}}) \tag{9.33}$$

and solve for the strain

$$\epsilon_{\bar{y}\bar{y}} = \frac{t_{\bar{y}}}{c_1} - c_2\epsilon_{\bar{x}\bar{x}} \tag{9.34}$$

The local stress tensor is given by

$$\bar{\boldsymbol{\sigma}} = \begin{pmatrix} \sigma_{\bar{x}\bar{x}} & \sigma_{\bar{x}\bar{y}} \\ \sigma_{\bar{y}\bar{x}} & \sigma_{\bar{y}\bar{y}} \end{pmatrix} \tag{9.35}$$

where

$$\sigma_{\bar{x}\bar{y}} = \sigma_{\bar{y}\bar{x}} = t_{\bar{x}} \tag{9.36}$$

with $t_{\bar{x}} = \mathbf{t} \cdot \mathbf{v}_{\bar{x}}$.

The global stress tensor $\boldsymbol{\sigma}$ can be obtained by the coordinate transformation

$$\boldsymbol{\sigma} = \mathbf{R}^{\mathrm{T}} \bar{\boldsymbol{\sigma}}\, \mathbf{R} \tag{9.37}$$

where the transformation matrix is given by

$$\mathbf{R} = \begin{pmatrix} \mathbf{v}_{\bar{x}}^{T} \\ \\ \mathbf{n}^{T} \end{pmatrix}. \tag{9.38}$$

The local strain tensor is given by

$$\bar{\boldsymbol{\epsilon}} = \begin{pmatrix} \epsilon_{\bar{x}\bar{x}} & \epsilon_{\bar{x}\bar{y}} \\ \epsilon_{\bar{y}\bar{x}} & \epsilon_{\bar{y}\bar{y}} \end{pmatrix} \tag{9.39}$$

where $\epsilon_{\bar{x}\bar{y}} = \frac{1}{2G}\sigma_{\bar{x}\bar{y}}$.

The global strain tensor $\boldsymbol{\epsilon}$ can be obtained by

$$\boldsymbol{\epsilon} = \mathbf{R}^{\mathrm{T}} \bar{\boldsymbol{\epsilon}}\, \mathbf{R}. \tag{9.40}$$

Finally the global derivatives of $\mathbf{u}$ are given by

$$\begin{aligned} \frac{\partial u_x}{\partial x} &= \epsilon_{xx} \quad , \quad \frac{\partial u_x}{\partial y} = \frac{1}{2}\epsilon_{xy} \\ \frac{\partial u_y}{\partial y} &= \epsilon_{yy} \quad , \quad \frac{\partial u_y}{\partial x} = \frac{1}{2}\epsilon_{yx} \end{aligned} \tag{9.41}$$

9.3.2 3-D Problems

For 3-D problems we establish a local orthogonal $\bar{x}, \bar{y}, \bar{z}$ coordinate system whose directions are defined as shown in Chap. 8. The local derivatives are computed by

$$\frac{\partial u_{\bar{x}}}{\partial \bar{x}} = \left[\frac{\partial \mathbf{u}}{\partial \xi} \cdot \mathbf{v}_{\bar{x}}\right] \frac{\partial \xi}{\partial \bar{x}} \tag{9.42}$$

$$\frac{\partial u_{\bar{y}}}{\partial \bar{y}} = \left[\frac{\partial \mathbf{u}}{\partial \xi} \cdot \mathbf{v}_{\bar{y}}\right] \frac{\partial \xi}{\partial \bar{y}} + \left[\frac{\partial \mathbf{u}}{\partial \eta} \cdot \mathbf{v}_{\bar{y}}\right] \frac{\partial \eta}{\partial \bar{y}} \tag{9.43}$$

$$\frac{\partial u_{\bar{x}}}{\partial \bar{y}} = \left[\frac{\partial \mathbf{u}}{\partial \xi} \cdot \mathbf{v}_{\bar{x}}\right] \frac{\partial \xi}{\partial \bar{y}} + \left[\frac{\partial \mathbf{u}}{\partial \eta} \cdot \mathbf{v}_{\bar{x}}\right] \frac{\partial \eta}{\partial \bar{y}} \tag{9.44}$$

$$\frac{\partial u_{\bar{y}}}{\partial \bar{x}} = \left[\frac{\partial \mathbf{u}}{\partial \xi} \cdot \mathbf{v}_{\bar{y}}\right] \frac{\partial \xi}{\partial \bar{x}} \tag{9.45}$$

where

$$\frac{\partial \mathbf{u}}{\partial \xi} = \sum_{k=1}^{K^u} \frac{\partial R_k^u(\xi_P, \eta_P)}{\partial \xi} \mathbf{u}_k^{ep} \quad \text{and} \quad \frac{\partial \mathbf{u}}{\partial \eta} = \sum_{k=1}^{K^u} \frac{\partial R_k^u(\xi_P, \eta_P)}{\partial \eta} \mathbf{u}_k^{ep}. \tag{9.46}$$

The local strains are given by:

$$\epsilon_{\bar{x}\bar{x}} = \frac{\partial u_{\bar{x}}}{\partial \bar{x}}, \qquad \epsilon_{\bar{y}\bar{y}} = \frac{\partial u_{\bar{x}}}{\partial \bar{y}}, \qquad \epsilon_{\bar{x}\bar{y}} = \frac{\partial u_{\bar{y}}}{\partial \bar{x}} + \frac{\partial u_{\bar{x}}}{\partial \bar{y}}. \tag{9.47}$$

The derivatives of ξ and η have been presented in Chap. 8. The stress in the $\bar{z}$-direction is computed by

$$\sigma_{\bar{z}\bar{z}} = c_1(c_2\epsilon_{\bar{x}\bar{x}} + c_2\epsilon_{\bar{y}\bar{y}} + \epsilon_{\bar{z}\bar{z}}) = t_{\bar{z}} \tag{9.48}$$

which can be solved for $\epsilon_{\bar{z}\bar{z}}$

$$\epsilon_{\bar{z}\bar{z}} = \frac{1}{c_1} t_{\bar{z}} - c_2\epsilon_{\bar{x}\bar{x}} - c_2\epsilon_{\bar{y}\bar{y}}. \tag{9.49}$$

Further strain components are

$$\epsilon_{\bar{x}\bar{z}} = \epsilon_{\bar{z}\bar{x}} = \frac{t_{\bar{x}}}{2G} \tag{9.50}$$

$$\epsilon_{\bar{y}\bar{z}} = \epsilon_{\bar{z}\bar{y}} = \frac{t_{\bar{y}}}{2G} \tag{9.51}$$

The local strain tensor is given by

$$\bar{\boldsymbol{\epsilon}} = \begin{pmatrix} \epsilon_{\bar{x}\bar{x}} & \epsilon_{\bar{x}\bar{y}} & \epsilon_{\bar{x}\bar{z}} \\ \epsilon_{\bar{y}\bar{x}} & \epsilon_{\bar{y}\bar{y}} & \epsilon_{\bar{y}\bar{z}} \\ \epsilon_{\bar{z}\bar{x}} & \epsilon_{\bar{z}\bar{y}} & \epsilon_{\bar{z}\bar{z}} \end{pmatrix}. \tag{9.52}$$

The global strain tensor $\boldsymbol{\epsilon}$ can be obtained by

$$\boldsymbol{\epsilon} = \mathbf{R}^{\mathrm{T}} \bar{\boldsymbol{\epsilon}}\, \mathbf{R} \tag{9.53}$$

where the transformation matrix is given by

$$\mathbf{R} = \begin{pmatrix} \mathbf{v}_{\bar{x}}^{\mathrm{T}} \\ \mathbf{v}_{\bar{y}}^{\mathrm{T}} \\ \mathbf{n}^{\mathrm{T}} \end{pmatrix} \tag{9.54}$$

Finally the global derivatives of $\mathbf{u}$ are given by

$$\begin{aligned}
\frac{\partial u_x}{\partial x} &= \epsilon_{xx} \quad , \quad \frac{\partial u_x}{\partial y} = \frac{1}{2}\epsilon_{xy} \quad , \quad \frac{\partial u_x}{\partial z} = \frac{1}{2}\epsilon_{xz} \\
\frac{\partial u_y}{\partial y} &= \epsilon_{yy} \quad , \quad \frac{\partial u_y}{\partial x} = \frac{1}{2}\epsilon_{yx} \quad , \quad \frac{\partial u_y}{\partial z} = \frac{1}{2}\epsilon_{yz} \\
\frac{\partial u_z}{\partial z} &= \epsilon_{zz} \quad , \quad \frac{\partial u_z}{\partial y} = \frac{1}{2}\epsilon_{zy} \quad , \quad \frac{\partial u_z}{\partial x} = \frac{1}{2}\epsilon_{zx}.
\end{aligned} \tag{9.55}$$

9.4 Examples

Here we present some examples of application that range from test examples that demonstrate the efficiency and accuracy of the IGABEM to some practical examples where the simulation effort is examined using the fast solution methods discussed in Sect. 6.6.

9.4.1 Circular Excavation in Infinite Domain

The first example relates to the simulation of a circular excavation in an infinite domain under plane strain conditions. The solution is a combination of two solutions: One without the excavation plus the change due to the excavation, i.e. the stress state is computed by:

$$\boldsymbol{\sigma} = \boldsymbol{\sigma}^{\mathrm{v}} + \triangle\boldsymbol{\sigma} \tag{9.56}$$

where $\boldsymbol{\sigma}^{\mathrm{v}}$ is the stress state before the excavation is made (virgin stress) and $\triangle\boldsymbol{\sigma}$ is the change in stress. The final state of the stress has to be such that the boundary is traction free. The traction at the boundary before excavation is:

$$t_i^{\mathrm{v}} = \sigma_{ij}^{\mathrm{v}} n_j \tag{9.57}$$

So to satisfy the condition that the excavation boundary is traction free, we have to apply a traction $\mathbf{t} = -\mathbf{t}^{\mathrm{v}}$ as a Neumann boundary condition to obtain the incremental stress state $\triangle\boldsymbol{\sigma}$. The method is shown in Fig. 9.1.

The virgin stress is normally specified as the value of the vertical stress σ^{v} and a factor k indicating the value of the horizontal stress as a proportion of the vertical stress (i.e. $\sigma_h = k\sigma^{\mathrm{v}}$).

There exists an analytical solution for this problem [129]. The normal stresses in radial (σ_r) and tangential (σ_θ) directions as well as the shear stresses ($\tau_{r\theta}$) of a circular excavation with the radius a are given for a point with the cylindrical coordinates r, θ by:

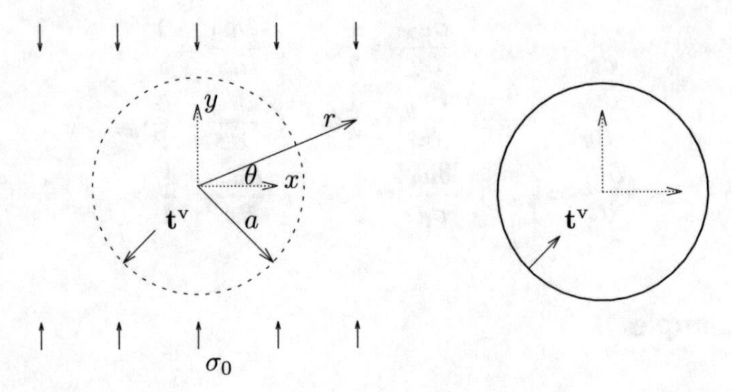

Fig. 9.1 Example 1: explanation of the simulation process for excavations: Left: stress state before excavation, Right: Neumann BC to be applied to obtain a zero traction condition on the boundary

$$\sigma_r = 0.5\sigma^{\mathrm{v}}\left[(1+k)(1-\frac{a^2}{r^2}) - (1-k)(1-\frac{4a^2}{r^2}+\frac{3a^4}{r^4})\right]\cos 2\theta \tag{9.58}$$

$$\sigma_r = 0.5\sigma^{\mathrm{v}}\left[(1+k)(1+\frac{a^2}{r^2}) + (1-k)(1+\frac{3a^4}{r^4})\right]\cos 2\theta \tag{9.59}$$

$$\tau_{r\theta} = 0.5\sigma^{\mathrm{v}}\left[(1-k)(1+\frac{2a^2}{r^2}-\frac{3a^4}{r^4})\right]\sin 2\theta \tag{9.60}$$

The displacements in radial and tangential directions are given by:

$$u_r = -\sigma^{\mathrm{v}}\frac{a^2}{4Gr}[1+k-(1-k)(2(1-2\nu)+(a/r)^2)\cos 2\theta] \tag{9.61}$$

$$u_t = -\sigma^{\mathrm{v}}\frac{a^2}{4Gr}[(1-k)(2(1-2\nu)+(a/r)^2)\sin 2\theta] \tag{9.62}$$

The discretisation of the circular excavation consists of 2 NURBS patches with basis function of order 2 and is shown in Fig. 9.2. It should be noted that the geometry of the circle is exactly described. The material properties assumed for the test example are $E = 1.0$ and $\nu = 0$ and the virgin stress is defined by $\sigma^{\mathrm{v}} = -1$ and $k = 0$. Since the geometry is exactly described, the outward normal is also exact and therefore the boundary condition is also exactly represented.

We simulate the problem without refinement (i.e. taking the same basis functions as are used for describing the geometry) and obtain the collocation points as shown in Fig. 9.2. We discover that the basis functions are able to represent the solution exactly, so no refinement is necessary. Indeed, the norm of the error of the solution obtained is 1.89×10^{-7} for the displacements and 1.82×10^{-6} for the stress. This is solely due to the specified error in the numerical integration and the precision of the input values and not due to the approximation of the unknowns. It has been verified that this accuracy can be obtained for any value of ν and k.

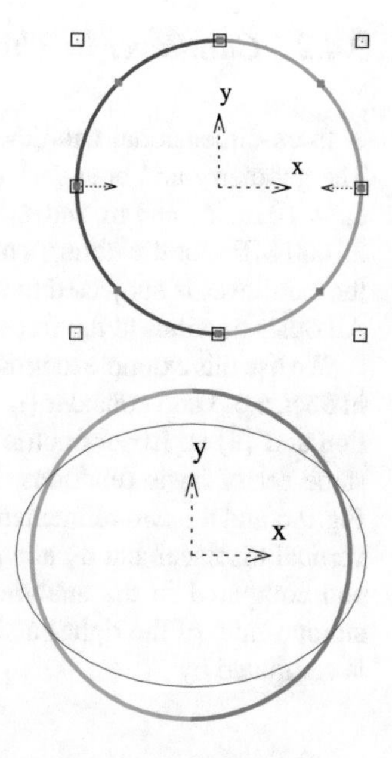

Fig. 9.2 Example 1: discretisation of circular excavation, control points are shown as hollow squares, collocation points as filled squares

Fig. 9.3 Example 1: deformed shape

The displaced shape is shown in Fig. 9.3. Next we compute the distribution of vertical stress along a horizontal line and compare the results with the analytical solution in Fig. 9.4. The values agree very well with the analytical solution even for points that are very close to the boundary, for the normal and regularised equations. However, the required number of Gauss points increases sharply for the normal integration, whereas the number remains nearly constant for the regularised integration.

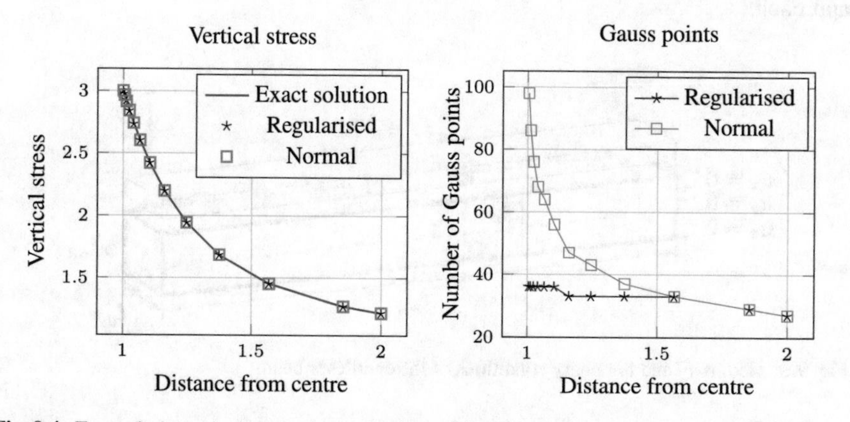

Fig. 9.4 Example 1: vertical stress along a horizontal line from the centre of the circle and number of Gauss points used

9.4.2 Cantilever in Three Dimensions

A three-dimensional cantilever beam with a constant vertical load is considered. The geometry and boundary conditions are shown in Fig. 9.5. The dimensions are $\ell_x = 10\,\text{m}$, $\ell_y = 1\,\text{m}$ and $\ell_z = 1\,\text{m}$, and the elastic material properties are $E = 29\,000\,\text{MPa}$ for the Young's modulus and $\nu = 0.0$ for the Poisson's ratio. At the top, the cantilever is subjected to a constant loading $t_z = -1\,\text{MPa}$ in vertical direction. All other tractions at the free surfaces are zero.

We use this example to illustrate an advantage of the refinement strategy described in Sect. 6.5.3 and consider (i) the proposed geometric independent field approximation and (ii) an isoparametric discretisation where all fields are represented by the same set of basis functions, i.e., $\hat{R}_k = \bar{R}_k = R_k$. The results are summarised in Fig. 9.6 and the two refinement cases are labelled by IFA and ISO, respectively. The vertical displacement u_z at a point $\boldsymbol{x}_{end}$ at the end of the cantilever beam is shown and compared to the analytical solution by Timoshenko [179]. Additionally, the storage-ratio of the right-hand side matrix due to the different refinement strategies is computed by

$$c_{\text{IFA}} = \frac{\mathsf{St}(\mathbf{R}_{\text{ISO}})}{\mathsf{St}(\mathbf{R}_{\text{IFA}})}. \tag{9.63}$$

The diagram on the left of Fig. 9.6 validates that geometry independent field approximation yields the same accuracy as an isoparametric approach. Furthermore, the graph on the right of the figure indicates that proposed refinement strategy saves storage, and as a consequence the computational effort to set-up the right-hand side of the system of equations. For the finest discretisation of this example, the value of $\mathsf{St}(\mathbf{R}_{\text{IFA}})$ is approximately 100 times smaller than as compared with an isoparametric approach.

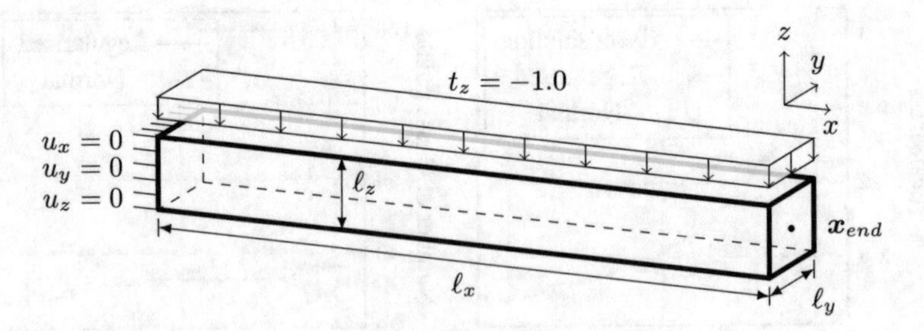

Fig. 9.5 Geometry and boundary conditions of the cantilever beam

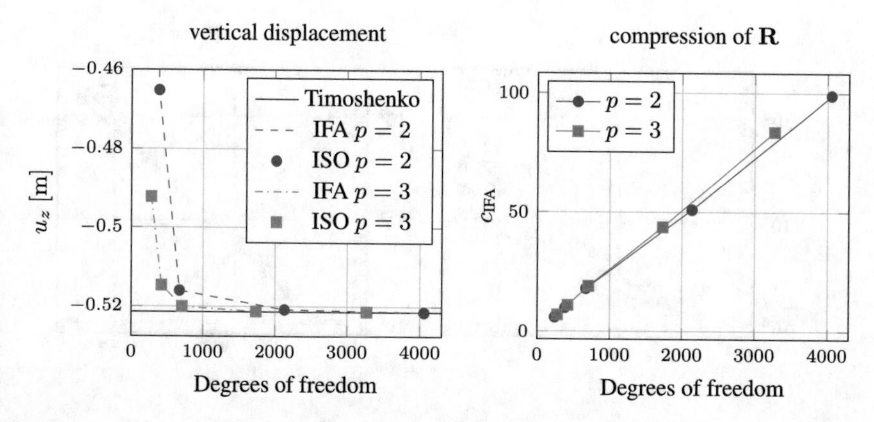

Fig. 9.6 Left: vertical displacement u_z of the cantilever beam at $\boldsymbol{x}_{end}$ for different orders p compared to the Timoshenko beam theory. Right: compression of the right-hand side matrix due to the application of geometric independent field approximation

9.4.3 Practical Example: Crankshaft

Finally, a practical example of a crankshaft is presented. This example is also used to show what compression rates can be achieved with $\mathcal{H}$-matrices (introduced in Sect. 6.6.4) for larger problems. The geometry is depicted in Fig. 9.7 and is described by basis functions of order 2. The material parameters are $E = 210$ GPa and $\nu = 0.25$. An arbitrary constant load is applied to the crank pins. The flywheel, as well as the axle on the other end of the crank, are fixed. A geometry independent field approximation is used. The approximation of the displacements was refined by knot insertion.

The calculation is performed using $\mathcal{H}$-matrices with a matrix approximation quality of $\epsilon_{\text{ACA}} = \{10^{-3}, 10^{-5}, 10^{-7}\}$. The minimal leaf size and the admissibility factor are set to $n_{min} = 30$ and $\eta = 1$, respectively.

Figure 9.8 shows the total compression rates c_{tot} for the left-hand side $\mathbf{L}$ and the right-hand side $\mathbf{R}$ of the system of equations with respect to the total number

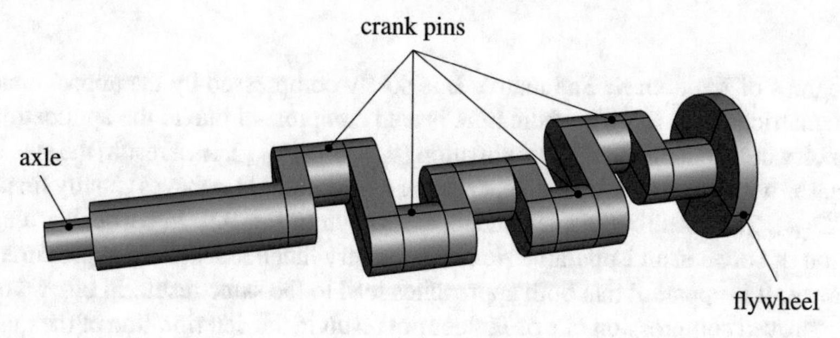

Fig. 9.7 Crankshaft geometry

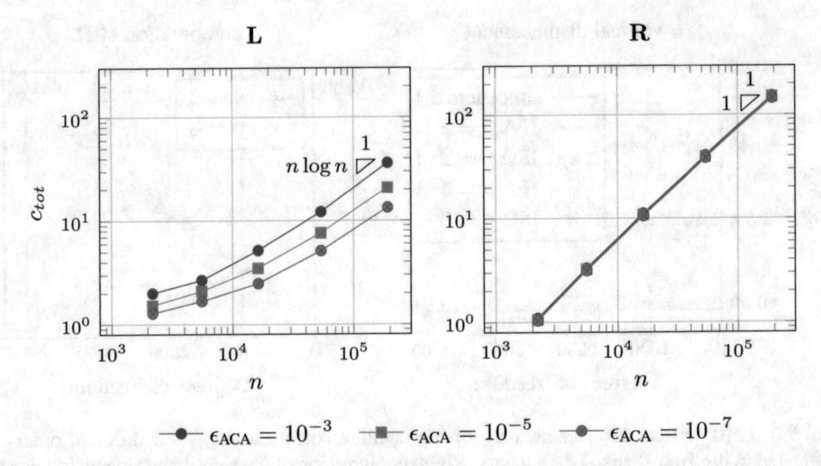

Fig. 9.8 Total compression rates c_{tot} for the system matrices of the crankshaft example with respect to the total degrees of freedom n

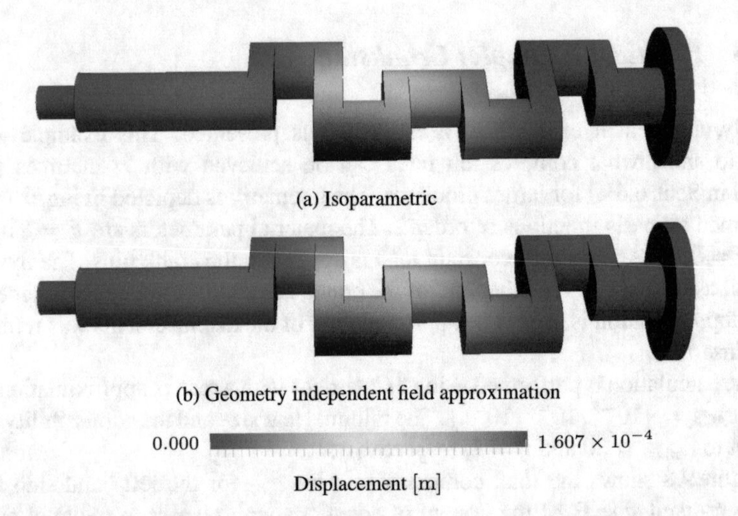

Fig. 9.9 The absolute displacement of the crankshaft example after three refinement steps due to an (a) isoparametric and a (b) geometry independent field discretisation

of degrees of freedom n. The matrix $\mathbf{L}$ is solely compressed by the approximation of $\mathcal{H}$-matrices ($c_{tot} = c_{\mathcal{H}}$), whereas $\mathbf{R}$ is also compressed due to the application of geometry independent field approximation ($c_{tot} = c_{\mathcal{H}}\, c_{\text{IFA}}$). As a result, the size of $\mathbf{L}$ increases with n nearly linearly, i.e, $\mathcal{O}(n \log n)$, whereas $\mathbf{R}$ behaves exactly linearly, i.e., $\mathcal{O}(n)$. The resulting displacements are shown in Fig. 9.9. In particular, a comparison of between an isoparametric and geometry independent field approximation is shown. It is apparent that both approaches lead to the same result. In other words, the improved compression rate of $\mathbf{R}$ does not result in the deterioration of the quality of the results.

9.5 Summary and Conclusions

In this chapter, we have presented the use of the IGABEM for problems in elasticity. Since the general solution procedures have already been outlined, we have concentrated here on specifics, especially the computation of results inside the domain and on the boundary.

On several examples, the power of the IGABEM has been demonstrated. The first example shows that in this case the exact result can be obtained without refinement, because the geometry and the variation of the unknown can be exactly represented by NURBS basis functions. On this example, it was also shown how very accurate results can be obtained very near or on the boundary using regularisation.

The other 3-D examples show how the solution converges rapidly with refinement, the advantage of the geometry independent field approximation and how the use of $\mathcal{H}$-matrices, introduced earlier, result in matrix compression rates that allows the computation of large problems with reasonable storage requirements and computation time.

Chapter 10
Simulation with Trimmed Models

So far we have discussed the simulation of steady state potential problems (Chap. 8) and linear elasticity problems (Chap. 9) on non-trimmed CAD geometries. That is, the entire parameter spaces of the patches were part of the domain of interest for the isogeometric analysis. In this chapter, we do not introduce a new problem type, but draw our attention to a more complex geometry representation, namely trimmed surfaces introduced in Sect. 4.2.3.

Geometries defined by trimmed surfaces impose several challenges on numerical simulations [114]. In a conventional analysis, they manifest themselves as severe problems during the model preparation and meshing process, as documented in numerous surveys and publications, see e.g., [61, 81, 162, 183]. Regarding isogeometric analysis, there are two fundamentally different philosophies for dealing with trimmed models. One seeks to resolve the deficiencies of trimmed models by a reconstruction of the geometric representation performed before the analysis. Since these procedures affect entire patches and their connectivity, we refer to them as *global* approaches. The other philosophy is to accept the numerical difficulties of trimmed models, implying that the analysis has to be adaptable enough to cope with them. This capability is accomplished by treating the occurring trimming situations on the knot span level. Hence, we label such techniques as *local* approaches. One aim of adapting trimmed models for simulation is to ensure that integration is only carried out inside the non-trimmed portion of the surface. Another aspect is the assembly of multiple patches. Furthermore, since we define the basis functions for the approximation of the unknowns on the untrimmed patch, we need to discuss the stabilisation of trimmed basis functions and the definition of their anchors.

Various aspects of isogeometric analysis with trimmed geometries will be discussed. The classification of global approaches is quite broad – in fact, classical meshing can be associated to this class. Isogeometric approaches include simulations based on T-splines, e.g., [8, 160, 197], subdivision surfaces, e.g., [42, 147, 191], and reconstruction schemes, e.g., [138, 182, 194]. Since discussing all of them would go beyond the scope of this book, we highlight the attributes of global approaches in Sect. 10.1 by a method proposed in [16] that reconstructs the domain of interest by ruled surfaces. The main focus of the remaining chapter will lie on local approach

G. Beer et al., *The Isogeometric Boundary Element Method*, Lecture Notes in Applied and Computational Mechanics 90,
https://doi.org/10.1007/978-3-030-23339-6_10

and the question how a simulation can be performed merely based on the basic data of a trimmed CAD model, i.e., boundary surfaces and trimming curves, defining the part of the surface to be visualised.

10.1 Double Mapping Method – A Global Approach

The following approach shines in simplicity and ease of implementation. Furthermore, it captures pretty well the goal of transforming trimmed surfaces such that they can subsequently be applied to "standard" isogeometric simulations. The basic idea is the following: the trimming curves $\boldsymbol{x}^t$ are used to specify a mapping $\mathcal{X}_t$ that fits a regular tensor basis into the valid area Ω_p of the corresponding trimmed patch $\boldsymbol{x}(\xi, \eta)$. To be precise, linear interpolation between two opposing trimming curves $\boldsymbol{x}^I$ and $\boldsymbol{x}^{II}$ defines $\mathcal{X}_t$. The geometrical mapping to the model space, however, is performed via the original trimmed patch. Hence, the approach is called *double mapping method*.

For the sake of notational simplicity, we assume that the regular basis functions $R_{ij}^{pq}(s,t)$ form a unit square, i.e., s and $t \in [0, 1]$, and both trimming curves are specified within the same parameter range $\zeta \in [a, b]$. Based on this setting, the intrinsic coordinate ζ can be linked to the boundaries of the regular basis at $t = 0$ and $t = 1$ by the coordinate transformations $f(s)$ and $g(s)$ given by

$$\zeta = f(s) = a + s(b - a) \quad \text{and} \quad \zeta = g(s) = b + s(a - b). \tag{10.1}$$

Note that $f(0) = g(1) = a$. Consequently this allows formulating $\mathcal{X}_t$ by

$$\boldsymbol{x}^{\Omega_p}(s,t) = (1 - t)\, \boldsymbol{x}^I(f(s)) + t\, \boldsymbol{x}^{II}(g(s)). \tag{10.2}$$

From a CAGD point of view, the mapping (10.2) is equivalent to the one of a ruled surface (4.22), but constructed in the parameter space. The geometric mapping $\mathcal{X}$ due to the trimmed patch determines the final representation in model space used for analysis. Figure 10.1 summarises the mappings of the double mapping approach.

The basis functions $R_{ij}^{pq}(s,t)$ define the approximation of the physical fields and the elements for the simulation. Note that this is equivalent to a regular (un-trimmed) patch and hence, the same integration and assembly concepts can be applied. The mapping (10.2) transforms the basis so that its elements cover the visible area Ω_p. The corresponding Jacobian adds a further component to the numerical integration (e.g., in Eq. (7.20)). Another distinguishing feature of the double mapping method is that the Gauss point coordinates are specified in the s, t map, requiring a further transformation to Gauss point coordinates $\bar{\xi}, \bar{\eta}$. Further details can be found in [16].

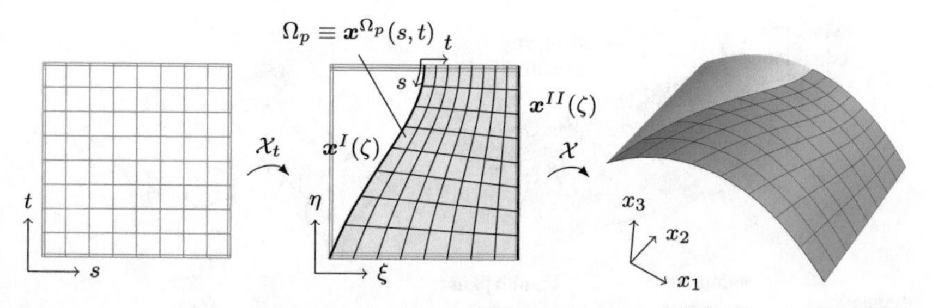

Fig. 10.1 Double mapping scheme to fit a regular tensor product surface to a trimmed patch. The first mapping $\mathcal{X}_t$ specifies the transformation to the valid area Ω_p of the trimmed parameter space, while the geometric mapping is denoted by $\mathcal{X}$. The trimming curves $\boldsymbol{x}^I(\zeta)$ and $\boldsymbol{x}^{II}(\zeta)$ are illustrated by thick lines

10.1.1 Numerical Example

The following example considers the deformation of a tunnel branch due to the excavation process. Figure 10.2 illustrates the CAD model of two intersecting cylinders with radii 7.5 m and 10 m, respectively, and the related trimmed patches.

For the simulation, we exported a quarter of this geometry as shown in Fig. 10.3(a). Two planes of symmetry (about the x–y and the x–z planes) are applied as boundary conditions. Furthermore, the tunnel ends are discretised using infinite plane strain patches (see Sect. 6.5.1.3) to simulate infinitely long tunnel tubes. The tunnel is assumed to be excavated in one step in an elastic pre-stressed ground (Poisson ratio $\nu = 0.0$ and Young's modulus $E = 1000$ MPa) with a virgin stress only in z-direction of $\sigma_{zz} = -1.00$ MPa. The excavation problem was solved by applying tractions as discussed in Sect. 9.4.1. Using a geometry independent field approximation (as introduced in Sect. 6.5.3), the basis functions for the definition of the unknown were refined using k-refinement until convergence was achieved. Continuous collocation

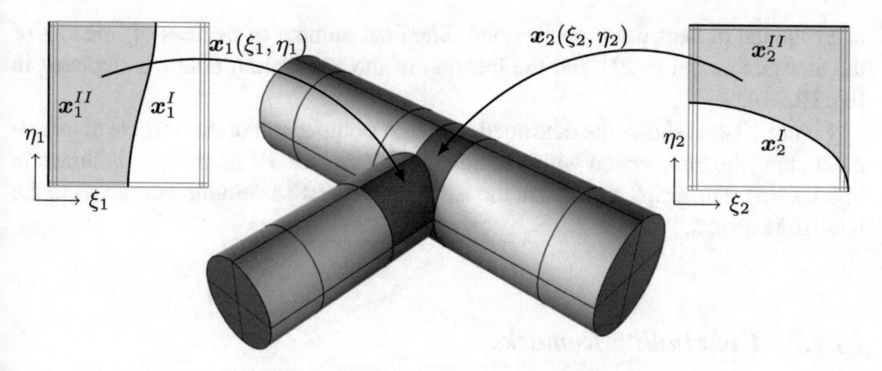

Fig. 10.2 The CAD model of the tunnel branch example together with the parameter space information of the two trimmed patches $\boldsymbol{x}_1(\xi_1, \eta_1)$ and $\boldsymbol{x}_2(\xi_2, \eta_2)$ that are employed in the simulation

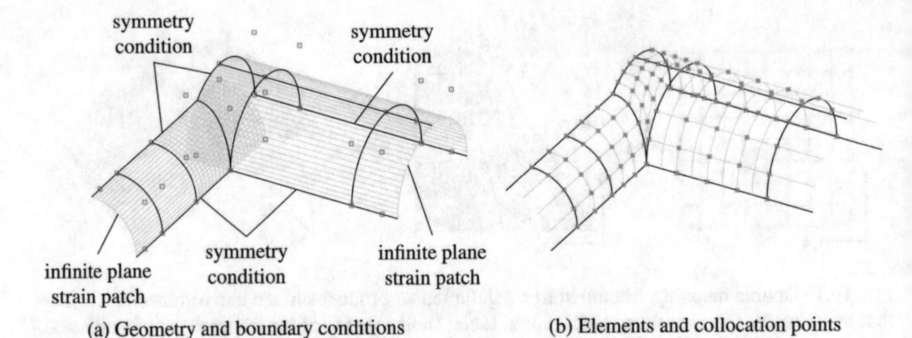

Fig. 10.3 Analysis model including the two trimmed surfaces, showing (a) the control points (red squares) of the geometry and the boundary conditions, and (b) the collocation points (red stars) and integration regions

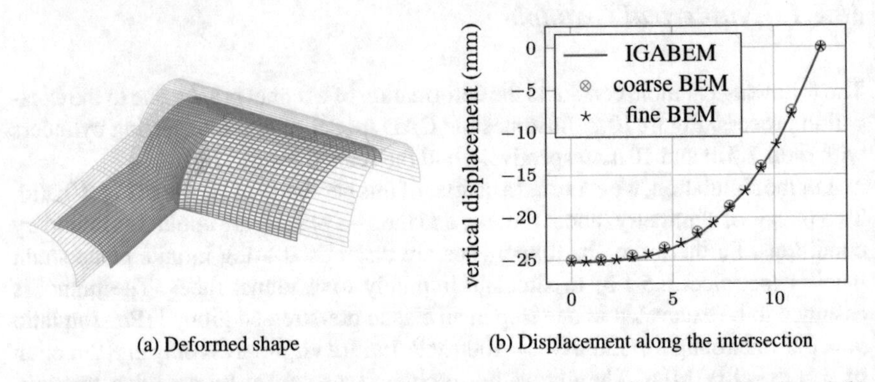

Fig. 10.4 Displacement due to the tunnel excavation: (a) the overall deformed shape and (b) the vertical displacement along the intersection of the two trimmed patches. In (b), the result of the double mapping method (IGABEM) is compared to results obtained with conventional BEM simulations using two different fine meshes

as explained in Sect. 6.5.4 was applied. The final number of degrees of freedom of the analysis model is 291 and the location of the collocation points is depicted in Fig. 10.3(b).

Figure 10.4(a) shows the deformed tunnel. A comparison of the vertical displacement along the intersection with a standard isoparametric BEM analysis is shown in Fig. 10.4(b). The graph shows that the isoparametric BEM solution converges to the IGABEM results.

10.1.2 Concluding Remarks

The presented global approach employs a simple strategy to fit a regular patch into the visualised area of the trimmed patch. This regular surface, together with the geometric mapping of the trimmed patch, defines the elements for the simulation. Hence, the actual analysis has not to deal with the issues induced by trimming, which is the distinguishing feature of a global scheme.

There are some limitations to the double mapping method worth to mention: first of all, the assumption that Ω_p can be represented by two opposing trimming curves limits its application to specific trimming situations. Furthermore, a four-sided nature of Ω_p is implied and thus, trimmed patches with more complex Ω_p have to be decomposed by an additional preprocessing step. Related concepts include cross fields or frame fields [25, 26], generalised Voronoi diagrams [79, 88] and decomposition based on characteristic points of the trimming curves [60, 82, 182]. Another potential issue is that elements may become distorted depending on the position of the trimming curves. Finally, the double mapping method works well for Bézier patches, but an integration issue arises as soon as the trimmed object is a spline surface with interior knots. Consider the situation depicted in Fig. 10.5. Note that the parameter lines of the geometry representation propagate through the elements defined by the mapped regular parameterisation. Thus, these integration regions are not C^∞-continuous but contain an arbitrarily located C^{p-m}-continuity which, in an extreme case, may even be an edge or kink. In other words, a proper distribution of quadrature points would require the subdivision of the elements along the non-smooth edges. A recent study [172] showed that also the continuity of the trimming curves has to be taken into account to achieve optimal high-order convergence rates with a conformal decomposition of trimmed patches.

Overall, the double mapping method is a simple, yet powerful, solution for trimmed (Bézier) patches and the concept can be used in various settings as will be shown in the remainder of this book. A natural extension of this approach is to define the mapping $\mathcal{X}_t$ to a trimmed parameter space by Coons patches. In contrast to the ruled surface interpolation, $\mathcal{X}_t$ takes four boundary curves into account (see Eq. (4.24)). Randrianarivony [138, 139] developed such an approach, including the decomposition of complex Ω_p, which was applied to wavelet Galerkin BEM in [80].

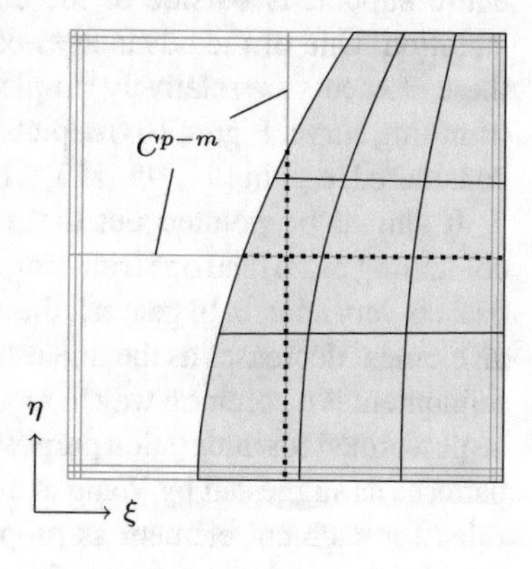

Fig. 10.5 Double mapping method for a B-spline patch. The dotted lines indicate parameter curves that are not C^∞-continuity within Ω_p

The implementation of this approach, however, is much more involved. In general, the complexity of global approaches raises rapidly with their capabilities, because they require a self-contained concept. This is indeed a drawback compared to local approaches which possess a modular structure, as will be discussed in the following.

10.2 Local Approaches

Local techniques do not modify the geometry, but the analysis has to deal with the deficiencies of trimmed solid models. Thereby, the trimmed parameter space is used as background parameterisation for the simulation while the trimming curves determine the domain of interest. Hence, the analysed area is embedded in a regular grid of knot spans which consists of *interior*, *exterior*, and *cut* elements. Considering geometry independent field approximation (Sect. 6.5.3), these elements correspond to the basis functions for the approximation of the physical fields, despite the fact that the trimming curves originate from the geometry representation.

In general, local approaches consists of the following distinct tasks: (i) the detection of the different element sets, (ii) the integration of cut elements, (iii) the treatment of multi-patch geometries along intersections, and (iv) the stability of a trimmed basis. If the known and unknown fields are approximated with different bases, these tasks are applied to both separately.

10.2.1 Element Detection

In an initial step, the various element types and their position within the trimmed basis need to be identified. Interior elements are defined by non-zero knot spans that are completely within the valid domain Ω_p. They can be treated as in regular isogeometric analysis. Exterior ones, on the other hand, can be ignored since their entire support is outside of the domain of interest. Cut elements require special attention. One of the advantages of local approaches is that the cutting patterns of these elements are relatively simple compared to the overall complexity of the entire trimming curve. Figure 10.6 depicts topological cases of cut elements that are usually considered, e.g. in [94, 95, 113, 116, 158].

It should be pointed out that other cases may exist as well, e.g., an element containing more than one trimming curve. These situations occur especially when the basis is very coarse. In general, the complexity of a trimming curve's topology within an element decreases as the fineness of the parameter space increases. Hence, (local) refinement is a common way to resolve invalid cutting patterns. This refinement may be performed for integration purpose only. An alternative is to extend the valid cutting patterns as suggested by Wang et al. [187] or the construction of tailored integration rules for each cut element as proposed by Nagy and Benson [123]. However, the benefit of a restricted number of trimming cases facilitates the subsequent integration process.

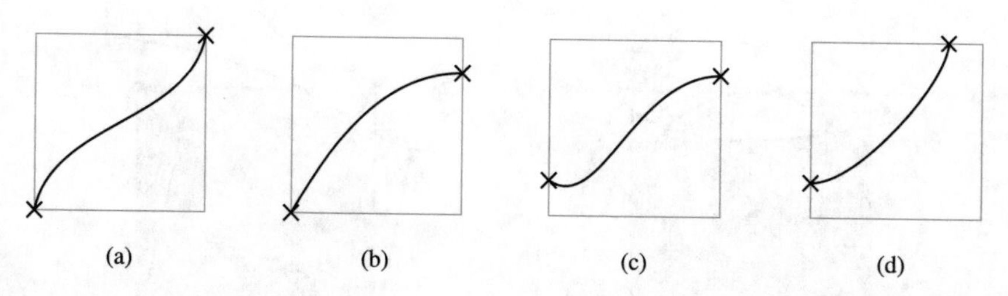

Fig. 10.6 Illustration of the most common valid cutting patterns of a single knot span. The actual element type is determined by the direction of the trimming curve. The crosses highlight the intersection points of the trimming curve with the element

Figure 10.7 illustrates a trimmed parameter space and the related element types. Note that cut elements are labelled based on their number of edges, whereas interior and exterior elements are referred to as elements of type 1 and –1, respectively. The knot span in the upper right corner is an example of an invalid case since smooth element edges are usually assumed for the numerical integration. Possible strategies to deal with this element include subdivision into several integration regions, treatment as a type 4 element with 2 curved edges, or knot refinement through the kink of the curve. Usual sources for invalid cases are kinks and straight trimming curves aligned with parameter lines.

The overall element detection task consists of the classification of knot span with respect to the trimming curves and the determination of trimming curve portions related to cut elements. Kim et al. [94, 95] and Schmidt et al. [158] presented two different approaches. The procedure suggested by the former can be summarised by:

(i) Compute the minimal signed distance $d_{i,j}$ of the centre of each non-zero knot span to all trimming curves to separate *interior* and *exterior* elements.
(ii) Identify *cut* elements by comparing $|d_{i,j}|$ with the radii $r_{i,j}^{in}$ and $r_{i,j}^{out}$ of the inscribed and circumscribed circles of the element. If $r_{i,j}^{in} \leqslant |d_{i,j}| < r_{i,j}^{out}$, the

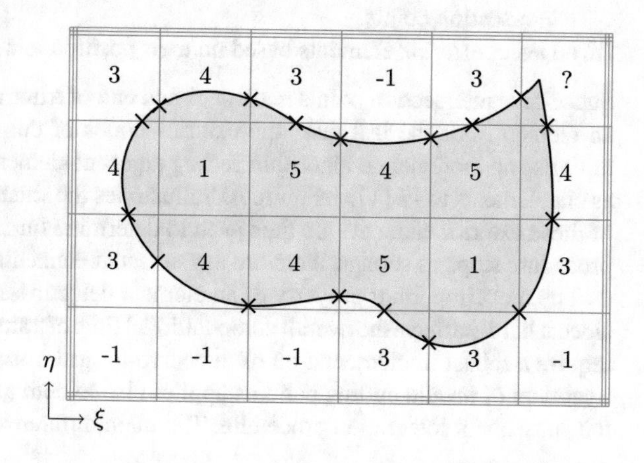

Fig. 10.7 Elements of a trimmed parameter space: 1 indicates untrimmed knot spans whereas –1 denotes knot spans which are outside of the domain. Cut elements are labelled by the number of interior edges (3, 4, or 5). The question mark indicates a special case. Crosses mark the intersections of the trimming curve with the parameter lines

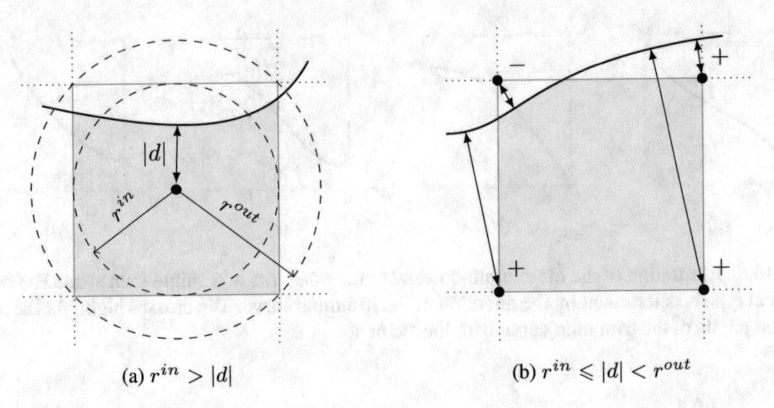

Fig. 10.8 Detection of cut elements according to Kim et al. [94, 95]: (a) first assessment based on the inscribed and circumscribed circles of the element and if necessary, (b) further comparison of the signed distance of the element corners

signed distance of the element corner nodes to the trimming curve are computed and compared as well.

(iii) Compute *intersection points* for each element cut by the trimming curve.

Both cases of the second step which specify cut elements are illustrated in Fig. 10.8. In Fig. 10.8(a), the distance of the element's centre to the trimming curve is smaller than the radius of the inscribed circle, whereas in Fig. 10.8(b), the cut element is identified since the signed distances of its corner nodes are positive and negative, where the sign indicates if the point is (+) inside or (−) outside of the visible domain Ω_p.

The other strategy by Schmidt et al. [158] starts by labelling all non-zero knot spans as interior elements and proceeds by:

(i) Determine all *intersections* of the trimming curve $\boldsymbol{x}^t$ with the tensor product grid and sort them in non-decreasing order with respect to the related parametric values of $\boldsymbol{x}^t$.
(ii) Assign the element type of *cut* elements based on the position of successive intersection points.
(iii) Detect *exterior* elements based on their position relative to the cut elements.

Successive intersection points mark start and end of trimming curve portions within an element. For the last task, the exterior nodes of cut elements can be used to initialise an incremental algorithm setting adjacent elements which are not labelled as cut elements to –1 [173]. Figure 10.9 illustrates the situation described. The nodes of these exterior elements are then used to determine further exterior elements. The procedure stops as soon as there are any adjacent elements of type 1 left.

The most important property of an element detection algorithm is its robustness since it hardly affects the overall computational time of a simulation. Both approaches require a robust implementation of the curve-to-grid intersection computation. The treatment of invalid cutting patterns applies also to both algorithms and depends on the subsequent integration procedure. The main difference between the approaches

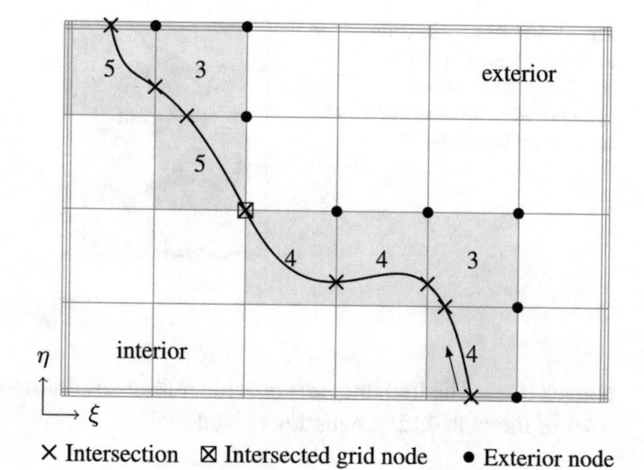

Fig. 10.9 Starting point for the separation of interior and exterior elements following the procedure of Schmidt et al. [158]. White knot spans are not classified yet. The arrow indicates the direction of the trimming curve

is that the former relies on a robust implementation of a point projection algorithm in order to determine the signed distance of a point to the trimming curve, while the latter requires a robust technique for detecting exterior elements based on the intersection information.

10.2.2 Integration Schemes

There is a vast body of literature proposing strategies to perform a proper integration of cut elements; either in the context of trimmed geometries or other simulation schemes such as fictitious domain methods and extended finite element methods. We focus on techniques presented for trimmed surfaces and divide them into schemes that take the trimming curve into account in an approximate or exact manner.

10.2.2.1 Approximated Trimming Curve

The following two schemes approximate the trimming curve $\boldsymbol{x}^t$ to define proper integration points within the parameter space. One approach uses a linear approximation of $\boldsymbol{x}^t$ to set up a tailored integration rule, whereas the other applies an adaptive subdivision to approximate the domain of the cut element.

Tailored integration rules. In [123, 188, 189], it is proposed to derive an integration rule for each cut element υ. The control polygon of $\boldsymbol{x}^t$ is used to represent υ by a polygon $\tilde{\rho}$ as shown in Fig. 10.10. The integral over $\tilde{\rho}$ can be reduced to a sum of line integrals over the edges of $\tilde{\rho}$ using Lasserre's theorems [104]. Therefore, the integration domain $\Omega_{\tilde{\rho}}$ has to be convex. In case of a non-convex region, a preprocessing step is applied so that the region is represented by a combination of

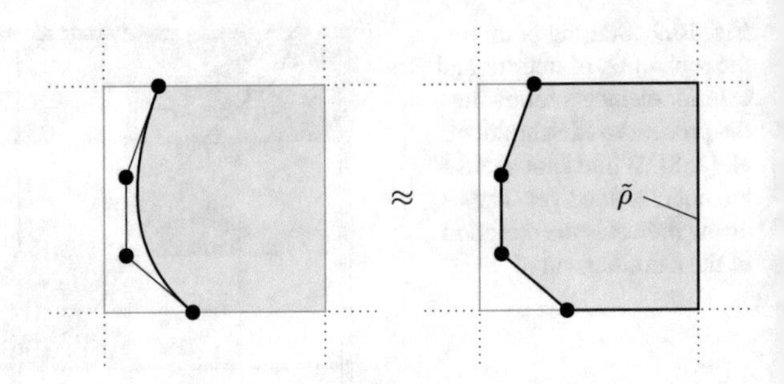

Fig. 10.10 Approximation of the cut element by a polygon $\tilde{\rho}$. The control points of the trimming curve are marked by circles

convex ones. The line integrals provide reference solutions for the right-hand side of a set of moment-fitting equations given by

$$\sum_{i=1}^{m} f_j(\xi_i, \eta_i) W_i = \int_{\Omega_{\tilde{\rho}}} f_j(\xi, \eta) d\Omega_{\tilde{\rho}}, \qquad j = 1, \ldots, n. \tag{10.3}$$

They are used to computed a tailored quadrature rule, i.e., points $\boldsymbol{q}_i = (\xi_i, \eta_i)^{\mathrm{T}}$ and weights W_i, for all functions f_j of the desired functions space, e.g., monomials up to a certain order. The goal is to find the lowest number of $\boldsymbol{q}_i$ so that the Eqs. (10.3) are satisfied up to a certain tolerance, for each v or rather $\tilde{\rho}$. The algorithm proposed in [123] starts with an initial set of $\boldsymbol{q}_i$ and successively eliminates one superfluous point after another. In each step, the reduced set of points is used to solve the system of Eqs. (10.3) in the least squares sense.

The construction of a tailored quadrature has two main benefits: (i) the number of integration points per v is optimised and (ii) *all* cutting patterns, including cases which had been labelled as invalid in Sect. 10.2.1, are covered by a single technique. This comes at the price of a more involved pre-processing phase since every cut element has to be treated individually. Furthermore, an error is introduced due to the approximation of v by $\tilde{\rho}$. This error can be reduced by refinement of the trimming curve since the control polygon converges to it. Still, the smooth high-order representation is removed by a linear one. Finally, it should be pointed out that the reduction of integration points does not take the smoothness of the basis into account, as it is done in case of optimised quadrature rules for regular splines, see e.g., [84].

Adaptive subdivision. Cut elements may be integrated by adaptive subdivision schemes as proposed in [76, 140, 151, 152]. The basic idea is to use a composed Gauss quadrature that aggregates integration points along the trimming curve. Therefore, a cut element v is decomposed into axis-aligned sub-cells $v^{\boxplus}$ based on a tree-structure, i.e., a quadtree in two dimensions. Starting from the initial cut element, each sub-cell is further subdivided into equally spaced sub-cells if it contains the trimming curve as displayed in Fig. 10.11(a). This recursive procedure is performed up to a user-defined maximal depth. The integral I^c over the entire cut element is defined as

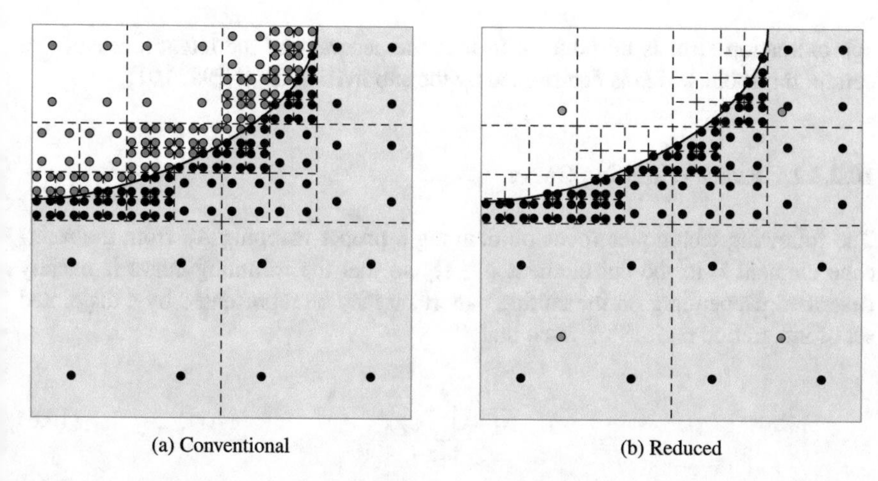

Fig. 10.11 Sub-cell structure of a single cut element: (a) conventional approach with quadrature points distributed within the valid (black points) and exterior (green points) domain and (b) reduced approach integrating over the whole element (orange points) and the valid domain (black points). The sub-cells are indicated by dashed lines

$$I^c = \sum_{i=1}^{I} I^{\text{v}}_{v_i^{\boxplus}}(\alpha^{\text{v}}) + \sum_{j=1}^{J} I^{-}_{v_j^{\boxplus}}(\alpha^{-}). \tag{10.4}$$

The factors $I^{\text{v}}_{v_i^{\boxplus}}$ and $I^{-}_{v_j^{\boxplus}}$ are the integrals over the valid domain Ω_p and the complementary exterior domain Ω_p^-, respectively. Integration points in the interior of Ω_p are multiplied by $\alpha^{\text{v}} = 1$, whereas exterior integration points are multiplied by a value that is almost zero, e.g., $\alpha^- = 10^{-14}$ as suggested in [151]. The integration procedure can be improved with respect to the number of quadrature points by

$$I^c = I^-_v(\alpha^-) + \sum_{i=1}^{I} I^{\text{v}}_{v_i^{\boxplus}}(\alpha^{\text{v}} - \alpha^-) \tag{10.5}$$

where $I^-_v(\alpha^-)$ represents the integral over the whole cut element without taken the trimming curve into account. The integration over the valid domain is performed as before by the composite quadrature, yet with another weighting factor, i.e., $(\alpha^{\text{v}} - \alpha^-)$. Such an improved sub-cell integration is illustrated in Fig. 10.11(b).

The key advantages of this approach are its simplicity and generality. The definition of integral transformation $\mathcal{X}_r$ from the reference element to the integration region and its Jacobian is straightforward, due to the axis-aligned shape of the sub-cells. Again, a single algorithm deals with *all* cut element types including invalid ones. Moreover, the extension to higher dimensions is straightforward. However, the downside is that the trimming curve is not represented exactly, and thus an additional approximation error is introduced. In fact, the accuracy of the integral ceases at a certain threshold which is determined by the subdivision depth [99, 101].

10.2.2.2 Exact Trimming Curve

The following techniques focus on defining a proper mapping $\mathcal{X}_r$ from the reference element $\hat{v}$ to the cut element $v \in \Omega_p$ so that the trimming curve is exactly described. Depending on the cutting pattern, v may be represented by a disjointed set of integration regions $v^{\boxdot}$ such that

$$v = \bigcup_{i=1}^{I} v_i^{\boxdot}. \tag{10.6}$$

In contrast to the sub-cells of the adaptive subdivision schemes outlined previously, the regions $v^{\boxdot}$ are usually not aligned with the axes of the parameter space and at least one $v^{\boxdot}$ has an edge which is defined by the portion of the trimming curve $\boldsymbol{x}^t$ within v.

There are various ways to specify $\mathcal{X}_r$. Ruled surface (4.22) and Coons patch (4.27) interpolations may be applied, where the portion of the trimming curve within v is considered for the construction [113, 187]. An example of local ruled surface mappings for various element types are shown in Fig. 10.12(a). It is worth noting that approaches based on the *blending function method* [64, 99, 100] can be included into this category. In the *nested Jacobian approach*, the integral transformation is also defined by a local NURBS surface combined with a nested subdivision [36, 131]. Thus, $\mathcal{X}_r$ consists of the local surface mapping and an additional transformation to the subregion. A corresponding distribution of quadrature points is shown in Fig. 10.12(b). In contrast to both previous references, i.e., [113, 187], type 5 elements are not decomposed into three triangular ones, but a bisection of the knot span is performed. An *adaptive Gaussian integration procedure* has been proposed in [37]. This variation of the nested Jacobian approach defines the local surface parameterisation within the reference space instead of the trimmed parameter space as illustrated in Fig. 10.12(c). Therefore, the trimming curve is transformed to the reference space by scaling and rotation. The integration points and their weights are adapted by scaling the $\bar{\eta}$-direction such that the points are located within the region described by the transformed trimming curve. The motivation for the adaptive Gaussian integration procedure is to treat the various cutting patterns by a single approach.

Another very common strategy is to adopt the integration concept developed in the context of the *NURBS-enhanced finite element method*, see e.g., [94, 95]. Using this scheme, every cut element is subdivided into a set of triangles. Those triangles that only consist of straight edges are subjected to conventional integration rules for triangles. The other triangles are treated by a series of mappings that take the curved edge into account

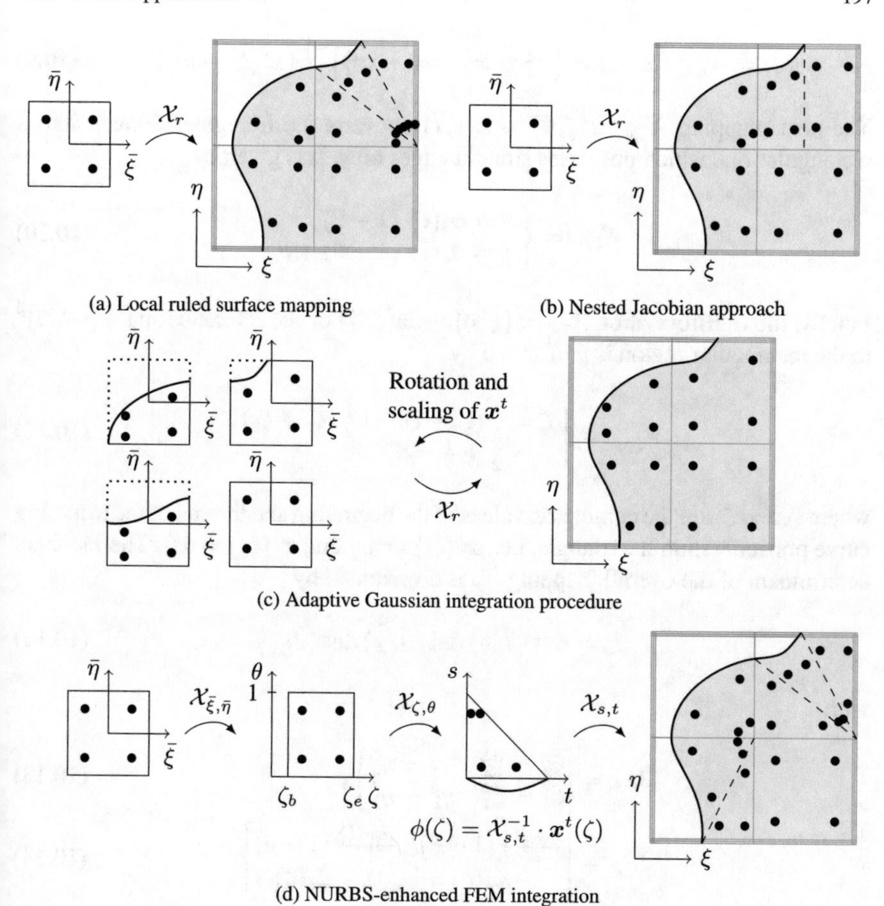

Fig. 10.12 Distribution of quadrature points due to various approaches that represent the trimming curve exactly. Dashed lines indicate the integration regions of a cut element

$$\mathcal{X}_r := \mathcal{X}_{s,t}\big(\mathcal{X}_{\zeta,\theta}\big(\mathcal{X}_{\bar{\xi},\bar{\eta}}(\bar{\xi},\bar{\eta})\big)\big). \tag{10.7}$$

Figure 10.12(d) displays the components of this series. Suppose the corner nodes of the triangle in the trimmed parameter space are labeled $\boldsymbol{x}_1^{\Delta}$ to $\boldsymbol{x}_3^{\Delta}$, where the beginning and the end of the trimming curve portion within the considered triangle are denoted by $\boldsymbol{x}_2^{\Delta}$ and $\boldsymbol{x}_3^{\Delta}$, respectively. The transformation $\mathcal{X}_{s,t}\colon \boldsymbol{x}(s,t) \mapsto \boldsymbol{x}(\xi,\eta)$ describes the mapping of a linear three node element

$$\mathcal{X}_{s,t} := \boldsymbol{x}(\xi,\eta) = t\,\boldsymbol{x}_1^{\Delta} + (1-s-t)\,\boldsymbol{x}_2^{\Delta} + s\,\boldsymbol{x}_3^{\Delta}. \tag{10.8}$$

To address the curved edge, the trimming curve $\boldsymbol{x}^t(\zeta)$ is transformed into the s,t-coordinate system by

$$\phi(\zeta) = \mathcal{X}_{s,t}^{-1} \cdot \boldsymbol{x}^t(\zeta) = \left[\boldsymbol{x}_3^{\triangle} - \boldsymbol{x}_2^{\triangle} \quad \boldsymbol{x}_1^{\triangle} - \boldsymbol{x}_2^{\triangle}\right]^{-1} \left(\boldsymbol{x}^t(\zeta) - \boldsymbol{x}_2^{\triangle}\right). \tag{10.9}$$

The next mapping $\mathcal{X}_{\zeta,\theta}$: $\boldsymbol{x}(\zeta,\theta) \mapsto \boldsymbol{x}(s,t)$ converts the triangular domain into a rectangular one which possesses straight edges only. It is given by

$$\mathcal{X}_{\zeta,\theta} := \begin{cases} s = \phi_s(\zeta)\,(1-\theta)\,, \\ t = \phi_t(\zeta)\,(1-\theta) + \theta. \end{cases} \tag{10.10}$$

Finally, the transformation $\mathcal{X}_{\bar{\xi},\bar{\eta}}$: $\boldsymbol{x}(\bar{\xi},\bar{\eta}) \mapsto \boldsymbol{x}(\zeta,\theta)$ of the reference space $[-1,1]^2$ to the rectangular region is performed by

$$\mathcal{X}_{\bar{\xi},\bar{\eta}} := \begin{cases} \zeta = \frac{\bar{\xi}}{2}\,(\zeta_e - \zeta_b) + \frac{1}{2}\,(\zeta_e + \zeta_b)\,, \\ \theta = \frac{\bar{\eta}}{2} + \frac{1}{2} \end{cases} \tag{10.11}$$

where ζ_b and ζ_e are the parametric values of the beginning and the end of the trimming curve portion within the triangle, i.e., $\boldsymbol{x}^t(\zeta_b) = \boldsymbol{x}_2^{\triangle}$ and $\boldsymbol{x}^t(\zeta_e) = \boldsymbol{x}_3^{\triangle}$. The Jacobian determinant of the overall mapping $\mathcal{X}_r$ is determined by

$$J_{\hat{\upsilon}} = \det\left(\mathbf{J}_{s,t}\right)\det\left(\mathbf{J}_{\zeta,\theta}\right)\det\left(\mathbf{J}_{\bar{\xi},\bar{\eta}}\right) \tag{10.12}$$

with

$$\mathbf{J}_{s,t} = \begin{bmatrix} \xi_3^{\triangle} - \xi_2^{\triangle} & \eta_3^{\triangle} - \eta_2^{\triangle} \\ \xi_1^{\triangle} - \xi_2^{\triangle} & \eta_1^{\triangle} - \eta_2^{\triangle} \end{bmatrix}, \tag{10.13}$$

$$\mathbf{J}_{\zeta,\theta} = \begin{bmatrix} \frac{\partial \phi_s(\zeta)}{\partial \zeta}\,(1-\theta) & \frac{\partial \phi_t(\zeta)}{\partial \zeta}\,(1-\theta) \\ -\phi_s(\zeta) & 1-\phi_t(\zeta) \end{bmatrix}, \tag{10.14}$$

$$\mathbf{J}_{\bar{\xi},\bar{\eta}} = \begin{bmatrix} \frac{1}{2}\,(\zeta_e - \zeta_b) & 0 \\ 0 & \frac{1}{2} \end{bmatrix}. \tag{10.15}$$

The coefficients $\xi_i^{\triangle}$ and $\eta_i^{\triangle}$ refer to the coordinates of the corner nodes $\boldsymbol{x}_i^{\triangle}$ and the derivative of the transformed trimming curve is calculated by

$$\frac{\partial \phi(\zeta)}{\partial \zeta} = \left[\boldsymbol{x}_3^{\triangle} - \boldsymbol{x}_2^{\triangle} \quad \boldsymbol{x}_1^{\triangle} - \boldsymbol{x}_2^{\triangle}\right]^{-1} \left(\frac{\partial \boldsymbol{x}^t(\zeta)}{\partial \zeta}\right). \tag{10.16}$$

Figure 10.12 summarises the different integration schemes discussed. Their common and most essential feature is that the trimmed element is exactly represented. The main difference between the strategies is the partitioning of a cut element υ into integration regions $\upsilon^{\square}$. In fact, the series of mappings (10.7) shown in Fig. 10.12(d) yields the same distribution of quadrature points over a triangular element as a ruled surface interpolation (4.22) illustrated in Fig. 10.12(a), if the trimming curve is a B-spline curve. In case of NURBS curves, on the other hand, different distributions are obtained. These two cases are compared in Fig. 10.13.

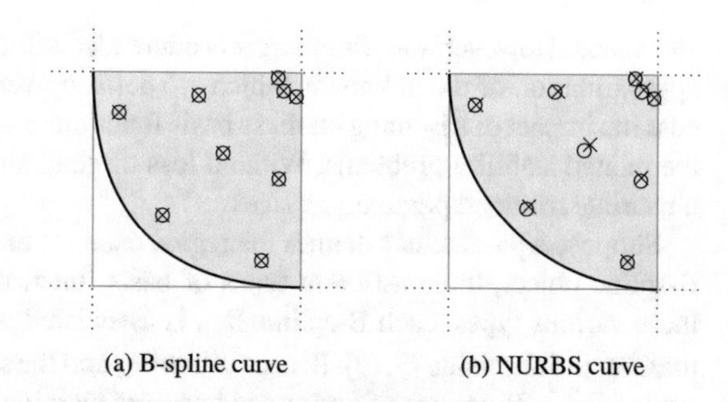

Fig. 10.13 Distribution of Gauss points within a cut element of type 3 based on the NURBS-enhanced FEM mapping (circles) and ruled surface parameterisation (crosses). The trimming curve is described by either (a) a B-spline curve or (b) a NURBS curve

In general, it seems that good results can be obtained with either of these concepts, especially for moderate orders. However, it has been demonstrated that the properties of coordinate mappings and the corresponding placement of interior nodes is crucial for the convergence behaviour of conventional high-order ($p > 3$) finite elements [124]. With this in mind, additional research might be useful to assess the quality of the mapping schemes presented with respect to their performance for higher orders.

10.2.3 Assembly of Multiple Patches

Special consideration has to be given for trimmed patches in the assembly of the system of equations. In order to ensure C^0-continuity, weak coupling strategies such as Lagrange multipliers [187], the penalty method [36, 37], and the Nitsche method [76, 97, 152] are often employed. A crucial element of all these techniques is the proper linking of the degrees of freedom of adjacent surfaces. This is usually quite challenging when dealing with trimmed patches since there is no explicit connectivity between adjacent trimmed surfaces. Instead, each surface has its independent approximation of the common intersection (see Sect. 4.2.3). In the case of BEM and the applications considered in this book, however, we are in the fortunate situation that continuity between patches is not required (see Sect. 6.5.4 on discontinuous collocation). In other words, we can treat geometries defined by multiple trimmed surfaces effectively by employing discontinuous collocation where no collocation point lies on an intersection of two surfaces. There is, however, the issue that collocation points may be located outside of the trimmed surface since the basis functions are defined on the entire patch. This aspect is addressed in the following subsection as part of the stabilisation concept.

10.2.4 Stabilisation

The integration schemes presented in Sect. 10.2.2 guarantee that integration is only carried out on the non-hidden part Ω_p of a trimmed patch, not over the whole parameter space. However, the trimming procedure also affects the basis functions for the approximation of the unknown, which are defined over cut elements. Here, we discuss the impact of trimming on these basis functions and present a concept to resolve the related stability problems. Without loss of generality, we are going to focus on univariate trimmed parameter spaces.

Suppose a parameter t defines the region used for the simulation Ω_p of a trimmed B-spline object, three different types of basis functions arise. In order to classify these various types, each B-spline $B_{i,p}$ is associated with an *anchor* $\tilde{\xi}_i$ defined by the Greville abscissae (3.19). Based on these $\tilde{\xi}_i$ and the support of the basis functions, $\text{supp}\,\{B_{i,p}\}$, B-splines of a trimmed basis are classified as:

- *Stable* if $\tilde{\xi}_i \in \Omega_p$
- *Degenerate* if $\tilde{\xi}_i \notin \Omega_p$ and $\text{supp}\,\{B_{i,p}\} \cap \Omega_p \neq \emptyset$
- *Exterior* if $\text{supp}\,\{B_{i,p}\} \cap \Omega_p = \emptyset$.

An example of these different types is shown in Fig. 10.14. Note that the given specification is by no means restricted to the univariate case; it can be easily applied to trimmed surfaces as well. Some authors identify the different B-splines of a trimmed parameter space based on the size of their support within Ω_p, see e.g., [86, 150]. The proposed scheme, however, provides a simple and fast separation of stable, degenerate, and exterior functions and it guarantees that anchors of stable B-splines, which will be used as collocation points for a simulation, are within the domain of interest.

The main purpose of the type classification is to detect degenerate B-splines. They are of particular interest, because their support may become arbitrary small due to the trimming procedure. In the example depicted in Fig. 10.14, this would be the case for $B_{6,3}$ as the trimming parameter t approaches the knot value 3. Trimmed basis functions with arbitrary small support can yield serious stability issues. This is demonstrated by an example of a bivariate basis due to uniform open knot vectors $\Xi_1 = \Xi_2 = (0, 0, 0, 0.25, 0.5, 0.75, 1, 1.25, 1.5, 1.75, 2, 2, 2)$ and $p = q = 2$. This basis is trimmed by a parameter t such that the domain of interest is $\Omega_p = [0, t]^2$. The condition numbers of the related spline interpolation matrix $\mathbf{A}$ and mass matrix $\mathbf{M}$ are computed (see Chap. 3 for details). The results for various

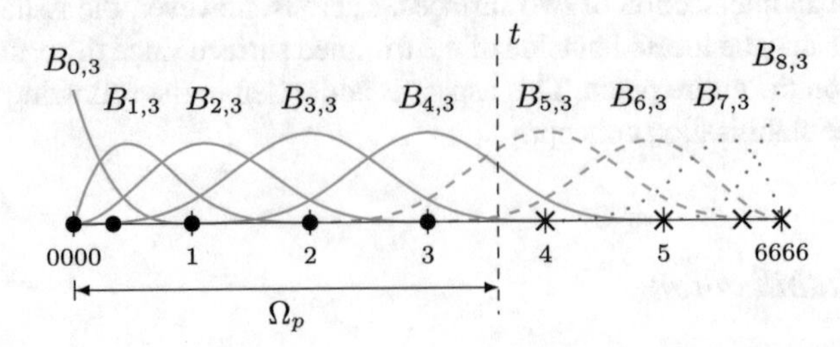

Fig. 10.14 Types of basis functions of a trimmed cubic B-spline basis: stable (solid), degenerate (dashed), and exterior (dotted). Circles mark anchors of basis functions that are within the valid domain Ω_p, while crosses indicate those outside the valid domain Ω_p

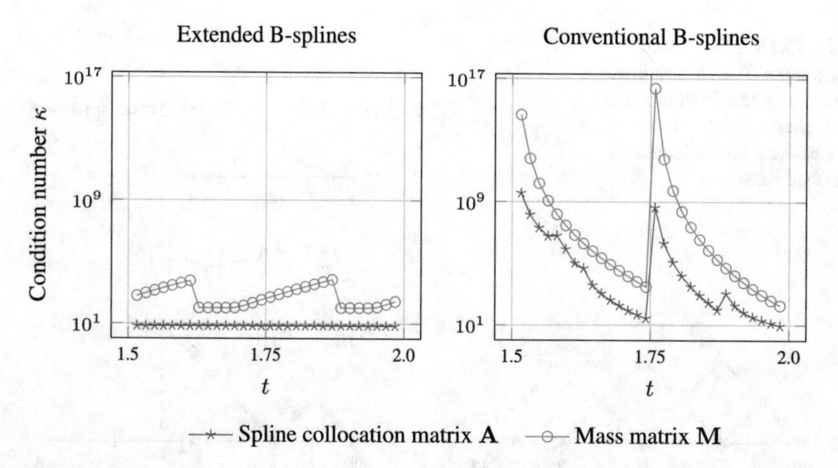

Fig. 10.15 Condition numbers of the spline collocation matrix $\mathbf{A}$ and mass matrix M of a trimmed bivariate basis with $\Omega_p = [0, t]^2$. The labels on the x-axis mark the knots of the uniform parameter space defined by an open knot vector. The graphs on the left correspond to the stabilised extended B-spline basis, whereas the performance of normal B-splines is shown on the right

positions of $t \in (1.5, 2.0)$ are summarised in Fig. 10.15 on the right. On the left-hand side, the same problems is analysed but with the stabilised extended B-spline basis which will be introduced in the following. Note that the extended B-spline basis yields moderate condition numbers, whereas drastic peaks occur in the graphs corresponding to conventional B-splines when t approaches a knot, e.g., the value 1.75.

10.2.4.1 Definition of Extended B-splines

The fundamental idea of extended B-splines $B^e_{i,p}$ is to substitute the polynomial segments of knot spans that contain degenerate functions by *extensions* of basis functions that are classified as stable. As a result, degenerate B-splines are eliminated from the basis and thus, the domain of interest Ω_p is described by stable basis functions only.

Recall that a B-spline consist of polynomial segments. Figure 10.16 illustrates an example of a univariate basis function and its three segments as well as their extensions indicated by the dashed lines. We label these segments by $\mathcal{B}^s_i$ where s denotes the knot span and i refers to the index of the corresponding B-spline $B_{i,p}$.

Considering univariate extended B-splines, the extensions for the stabilisation are provided by the polynomial segments $\mathcal{B}^s_i$ of the *closest* knot span with respect to the boundary of Ω_p that contains *only* stable B-splines. Figure 10.17 shows this concept and the components involved for a set of linear basis functions; exterior B-splines are omitted since they do not make any contribution to the basis. Noted that the supports of the resulting extended B-splines $B^e_{i,p}$ displayed in Fig. 10.17(b) can never become arbitrarily small. Thus, an extended B-spline basis is stable.

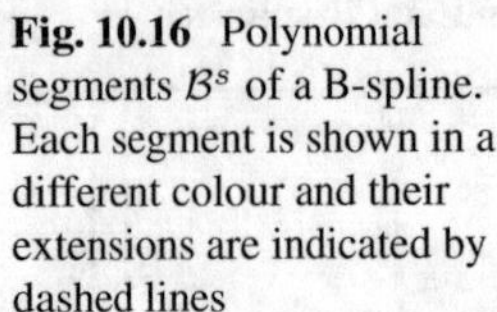

Fig. 10.16 Polynomial segments $\mathcal{B}^s$ of a B-spline. Each segment is shown in a different colour and their extensions are indicated by dashed lines

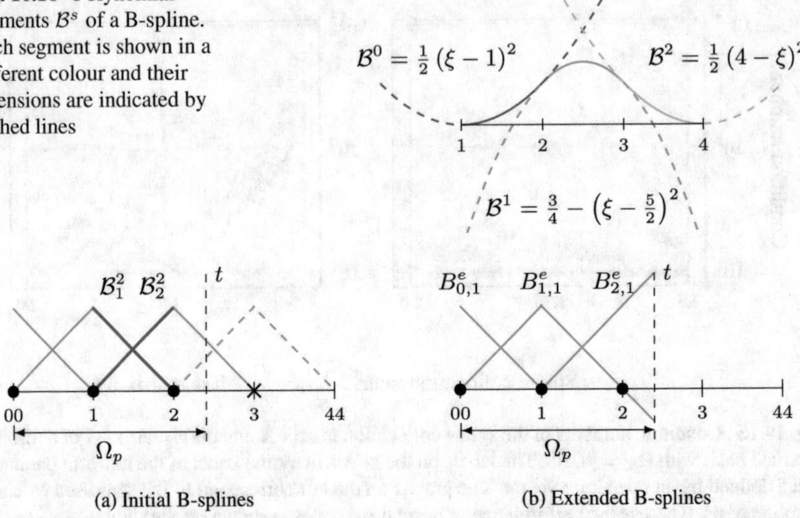

Fig. 10.17 Conversion of a potentially unstable B-spline basis to a stable extended B-spline basis: (a) initial stable and degenerate basis functions and (b) the resulting extended B-splines. In (a), the polynomial segments $\mathcal{B}_1^2$ and $\mathcal{B}_2^2$ that provide the extensions for the stabilisation are highlighted by thick red lines

Extended B-splines can be defined as a linear combination of the original B-splines. To be precise, the extended polynomial segments $\mathcal{B}_i^s$ of the non-trimmed knot span s are expressed by B-splines of the trimmed knot span t such that

$$\mathcal{B}_i^s(\xi) = \sum_{j=t-p}^{t} B_{j,p}(\xi)\, e_{i,j} \qquad \xi \in [\xi_t, \xi_{t+1}) . \tag{10.17}$$

This representation is exact since all $\mathcal{B}_i^s$ are in the same space of splines $\mathbb{S}_{\Xi,p}$. The main task is to determine the values of the *extrapolation weights* $e_{i,j}$. If the index j corresponds to a stable B-spline, the weights are trivial:

$$e_{i,i} = 1 \quad \Leftrightarrow \quad \mathcal{B}_i^s(\xi) \equiv B_{i,p}(\xi), \xi \in [\xi_s, \xi_{s+1}) \,, \forall i \in \{s-p, \ldots, s\}, \tag{10.18}$$

$$e_{i,j} = 0 \quad \Leftrightarrow \quad \mathcal{B}_i^s(\xi) \neq B_{j,p}(\xi), \xi \in [\xi_s, \xi_{s+1}) \,, \forall j \in \{s-p, \ldots, s\} \backslash i. \tag{10.19}$$

The remaining extrapolation weights related to degenerate B-splines can be computed by

$$e_{i,j} = \frac{1}{p!} \sum_{k=0}^{p} (-1)^k \, (p-k)! \, \beta_{p-k} \, k! \, \tilde{\alpha}_k \tag{10.20}$$

with the coefficients $\tilde{\alpha}$ and β denoting the constants of the polynomials

$$\mathcal{B}_i^s(\xi) = \sum_{k=0}^{p} \tilde{\alpha}_k \, \xi^k \quad \text{and} \quad \psi_{j,p}(\xi) = \prod_{m=1}^{p} (\xi - \xi_{j+m}) = \sum_{k=0}^{p} \beta_k \, \xi^k. \tag{10.21}$$

The former is the power basis form of the extended polynomial segment, which may be derived by Taylor expansion, i.e.,

$$\mathcal{B}_i^s(\xi) = \sum_{k=0}^{p} \frac{d^k}{d\xi^k} B_{i,p}(\tilde{\xi}) \, \frac{(\xi - \tilde{\xi})^k}{k!} = \sum_{k=0}^{p} \alpha_k \, (\xi - \tilde{\xi})^k, \quad \tilde{\xi} \in [\xi_s, \xi_{s+1}) \tag{10.22}$$

where the point $\tilde{\xi}$ is within the corresponding knot span s. Based on Eq. (10.22), the coefficients $\tilde{\alpha}_k$ are determined by

$$\tilde{\alpha}_k = \sum_{m=k}^{p} \binom{m}{k} \alpha_m \left(-\tilde{\xi}\right)^{m-k} \quad \text{with} \quad \binom{m}{k} := \frac{m!}{(m-k)! \, k!} . \tag{10.23}$$

The Newton polynomials $\psi_{j,p}$ result from a quasi interpolation procedure called *de Boor–Fix* functional [28, 29]. The corresponding coefficients β_k can be obtained by

$$\beta_k = (-1)^{p-k} \sum_{\ell=1}^{L} \prod_{m \in \mathbb{T}_{p-k,\ell}} \xi_m \qquad \text{with} \qquad L = \frac{p!}{(p-k)! \, k!} \tag{10.24}$$

where the sum over $\mathbb{T}_{n,\ell}$ denotes all n-combinations with repetition of the knots appearing in the definition of $\psi_{j,p}$, i.e., $\xi_{j+1}, \ldots, \xi_{j+p}$. For example, $\mathbb{T}_{n,\ell}$ for $p = 3$ would be given by

$$\begin{aligned}
\mathbb{T}_{3,1} = \{\xi_{j+1}, \xi_{j+2}, \xi_{j+3}\}, \quad & \mathbb{T}_{2,1} = \{\xi_{j+1}, \xi_{j+2}\}, \quad && \mathbb{T}_{1,1} = \{\xi_{j+1}\}, \\
& \mathbb{T}_{2,2} = \{\xi_{j+2}, \xi_{j+3}\}, && \mathbb{T}_{1,2} = \{\xi_{j+2}\}, \\
& \mathbb{T}_{2,3} = \{\xi_{j+1}, \xi_{j+3}\}, && \mathbb{T}_{1,3} = \{\xi_{j+3}\}.
\end{aligned}$$

By taking the trivial extrapolations weights (10.18) and (10.19) into account, an extended B-spline is defined as

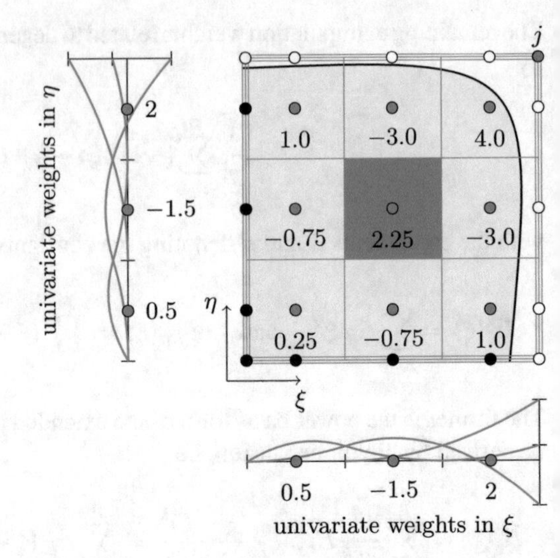

Fig. 10.18 The construction of bivariate extrapolation weights $e_{i,j}$. The anchors of stable B-splines are marked by black and green circles. The shown values of $e_{i,j}$ are related to the degenerate basis function marked by the red circle in the upper right corner of the parameter space. The univariate components of these $e_{i,j}$ are indicated by the splines displayed outside of the bivariate parameter space. The closest non-trimmed knot span and the related anchors are indicated in green

$$B^e_{i,p} = B_{i,p} + \sum_{j \in \mathbb{J}_i} e_{i,j} B_{j,p} \tag{10.25}$$

where $B_{i,p}$ is the stable B-spline from which the extension originates, and $\mathbb{J}_i$ is the index-set of all degenerate B-splines related to $B^e_{i,p}$. Definition (10.25) also applies to bivariate basis functions of B-spline surfaces. The corresponding extrapolation weights are determined by the tensor product of the univariate counterparts computed for each parametric direction. The construction procedure is visualised in Fig. 10.18 and examples of different bivariate extended B-splines are shown in Fig. 10.19. The closest non-trimmed knot span can be detected by measuring the distance between the anchor of the degenerate B-spline and the midpoints of the surrounding knot spans.

In general, a degenerate B-spline is distributed to $(p+1)(q+1)$ stable ones. Further, several degenerate B-splines $B_{j,p}$ may be associated to a single stable B-spline $B_{i,p}$. In particular, the number of $B_{j,p}$ related to $B_{i,p}$ is determined by the cardinality of the corresponding index-set $\#\mathbb{J}_i$. However, the extension procedure is restricted to those basis functions that are close to the trimming curve. For instance, the basis function shown in Fig. 10.19(a) is a conventional B-spline since it is far enough away from the trimming curve. The actual size of the affected region depends on the fineness of the parameter space and the order of its basis functions. For more information regarding the properties of extended B-splines, the interested reader is referred to [86, 116].

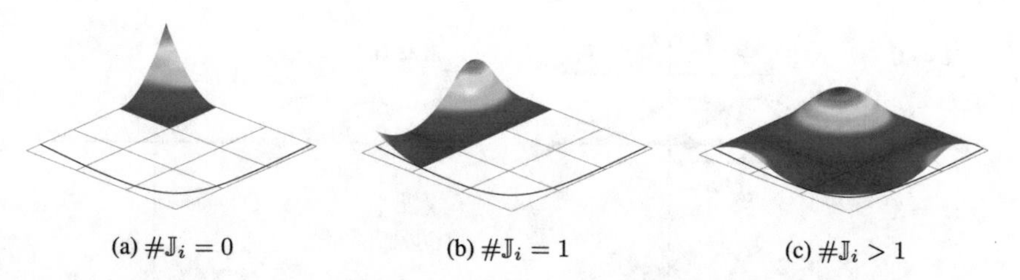

Fig. 10.19 Bivariate extended B-splines $B^e_{i,p}$ with various cardinalities of the index-set $\mathbb{J}_i$ which indicates the number of related degenerate B-splines

10.2.4.2 Extended THB-splines

The definition of extended B-splines (10.25) is applicable to any order p. However, the order has an impact on the extrapolation process. That is, the area represented by extensions, rather than original B-splines, increases with p due to the fact that supports of degenerate B-splines propagate further into Ω_p as indicated in Fig. 10.20. We specify the related *extrapolation length*, d_e, by the shortest distance of the boundary of Ω_p to the centre of the knot span that provides the extensions. A large extrapolation length may increase the approximation error in the vicinity of the trimming curve. It is worth noting that this error is only introduced in the representation of the physical fields of the problem considered and does not affect the geometry description.

Local refinement provides a means to gain control over the extrapolation length. In the following, the complementary features of extended B-splines and THB-splines (see Sect. 3.6.4) are combined to obtain a more powerful stabilisation technique for isogeometric analysis of trimmed geometries. While extended B-splines resolve the stability issue introduced by degenerate B-splines of a trimmed parameter space, truncated hierarchical refinement provides control over the extrapolation length d_e, thereby making the stabilisation independent of the order of the basis.

Fig. 10.20 Correlation between the polynomial order p and the extrapolation length d_e by comparing a quadratic basis (top) with its cubic counterpart (bottom)

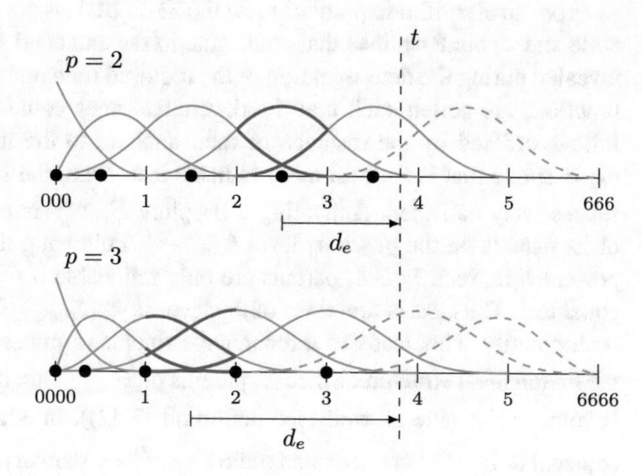

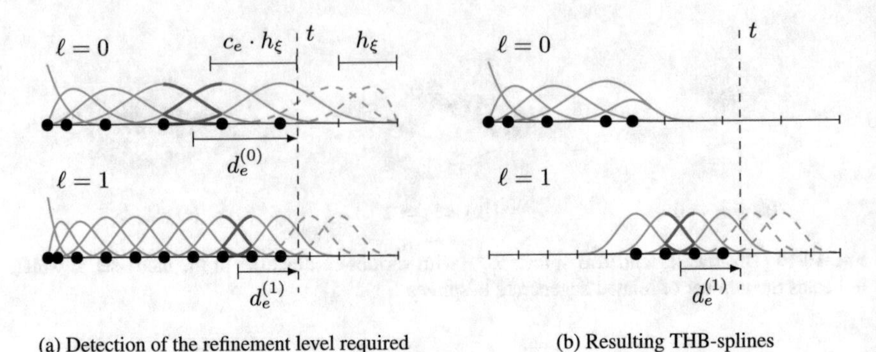

(a) Detection of the refinement level required (b) Resulting THB-splines

Fig. 10.21 Setting up of an extended THB-spline basis: (a) determination of the refinement level $\ell_{\max}$ required to fulfil the criterion $d_e < c_e \cdot h_\xi$, and (b) activation of the basis function for the extrapolation on the finest level

Following [117], these two components are united based on an admissibility criterion for the extrapolation length specified by

$$d_e < c_e \cdot h_\xi \tag{10.26}$$

where h_ξ refers to the average knot span size of the initial level 0 of the THB-basis and c_e is a user-defined constant. A guideline on how to chose c_e is given later on during the discussion of the numerical results in Sect. 10.2.5. Refinement levels ℓ are added until the related extrapolation length $d_e^{(\ell)}$ complies with condition (10.26). Figure 10.21(a) illustrates the identification of the required level. Noted that $d_e^{(\ell)}$ is checked for each level separately. Once the number of hierarchical levels is determined, the THB-basis is initiated by activating those basis functions of the finest level that are involved in the stabilisation. The rest of the space is spanned by basis functions of the previous levels as indicated in Fig. 10.21(b).

From an algorithmic point of view the THB-basis is set up as follows: the degenerate and stable B-splines that contribute to the extended B-spline construction are revealed during the determination of the required refinement level $\ell_{\max}$. These basis functions are sorted such that the degenerate ones come first and the stable ones follow, ordered by the distance of their anchors to the trimming curve such that the B-spline that is furthest away is listed last. Next, the sorted basis functions are successively activated. Activating a B-spline $B_i^{(\ell_{\max})}$ is performed by refining *all* of its parents on the previous level $\ell_{\max} - 1$. Following the refinement rule (3.42) presented in Sect. 3.6.3.4, parents are only refineable if their subdivision weight is equal to 1. Thus, basis functions of the levels $\ell < (\ell_{\max} - 1)$ may have to be refined preliminarily. This leads to a recursive refinement process which evolves through the hierarchical structure. Since all parents of $B_i^{(\ell_{\max})}$ are refined, the function itself becomes refineable as well (see definition (3.42)). In addition, B-splines $B_j^{(\ell_{\max})}$ adjacent to $B_i^{(\ell_{\max})}$ are activated indirectly, if they share a parent function. The overall refinement process stops once all basis functions of the sorted set are either directly or indirectly activated. Due to the initial sorting of the functions, the evolution of the THB-basis begins outside of Ω_p and gradually propagates into it. Hence, the refinement ends before it affects basis functions on level $\ell_{\max}$ that are not related to the extended B-splines. In the following, the term *extended THB-splines* will be used to refer to the overall stabilisation procedure described.

It is underscored that extended B-splines are restricted to the finest level of the hierarchical basis; all other levels consists of conventional THB-splines only. Thus, the actual stabilisation, i.e., the determination of the extrapolation weights, remains unchanged. Moreover, no superfluous basis functions are included to the THB-basis since all stable B-splines of the finest level have at least one degenerate B-spline associated to them.

10.2.4.3 Application to a System of Equations

A very convenient feature of extended B-splines is that they do *not* need to be explicitly evaluated during an analysis. Suppose we have a linear system of n equations, one for each stable B-spline, set up by all basis functions m which are at least partially inside Ω_p. This yields a system of equations

$$\mathbf{K}\mathbf{u} = \mathbf{f} \quad \text{where} \quad \mathbf{u} \in \mathbb{R}^m, \mathbf{f} \in \mathbb{R}^n \quad \text{and} \quad \mathbf{K} \in \mathbb{R}^{n\times m} \quad \text{with} \quad m > n. \tag{10.27}$$

Up to this point, only conventional B-splines have been used to compute the system matrix $\mathbf{K}$. In order to get a stabilised square system matrix, $\mathbf{K}$ is multiplied by an *extension matrix* $\mathbf{E} \in \mathbb{R}^{m\times n}$ [86]. The entries of $\mathbf{E}$ are the extrapolation weights $e_{i,j}$ of all extended B-splines. The trivial weights $e_{i,i} = 1$ are stored as well, even if, a stable B-spline has no related degenerate ones. The matrix entries of $\mathbf{E}$ are assembled such that columns of the ith row of $\mathbf{K}$ are distributed according to the definition (10.25) of the associated extended B-spline $B^e_{i,p}$. In case of multi-patch models, the indices refer to the global degree of freedom. The stable system of equations due to the extended B-spline basis is given by

$$\mathbf{K}_{st}\mathbf{u}_{st} = \mathbf{f} \qquad \text{with} \qquad \mathbf{K}_{st} = \mathbf{K}\mathbf{E}, \qquad \mathbf{K}_{st} \in \mathbb{R}^{n\times n}. \tag{10.28}$$

The solution vector $\mathbf{u}_{st} \in \mathbb{R}^n$ corresponds to the extended B-spline basis. Its relation to the original basis can also be expressed by the extension matrix as $\mathbf{u} = \mathbf{E}\mathbf{u}_{st}$.

Let us consider a simple interpolation problem (3.2) to demonstrate the application of this concept: a univariate basis defined by order $p = 2$ and $\Xi = (1, 1, 1, 2, 3, 4, 4, 4)$ is trimmed at a point $t \in (1, 1.5)$, as illustrated in Fig. 10.22(a). The related interpolation sites are given by the Greville abscissae $\tilde{\xi}_i = \{1, 1.5, 2.5, 3.5, 4\}$. Note that $\tilde{\xi}_0$ is outside of the domain Ω_p and thus, cannot be used as interpolation site. Hence, the (unstable) spline collocation matrix $\mathbf{K} \in \mathbb{R}^{4\times 5}$ reads

$$\mathbf{K} = \begin{pmatrix} 0.25 & 0.625 & 0.125 & & \\ & 0.125 & 0.750 & 0.125 & \\ & & 0.125 & 0.625 & 0.25 \\ & & & & 1.00 \end{pmatrix}. \tag{10.29}$$

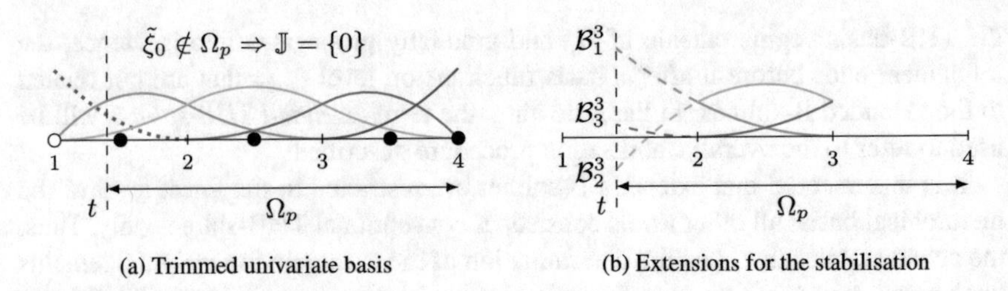

(a) Trimmed univariate basis (b) Extensions for the stabilisation

Fig. 10.22 Illustration of the interpolation examples: (a) the trimmed unstable B-spline basis and the corresponding Greville abscissae and (b) the extensions of the extended B-splines that provide the stable functions for the evaluation of the trimmed knot span

The extensions for the stabilisation are illustrated in Fig. 10.22(b) and can be obtained by (10.22) and (10.23). They are given by

$$\begin{aligned} \mathcal{B}_1^3 &= 0.5\,\xi^2 - 3\,\xi + 4.5, \\ \mathcal{B}_2^3 &= -1\,\xi^2 + 5\,\xi - 5.5, \\ \mathcal{B}_3^3 &= 0.5\,\xi^2 - 2\,\xi + 2. \end{aligned} \tag{10.30}$$

The Newton polynomial for setting up the extrapolation weights $e_{i,0}$ is

$$\psi_{0,p}(\xi) = 1\,\xi^2 - 2\,\xi + 1. \tag{10.31}$$

Equation (10.20) provides the final weights to represent the extensions (10.30) computed by

$$\begin{aligned} e_{1,0} &= \frac{1}{2}\left[\quad 2\cdot 0.5\cdot 1\cdot 1 - 1\cdot(-3)\cdot 1\cdot(-2) + 1\cdot 4.5\cdot 2\cdot 1 \quad\right] = 2, \\ e_{2,0} &= \frac{1}{2}\left[\quad 2\cdot(-1)\cdot 1\cdot 1 - 1\cdot 5\cdot 1\cdot(-2) + 1\cdot(-5.5)\cdot 2\cdot 1 \quad\right] = -1.5, \\ e_{3,0} &= \frac{1}{2}\left[\quad 2\cdot 0.5\cdot 1\cdot 1 - 1\cdot(-2)\cdot 1\cdot(-2) + 1\cdot 2\cdot 2\cdot 1 \quad\right] = 0.5. \end{aligned}$$

Since there is only one degenerate basis function, these are the only non-trivial entries of the extrapolation matrix $\mathbf{E} \in \mathbb{R}^{5\times 4}$ which is given by

$$\mathbf{E} = \begin{pmatrix} 2 & -1.5 & 0.5 & \\ 1 & & & \\ & 1 & & \\ & & 1 & \\ & & & 1 \end{pmatrix}. \tag{10.32}$$

Finally, the stable spline collocation matrix is computed by

$$\mathbf{K}_{st} = \mathbf{KE} = \begin{pmatrix} 1.125 & -0.250 & 0.125 & \\ 0.125 & 0.750 & 0.125 & \\ & 0.125 & 0.625 & 0.25 \\ & & & 1.00 \end{pmatrix}. \tag{10.33}$$

Note that most entries in (10.29) and (10.33) are the same since only at the first interpolation site $\tilde{\xi}_1$ the basis functions are not conventional B-splines. Furthermore, the non-zero coefficient in the first row of $\mathbf{K}_{st}$ are the values of the extensions (10.30) at $\tilde{\xi}_1$. In other words, we evaluated these functions by the extrapolation matrix, without the need of explicitly specifying extended B-splines.

10.2.4.4 Application to NURBS Models

The extended (TH)B-spline stabilisation is tailored to B-spline functions where it is exploited that the extensions of any polynomial segment $\mathcal{B}_i^s$ can be exactly represented by a linear combination of basis functions of the trimmed knot span. In case of NURBS, this property is not guaranteed due to the local influence of the NURBS weights which are associated to the basis functions. In order to apply extended B-splines to a trimmed NURBS CAD model, we propose the application of geometry independent field approximation as described in Sect. 6.5.3. This allows us to use B-splines discretisation for the field variables over NURBS geometries.

10.2.5 Numerical Results

In this section, simulations of potential and elasticity problems with trimmed B-spline and NURBS geometries are preformed. In particular, the local ruled surface mapping outlined in Sect. 10.2.2.2 is utilised for the regular integration over cut elements and extended THB-splines presented in Sect. 10.2.4 are used to stabilise the trimmed bases.

10.2.5.1 Trimmed Cube

Exterior Neumann problems for the Laplace equation and elasticity are considered. The numerical example compares direct simulations of trimmed objects with analyses of equivalent geometries defined by non-trimmed surfaces. In particular, a unit cube ($\ell_x = \ell_y = \ell_z = 1.0$) specifies the boundary Γ of the domain of interest $\Omega = \mathbb{R}^3 \setminus \Omega^-$ where Ω^- refers to the void. The two different geometry models are depicted in Fig. 10.23: (a) six regular patches with matching parameterisation provide the basis of the reference solutions, and (b) a model with two regular and four trimmed sides. The invisible area of the trimmed patches is defined by $\ell_x = 1.70$. The prescribed Neumann data along Γ are given by

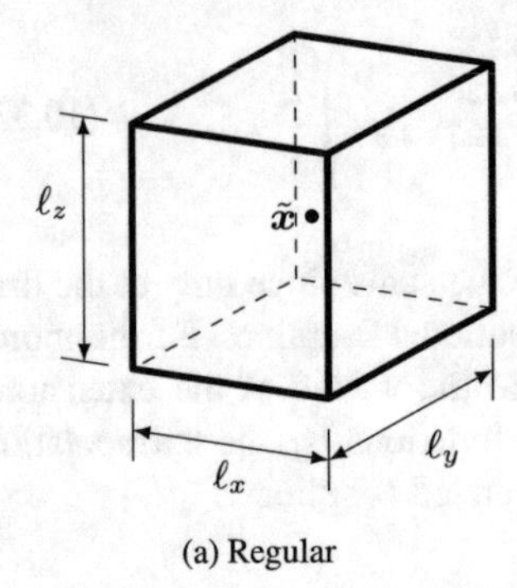

(a) Regular

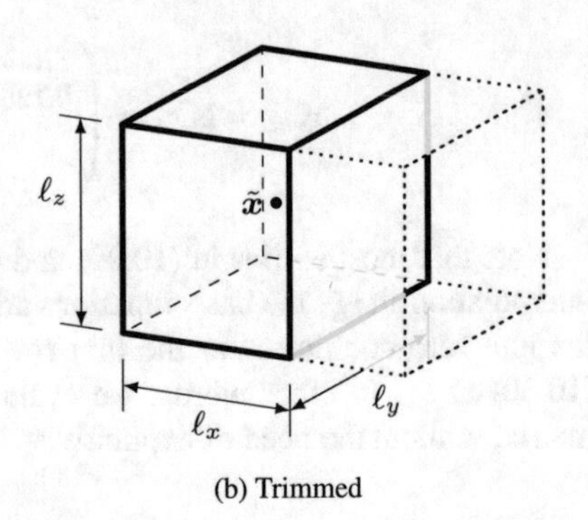

(b) Trimmed

Fig. 10.23 The unit cube of the exterior Neumann problem defined by (a) regular patches and (b) trimmed patches

$$t(\boldsymbol{x}) := \mathsf{T}(\tilde{\boldsymbol{x}}, \boldsymbol{x}) \qquad \boldsymbol{x} \in \Gamma,\ \tilde{\boldsymbol{x}} \in \Omega^- \tag{10.34}$$

where $\tilde{\boldsymbol{x}}$ is a source point located in the centre of the cube. The exact solution of the primary variable $u(\boldsymbol{x})$ is defined by the fundamental solution $\mathsf{U}(\tilde{\boldsymbol{x}}, \boldsymbol{x})$. Thus, the relative approximation error of a simulation can be calculated by

$$\epsilon_{rel} = \frac{u_h(\boldsymbol{x}) - \mathsf{U}(\tilde{\boldsymbol{x}}, \boldsymbol{x})}{\mathsf{U}(\tilde{\boldsymbol{x}}, \boldsymbol{x})} \qquad \boldsymbol{x} \in \Gamma,\ \tilde{\boldsymbol{x}} \in \Omega^- \tag{10.35}$$

and measured with respect to the L_2-norm $\|\epsilon_{rel}\|_{L_2}$, where $u_h(\boldsymbol{x})$ is the numerical solution of the boundary data.

Uniform knot refinement is used to improve $u_h(\boldsymbol{x})$; an additional knot insertion step is applied in the x-direction of the trimmed patches to get similar element sizes on all faces of the cube. The final THB-basis of the trimmed surfaces is determined by the admissibility criterion (10.26) with the knot span size h_ξ measured after the global refinement. The influence of this criterion is investigated by using different constants $c_e = \{p/2, 0.5, 10\}$. In the latter case, the allowed extrapolation length is very large and hence, no local refinement is performed. On the other hand, $c_e = 0.5$ is a very strict condition yielding a fine resolution along the trimmed edge. Finally, moderate local refinement is established by $c_e = p/2$. The study is repeated for various polynomial orders $p = \{2, 3, 4\}$.

Figure 10.24 provides an overview of the results obtained. For a better comparison of the regular and the trimmed settings, the convergence is plotted versus the number of degrees of freedom. The corresponding number of hierarchical refinement levels for $c_e = 0.5$ and $c_e = p/2$ are summarised in Table 10.1. There is no difference between the Laplace and the elasticity case, because the determination of the required level is independent of the problem type. In addition, the distribution of the relative error of the Laplace problem discretised by $p = 3$ and two knot insertion steps is shown for the simulation based on extended B-splines without local refinement ($c_e = 10$) and the proposed extended THB-spline approach ($c_e = p/2$) in Fig. 10.25.

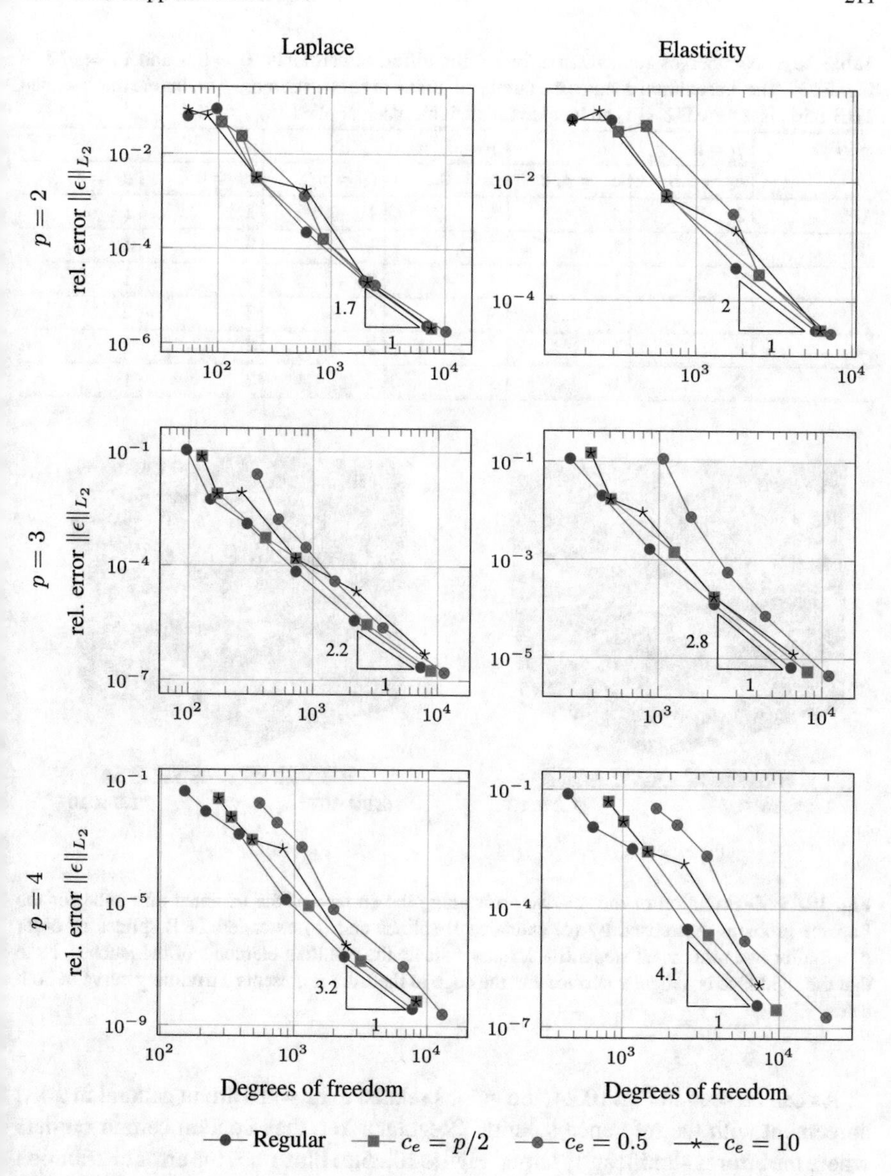

Fig. 10.24 Relative L_2-error of the exterior Neumann problem of the cube example. The left column refers to the Laplace problem, whereas the elasticity results are shown on the right. The rows correspond to the different orders. Each diagram contains results related to discretisations using regular patches (Regular), and discretisation based on trimmed ones which employ different refinement criteria c_e

Table 10.1 Number of refinement levels ℓ for different constants $c_e = 0.5$ and $c_e = p/2$ of Fig. 10.24. The first column denotes the number of uniform knot insertions, $\#ki$, before the extended THB-basis is set up. If $\ell = 1$, no local refinement has been applied

$\#ki$	$p = 2$		$p = 3$		$p = 4$	
	$c_e = 0.5$	$c_e = p/2$	$c_e = 0.5$	$c_e = p/2$	$c_e = 0.5$	$c_e = p/2$
0	2	2	3	1	3	1
1	2	2	3	1	3	1
2	3	1	3	2	3	1
3	2	2	3	1	3	2
4	2	1	3	2	4	2
5	3	1	3	2	3	1

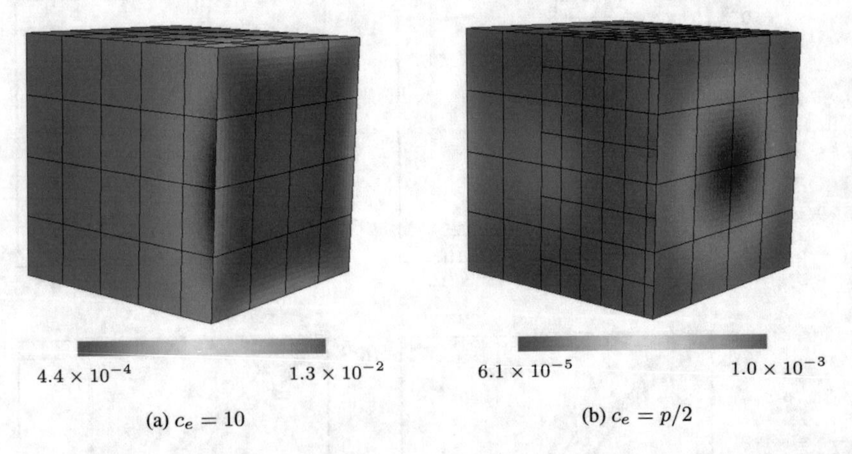

(a) $c_e = 10$ (b) $c_e = p/2$

Fig. 10.25 Distribution of the relative error along the surface of the trimmed unit cube for the Laplace problem discretised by (a) extended B-splines and (b) extended THB-splines of order $p = 3$ after two refinement steps. Black lines indicate the resulting elements of the patches. Note that the right face is a regular surface and the edge in the front represents a trimming curve in both cases

As can be seen in Fig. 10.24, the graphs related to $c_e = 10$ are in general in good agreement with the reference solution (Regular). Yet, they contain certain outliers where the error is significantly larger. Figure 10.25(a) illustrates the error distribution of such an outlier. Apparently, the inaccuracy along the trimmed edges diminishes the overall quality of the numerical solution. This is a manifestation of the potentially negative effect of an uncontrolled extrapolation length. The outliers do not correlate to a specific refinement step; they rather depend on the trimming situation and the corresponding set of degenerate B-splines. However, the amplitude of the outliers tend to decrease with the fineness of a discretisation since the contribution of the results along trimming curves becomes smaller with respect to the overall solution.

The results related to $c_e = 0.5$, on the other hand, clearly demonstrate that local refinement can become counterproductive if the admissibility condition is too strict.

That is, the number of degrees of freedom concentrated in the vicinity of the trimmed edges is much higher than needed. Setting c_e to a fixed value has also the disadvantage that the number of refinement levels increases with the order.

A good balance between accuracy and the number of degrees of freedom is obtained when the admissibility constant is set to $c_e = p/2$. Using this criterion, only a few refinement levels are introduced and in those cases where conventional extended B-splines already yield sufficient accuracy, no local refinement is applied at all. In fact, not more than one level is added in the examples considered and yet the results of $c_e = p/2$ are in an excellent agreement with the reference solution for all orders. Still, there is an offset between the graphs related to $p = 4$. This deviation may occur due to the fact that the local refinement procedure introduces basis functions also in the direction parallel to the trimmed edge. These B-splines are indeed superfluous for the given trimming situation which is aligned with one parametric direction, and their number increases with the order. Overall, it can be concluded that extended THB-splines make the stabilisation of trimmed parameter spaces more robust and the related numerical results more accurate.

10.2.5.2 Hollow Cylinder

The interior side of a solid hollow cylinder is subjected to an internal constant pressure p_c. The analytical solution of the corresponding displacement u_r in radial direction is given by

$$u_r(r) = \frac{p_c}{E} \frac{r_i^2}{r_o^2 - r_i^2} \left((1 - \nu)\, r + (1 + \nu) \frac{r_o^2}{r} \right) \tag{10.36}$$

with r_i and r_o denoting the inner and outer radius of the cylinder, respectively [160].

Figure 10.26 shows the geometry of the problem and its discretisation before local refinement is applied. Quadratic basis functions are used to define all patches of the model including the trimmed planar surfaces on the top and bottom of the cylinder. The material parameters are set to the Poisson ratio $\nu = 0$ and the Young's modulus $E = 1 \times 10^5$ MPa. The interior side of the cylinder is subjected to a constant pressure of $p_c = -100$ MPa which is applied to the model by prescribing the related displacement $u_r(r_i = 1)$ multiplied by the outward normal $\mathbf{n}$. The boundary traction of all other surfaces are set to zero. Two simulations with different constants for the admissibility criterion (10.26) are performed, i.e., $c_e = \{1.0, 0.8\}$. The former uses the factor $p/2$ based on the discussion of the previous numerical example, leading to an extended THB-basis with 2 levels, whereas the latter introduces an additional refinement level. The number of degrees of freedom of these discretisations are 1800 and 4056, respectively.

The resulting integration regions and the displacements obtained are depicted in Fig. 10.27. In addition, Fig. 10.28 illustrates a comparison of the radial displacement along the x-axis of each model with the analytical solution (10.36).

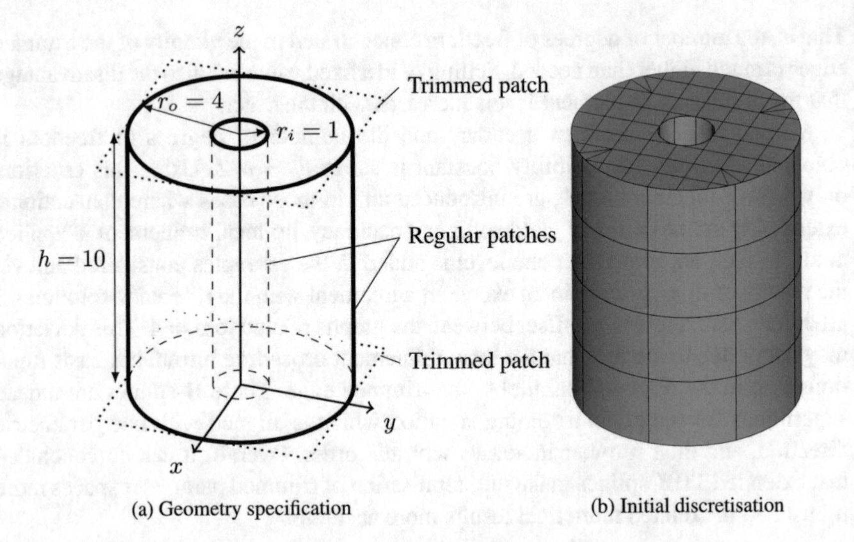

(a) Geometry specification (b) Initial discretisation

Fig. 10.26 Model of the hollow cylinder: (a) the interior and outer sides of the cylinder are described by regular surfaces, whereas trimmed patches represent its top and bottom; (b) the corresponding integration regions before local refinement is applied

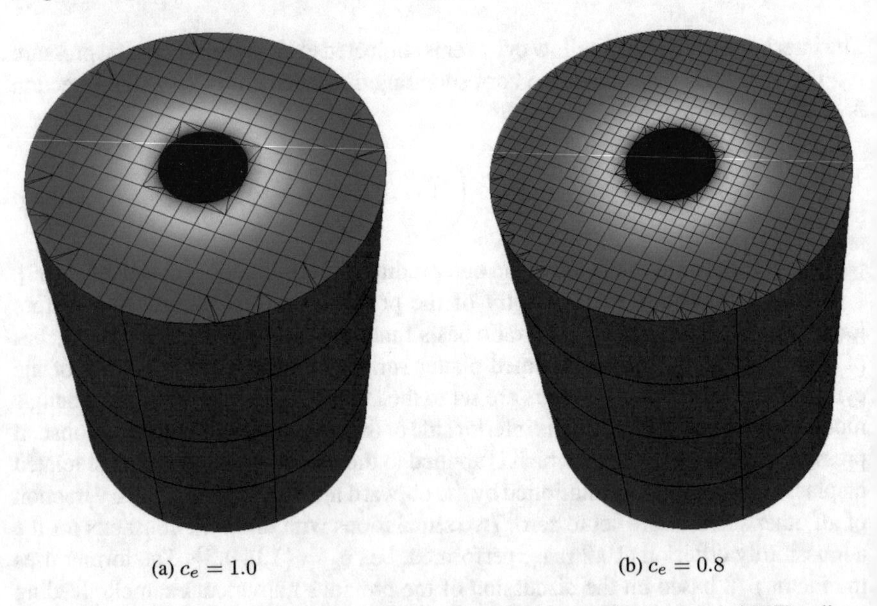

(a) $c_e = 1.0$ (b) $c_e = 0.8$

Fig. 10.27 Displacement of the hollow cylinder discretised by quadratic extended THB-splines. The local refinement is determined by (a) the proposed factor $c_e = p/2$ and (b) a stricter admissibility criterion. The same parameter range is used for both figures, i.e., 5.26×10^{-4} to 1.13×10^{-3}

Fig. 10.28 Displacement of the hollow cylinder along the x-axis

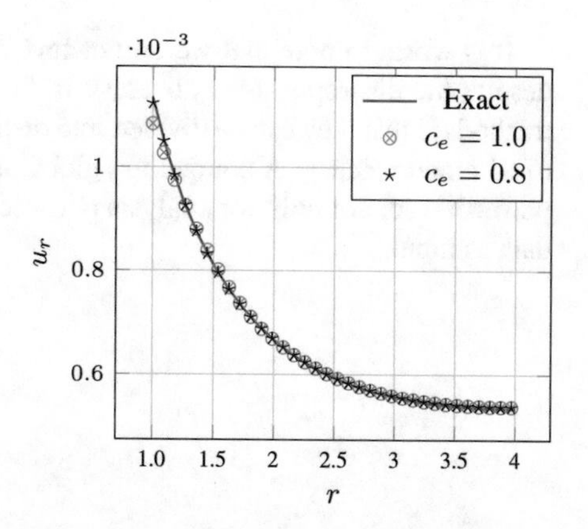

It is apparent that the axisymmetric behaviour of the solution is resolved very well in both cases, despite the fact that the trimmed parameter spaces are not aligned with the radial direction. The comparison with the reference solution verifies the quality of the numerical solutions and the choice of $c_e = p/2$ for a good balance between accuracy and numerical effort.

10.3 Summary and Conclusions

This chapter addresses the treatment of trimmed surfaces in the context of an isogeometric BEM analysis, which is a crucial obstacle for the interaction of CAD and simulations. Trimmed representations impose several challenges such as missing connectivity between patches, random element topology, and cut basis functions with arbitrary small support. In general, there are two different philosophies to treat trimmed CAD models: (i) global approaches aim to update the representation of the geometry such that the resulting patches can be analysed with hardly any adaptations to the analysis process, whereas (ii) local approaches do not alter the geometry but adapt the simulation routines. In Sect. 10.1, we present an example of a global method. However, the main focus is on local approaches. Here concepts for the proper definition of elements cut by a trimming curve are provided, and it is argued that robustness is the most important attribute to be considered. Based on the element detection, tailored integration schemes can be applied, and we presented a broad spectrum of different options. Using BEM, the treatment of multi-patch models is straightforward since discontinuities between patches are permitted. Finally, extended THB-splines are presented as a general approach to deal with the stability issues induced by cut basis functions.

It is worth to note that we do not favour the local strategy over the global one, despite the disproportional coverage in this chapter. An appealing feature of local methods is that they can be divided into distinct tasks, which simplifies the treatment of trimmed patches. A compelling global approach, on the other hand, would be a powerful tool not only for analysis purposes but any application that receives CAD data as input.

Chapter 11
Body Force Effects

11.1 Introduction

With boundary integral equations alone we cannot consider effects that occur inside the domain and therefore can only deal with linear problems and homogeneous domains, with effects on the boundary only. This restricts the application of the method to practical problems.

The restriction can be lifted by considering forces that occur inside the domain, also known as body forces. Body forces can be heat generated inside the domain, self-weight or centrifugal forces but they can also arise from material nonlinear behaviour and non-homogeneous conditions.

Consider the domain Ω in Fig. 11.1 with a subdomain Ω_0 where body forces $\mathbf{b}(\hat{\boldsymbol{x}})$ are present. We explain the theory on elasticity problems and apply the theorem by Betti, which has been used for deriving the original integral equations.

Remark The theory can also be applied to potential problems where the body force is heat generated inside the domain, the displacements are replaced by the potential and the tractions are replaced by the flow normal to the boundary.

If body forces are present, additional work is done in the domain Ω_0 by $\mathbf{b}(\hat{\boldsymbol{x}})$ times displacements $\mathsf{U}(\tilde{\boldsymbol{x}}_n, \hat{\boldsymbol{x}})$ and on the interface Γ_0 by tractions, arising from the body forces, $\mathbf{b}_n$ times displacements $\mathsf{U}(\tilde{\boldsymbol{x}}_n, \hat{\boldsymbol{x}})$. Therefore the regularised integral equation (6.11) has to be expanded by 2 terms

$$\int_{\Gamma} \mathsf{T}(\tilde{\boldsymbol{x}}_n, \hat{\boldsymbol{x}})(\mathbf{u}(\hat{\boldsymbol{x}}) - \mathbf{u}(\tilde{\boldsymbol{x}}_n))d\Gamma(\hat{\boldsymbol{x}}) - \mathbf{A}_n\mathbf{u}(\tilde{\boldsymbol{x}}_n) = \int_{\Gamma} \mathsf{U}(\tilde{\boldsymbol{x}}_n, \hat{\boldsymbol{x}})\mathbf{t}(\hat{\boldsymbol{x}})d\Gamma(\hat{\boldsymbol{x}})$$
$$+ \int_{\Omega_0} \mathsf{U}(\tilde{\boldsymbol{x}}_n, \hat{\boldsymbol{x}})\mathbf{b}(\hat{\boldsymbol{x}})d\Omega_0(\hat{\boldsymbol{x}}) + \int_{\Gamma_0} \mathsf{U}(\tilde{\boldsymbol{x}}_n, \hat{\boldsymbol{x}})\mathbf{b}_n(\hat{\boldsymbol{x}})d\Gamma_0(\hat{\boldsymbol{x}}). \tag{11.1}$$

G. Beer et al., *The Isogeometric Boundary Element Method*, Lecture Notes in Applied and Computational Mechanics 90,
https://doi.org/10.1007/978-3-030-23339-6_11

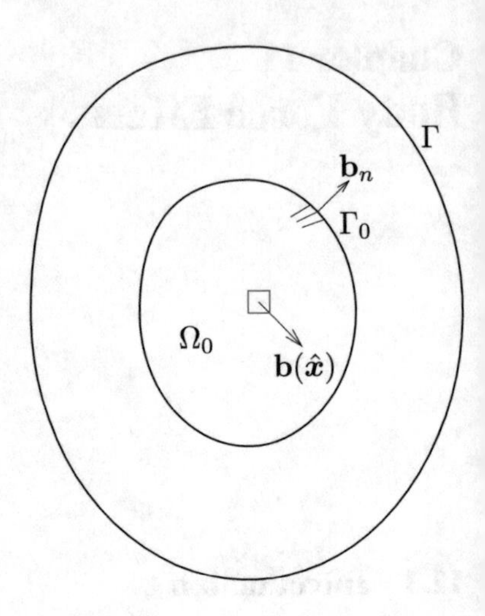

Fig. 11.1 Subdomain Ω_0 with body forces

The integral equation can be discretised as shown in Chap. 6 and the following system of equations obtained

$$\mathbf{Lx} = \mathbf{r} + \mathbf{r}_0 \tag{11.2}$$

where $\mathbf{r}_0 = \mathbf{r}_0^{\Omega} + \mathbf{r}_0^{\Gamma}$ and the coefficients of the vectors are given by:

$$\mathbf{r}_{0n}^{\Omega} = \int\limits_{\Omega_0} \mathsf{U}(\tilde{\boldsymbol{x}}_n, \hat{\boldsymbol{x}})\mathbf{b}(\hat{\boldsymbol{x}})d\Omega_0(\hat{\boldsymbol{x}}) \tag{11.3}$$

$$\mathbf{r}_{0n}^{\Gamma} = \int\limits_{\Gamma_0} \mathsf{U}(\tilde{\boldsymbol{x}}_n, \hat{\boldsymbol{x}})\mathbf{b}_n(\hat{\boldsymbol{x}})d\Gamma_0(\hat{\boldsymbol{x}}). \tag{11.4}$$

For the computation of $\mathbf{u}$ at a point $\boldsymbol{x}$ inside the domain, we need to expand Eq. (6.59) to

$$\begin{aligned}\mathbf{u}(\boldsymbol{x}) = &\int\limits_{\Gamma} \left[\mathsf{U}(\boldsymbol{x}, \hat{\boldsymbol{x}})\mathbf{t}(\hat{\boldsymbol{x}}) - \mathsf{T}(\boldsymbol{x}, \hat{\boldsymbol{x}})\,\mathbf{u}(\hat{\boldsymbol{x}})\right] d\Gamma(\hat{\boldsymbol{x}}) \\ &+ \int\limits_{\Omega_0} \mathsf{U}(\boldsymbol{x}, \hat{\boldsymbol{x}})\mathbf{b}(\hat{\boldsymbol{x}})d\Omega_0(\hat{\boldsymbol{x}}) + \int\limits_{\Gamma_0} \mathsf{U}(\boldsymbol{x}, \hat{\boldsymbol{x}})\mathbf{b}_n(\hat{\boldsymbol{x}})d\Gamma_0(\hat{\boldsymbol{x}}).\end{aligned} \tag{11.5}$$

The discretised form of Eq. (11.5) is with reference to Chap. 6

$$\mathbf{u}(\boldsymbol{x}) = \mathbf{Ex} + \bar{\mathbf{u}} + \mathbf{u}^{\Omega} + \mathbf{u}^{\Gamma} \tag{11.6}$$

where the additional terms are

$$\mathbf{u}^{\Omega} = \int_{\Omega_0} \mathsf{U}(\boldsymbol{x}, \hat{\boldsymbol{x}}) \mathbf{b}(\hat{\boldsymbol{x}}) d\Omega_0(\hat{\boldsymbol{x}}) \tag{11.7}$$

$$\mathbf{u}^{\Gamma} = \int_{\Gamma_0} \mathsf{U}(\boldsymbol{x}, \hat{\boldsymbol{x}}) \mathbf{b}_n(\hat{\boldsymbol{x}}) d\Gamma_0(\hat{\boldsymbol{x}}). \tag{11.8}$$

We can also compute derived values, for example the stress $\{\boldsymbol{\sigma}\}$ by expanding Eq. (6.61)

$$\begin{aligned} \{\boldsymbol{\sigma}\}(\boldsymbol{x}) = & \int_{\Gamma} [\mathsf{S}(\boldsymbol{x}, \hat{\boldsymbol{x}})\, \mathbf{t}(\hat{\boldsymbol{x}}) - \mathsf{R}(\boldsymbol{x}, \hat{\boldsymbol{x}})\, \mathbf{u}(\hat{\boldsymbol{x}})]\, d\Gamma(\hat{\boldsymbol{x}}) \\ & + \int_{\Omega_0} \mathsf{S}(\boldsymbol{x}, \hat{\boldsymbol{x}}) \mathbf{b}(\hat{\boldsymbol{x}}) d\Omega_0(\hat{\boldsymbol{x}}) + \int_{\Gamma_0} \mathsf{S}(\boldsymbol{x}, \hat{\boldsymbol{x}}) \mathbf{b}_n(\hat{\boldsymbol{x}}) d\Gamma_0(\hat{\boldsymbol{x}}). \end{aligned} \tag{11.9}$$

The discretised form of Eq. (11.9) is

$$\{\boldsymbol{\sigma}\} = \mathbf{S}\mathbf{x} + \{\bar{\boldsymbol{\sigma}}\} + \mathbf{d}^{\Omega} + \mathbf{d}^{\Gamma} \tag{11.10}$$

where the additional terms are

$$\mathbf{d}^{\Omega} = \int_{\Omega_0} \mathsf{S}(\boldsymbol{x}, \hat{\boldsymbol{x}}) \mathbf{b}(\hat{\boldsymbol{x}}) d\Omega_0(\hat{\boldsymbol{x}}) \tag{11.11}$$

$$\mathbf{d}^{\Gamma} = \int_{\Gamma_0} \mathsf{S}(\boldsymbol{x}, \hat{\boldsymbol{x}}) \mathbf{b}_n(\hat{\boldsymbol{x}}) d\Gamma_0(\hat{\boldsymbol{x}}). \tag{11.12}$$

In the following we distinguish between body forces that are constant inside the domain Ω and those that vary within the sub-domain Ω_0. In the first case, we can use Green's theorem to transfer the integral to a surface integral; in the second case, the volume integral has to be solved numerically.

11.2 Constant Body Forces

Here we deal with body forces that are constant over the whole domain Ω. Examples are gravity and centrifugal forces or a constant heat generated in the domain. Since $\mathbf{b}_n = \mathbf{t}$ and $\Omega_0 = \Omega, \Gamma_0 = \Gamma$ in this case, the integral equation reduces to:

$$\int_\Gamma \mathsf{T}(\tilde{\boldsymbol{x}}_n, \hat{\boldsymbol{x}})(\mathbf{u}(\hat{\boldsymbol{x}}) - \mathbf{u}(\tilde{\boldsymbol{x}}_n))\, d\Gamma(\hat{\boldsymbol{x}}) - \mathbf{A}_n \mathbf{u}(\tilde{\boldsymbol{x}}_n) = \int_\Gamma \mathsf{U}(\tilde{\boldsymbol{x}}_n, \hat{\boldsymbol{x}})\, \mathbf{t}(\hat{\boldsymbol{x}})\, d\Gamma(\hat{\boldsymbol{x}}) + \int_\Omega \mathsf{U}(\tilde{\boldsymbol{x}}_n, \hat{\boldsymbol{x}}) \mathbf{b}(\hat{\boldsymbol{x}}) d\Omega(\hat{\boldsymbol{x}}) \tag{11.13}$$

Using Green's theorem we can transform the integral over Ω to a surface integral by:

$$\int_\Omega \mathsf{U}(\tilde{\boldsymbol{x}}_n, \boldsymbol{x}) \mathbf{b}(\hat{\boldsymbol{x}}) d\Omega(\hat{\boldsymbol{x}}) = \int_\Gamma \mathsf{G}(\tilde{\boldsymbol{x}}_n, \hat{\boldsymbol{x}}) d\Gamma(\hat{\boldsymbol{x}}) \tag{11.14}$$

where $\mathsf{G}(\tilde{\boldsymbol{x}}_n, \hat{\boldsymbol{x}})$ is presented in the Appendix A.5 for elasticity problems. The solution of this problem now proceeds exactly like for homogeneous problems.

For the computation of primary internal values we have

$$\mathbf{u}(\boldsymbol{x}) = \int_\Gamma [\mathsf{U}(\boldsymbol{x}, \hat{\boldsymbol{x}}) \mathbf{t}(\hat{\boldsymbol{x}}) - \mathsf{T}(\boldsymbol{x}, \hat{\boldsymbol{x}})\, \mathbf{u}(\hat{\boldsymbol{x}})]\, d\Gamma(\hat{\boldsymbol{x}}) + \int_\Gamma \mathsf{G}(\boldsymbol{x}, \hat{\boldsymbol{x}}) d\Gamma(\hat{\boldsymbol{x}}) \tag{11.15}$$

and for the stress we have

$$\{\boldsymbol{\sigma}\}(\boldsymbol{x}) = \int_\Gamma [\mathsf{S}(\boldsymbol{x}, \hat{\boldsymbol{x}})\, \mathbf{t}(\hat{\boldsymbol{x}}) - \mathsf{R}(\boldsymbol{x}, \hat{\boldsymbol{x}})\, \mathbf{u}(\hat{\boldsymbol{x}})]\, d\Gamma(\hat{\boldsymbol{x}}) + \int_\Gamma \hat{\mathsf{G}}(\boldsymbol{x}, \hat{\boldsymbol{x}}) d\Gamma(\hat{\boldsymbol{x}}) \tag{11.16}$$

where $\hat{\mathsf{G}}$ is listed in Appendix A.5.3.

11.3 General Body Forces

11.3.1 Types of Body Forces

Apart from general body forces, that are not constant over the domain, there are basically three scenarios where body forces have to be considered as will be explained in the subsequent chapters:

- Inclusions with different material properties
- Non-linear material behaviour
- Sources inside part of domain

As already mentioned the arising additional integrals

$$\mathbf{r}_{0n}^{\Omega} = \int_{\Omega_0} \mathsf{U}(\tilde{\boldsymbol{x}}_n, \hat{\boldsymbol{x}}) \mathbf{b}(\hat{\boldsymbol{x}}) d\Omega_0(\hat{\boldsymbol{x}}) \tag{11.17}$$

$$\mathbf{r}_{0n}^{\Gamma} = \int_{\Gamma_0} \mathsf{U}(\tilde{\boldsymbol{x}}_n, \hat{\boldsymbol{x}}) \mathbf{b}_n(\hat{\boldsymbol{x}}) d\Gamma_0(\hat{\boldsymbol{x}}) \tag{11.18}$$

have to be evaluated numerically and we need to discretise the domain Ω_0. There are two possibilities: discretise the domain into cells or use isogeometric methods.

11.3.2 Internal Cell Method

This method has been used traditionally in the past [65, 177, 178] and involves discretising the domain Ω_0 into C cells. Equations (11.17) and (11.18) are now replaced by

$$\mathbf{r}_{0n}^{\Omega} = \sum_{c=1}^{C} \int_{\Omega_c} \mathsf{U}(\tilde{\boldsymbol{x}}_n, \hat{\boldsymbol{x}}) \mathbf{b}(\hat{\boldsymbol{x}}) d\Omega_c(\hat{\boldsymbol{x}}) \tag{11.19}$$

$$\mathbf{r}_{0n}^{\Gamma} = \sum_{c_s=1}^{C_s} \int_{\Gamma_{c_s}} \mathsf{U}(\tilde{\boldsymbol{x}}_n, \hat{\boldsymbol{x}}) \mathbf{b}_n(\hat{\boldsymbol{x}}) d\Gamma_{c_s}(\hat{\boldsymbol{x}}) \tag{11.20}$$

where Ω_c specifies the volume of cell c , C_s the cells that lie on Γ_0 and Γ_{c_s} the boundary of cells that coincide with Γ_0. We approximate the body forces within a cell by

$$\mathbf{b}(\boldsymbol{x}) = \sum_{i=1}^{I} N_i \mathbf{b}_i \tag{11.21}$$

where $\mathbf{b}_i$ are nodal point values of the body force and N_i are suitable shape functions. Cells may either be of linear, triangular or tetrahedra shape, very similar to isoparametric finite elements. The numerical integration over a cell is performed using Gauss Quadrature.

The disadvantage of the cell method is that additional mesh generation effort is necessary and that values of the body force have to be approximated inside the domain.

11.3.3 Isogeometric Mapping Method

In this method, we use NURBS to describe the domain Ω_0 and map from the global coordinate system to a local one, where all operations such as integration and differentiation are carried out. In the following, we present the mapping from local to global coordinates as well as the Jacobi matrix and the Jacobian. The mapping methods depend on the type of subdomain. We distinguish between domains:

1. Very thin domains (for example reinforcement, rock bolts etc.)
2. Moderately thin domains (for example geological features)
2. Domains of general shape.

We discuss the different mapping methods next.

11.3.3.1 Mapping Method 1

This is for domains that are very thin and where the variation of the body force can be assumed constant across the thickness.

Plane domains. Here the domain is defined as a curve with a thickness w. We establish a local coordinate system $s = [0, 1]$ as shown in Fig. 11.2, perform all computations such as integration and differentiation in this system and then map it to the global $\boldsymbol{x}$-system.

The global coordinates of a point $\boldsymbol{x}$ with the local coordinates s are given by

$$\boldsymbol{x}(s) = \sum_{k=1}^{K} R_k(s)\, \boldsymbol{x}_k \tag{11.22}$$

where K is the number of control points, $R_k(s)$ are NURBS basis functions and $\boldsymbol{x}_k$ are control point coordinates. The derivative is given by

$$\frac{\partial \boldsymbol{x}(s)}{\partial s} = \sum_{k=1}^{K} \frac{\partial R_k(s)}{\partial s}\, \boldsymbol{x}_k = \mathbf{v}. \tag{11.23}$$

The Jacobian is

$$J = \mathrm{w}\sqrt{\mathrm{v}_x^2 + \mathrm{v}_y^2}. \tag{11.24}$$

For the numerical integration we have to transform from $s = [0, 1]$ to the Gauss coordinate system $\xi = [-1, 1]$

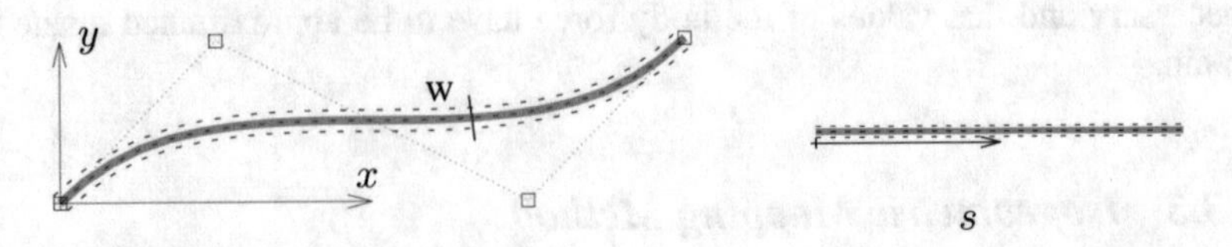

Fig. 11.2 Mapping of plane linear inclusion with width w showing the NURBS curve and the associated control points defining the inclusion: left in global $\boldsymbol{x}$, right in local s space

$$s = \frac{1}{2}(1 + \bar{\xi}). \tag{11.25}$$

The Jacobian of this transformation is $J_{\bar{\xi}} = 0.5$. The numerical integration of the volume term is given by

$$\mathbf{r}_{0n}^{\Omega} = \sum_{g=1}^{G} \left[\mathsf{U}(\tilde{\boldsymbol{x}}_n, \hat{\boldsymbol{x}}(s_g)) \mathbf{b}(s_g) J(s_g) \; J_{\bar{\xi}} \; \mathrm{w} \; W_g \right] \tag{11.26}$$

where s_g are Gauss point coordinates, W_g are weights and G is the number of Gauss points dependent on the proximity of $\tilde{\boldsymbol{x}}_n$ to the integration region.

3-D domains. Here the domain is defined as a curve with a circular cross-section of radius r and the body force is assumed constant across it. The definition of the curve and the local derivative is the same as before but the global system is three-dimensional. The Jacobian is now

$$J = r^2 \pi \sqrt{\mathrm{v}_x^2 + \mathrm{v}_y^2}. \tag{11.27}$$

The numerical integration of the volume term is given by

$$\mathbf{r}_{0n}^{\Omega} = \sum_{g=1}^{G} \left[\mathsf{U}(\tilde{\boldsymbol{x}}_n, \hat{\boldsymbol{x}}(s_g)) \mathbf{b}(s_g) J(s_g) \; J_{\bar{\xi}} \; r^2 \pi \; W_g \right]. \tag{11.28}$$

11.3.3.2 Mapping Method 2

This method is for domains that are moderately thin, but where the variation of the body force across it can no longer be assumed constant.

Plane domains. For plane domains, we establish a local coordinate system $\boldsymbol{s} = (s, t)^{\mathrm{T}} = [0, 1]^2$ as shown in Fig. 11.3. Note that there is a one to one mapping between the coordinate s and the local coordinate ξ of the red and green NURBS curve in Fig. 11.3. This means that if we use the local coordinate s of the inclusion in the equations we refer at the same time to the local coordinate ξ of the curve. The global coordinates of a point $\boldsymbol{x}$ with the local coordinates $\boldsymbol{s}$ are given by

$$\boldsymbol{x}(s, t) = (1 - t) \, \boldsymbol{x}^{I}(s) + t \, \boldsymbol{x}^{II}(s) \tag{11.29}$$

where

$$\boldsymbol{x}^{I}(s) = \sum_{k=1}^{K^I} R_k^I(s) \, \boldsymbol{x}_k^I \qquad \text{and} \qquad \boldsymbol{x}^{II}(s) = \sum_{k=1}^{K^{II}} R_k^{II}(s) \, \boldsymbol{x}_k^{II}. \tag{11.30}$$

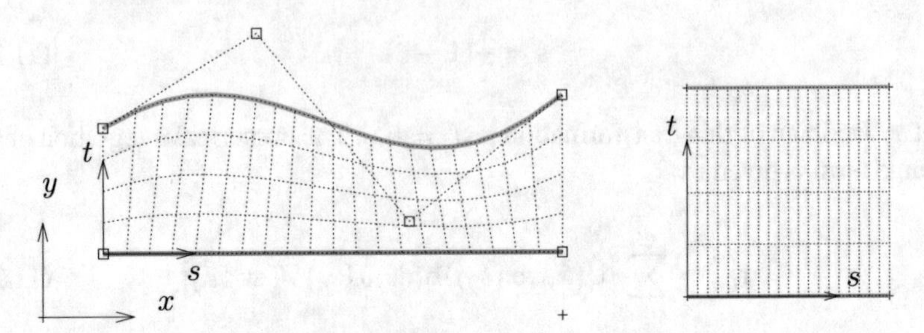

Fig. 11.3 Mapping of plane inclusion showing the bottom and top NURBS curves and the associated control points defining the inclusion: left in global $\boldsymbol{x}$, right in local $\boldsymbol{s}$ space

The superscript I relates to the bottom (red) curve and II to the top (green) curve and $\mathbf{x}_k^I$, $\mathbf{x}_k^{II}$ are control point coordinates. K^I and K^{II} are the number of control points, $R_k^I(s)$ and $R_k^{II}(s)$ are NURBS basis functions. The derivatives are given by

$$\frac{\partial \boldsymbol{x}(s,t)}{\partial s} = (1-t)\,\frac{\partial \boldsymbol{x}^I(s)}{\partial s} + t\,\frac{\partial \boldsymbol{x}^{II}(s)}{\partial s} \tag{11.31}$$
$$\frac{\partial \boldsymbol{x}(s,t)}{\partial t} = -\boldsymbol{x}^I(s) + \boldsymbol{x}^{II}(s)$$

where

$$\frac{\partial \boldsymbol{x}^I(s)}{\partial s} = \sum_{k=1}^{K^I} \frac{\partial R_k^I(s)}{\partial s}\,\boldsymbol{x}_k^I \tag{11.32}$$
$$\frac{\partial \boldsymbol{x}^{II}(s)}{\partial s} = \sum_{k=1}^{K^{II}} \frac{\partial R_k^{II}(s)}{\partial s}\,\boldsymbol{x}_k^{II}.$$

The Jacobi matrix of this mapping is

$$\mathbf{J} = \begin{pmatrix} \frac{\partial \boldsymbol{x}}{\partial s} \\ \frac{\partial \boldsymbol{x}}{\partial t} \end{pmatrix} \tag{11.33}$$

and the Jacobian is $J = |\mathbf{J}|$.

For the Gauss integration, we have to transform form $\boldsymbol{s} = (s,t)^{\mathrm{T}} = [0,1]^2$ to $\bar{\boldsymbol{\xi}} = (\bar{\xi},\bar{\eta})^{\mathrm{T}} = [-1,1]^2$

$$s = \frac{1}{2}(1+\bar{\xi}), \quad t = \frac{1}{2}(1+\bar{\eta}). \tag{11.34}$$

The Jacobian of this transformation is $J_{\bar{\xi}} = 0.25$.

The numerical integration of the volume term is given by

$$\mathbf{r}_{0n}^{\Omega} = \sum_{g_s=1}^{G_s} \sum_{g_t=1}^{G_t} \left[\mathsf{U}(\tilde{\boldsymbol{x}}_n, \hat{\boldsymbol{x}}(s_{g_s}, t_{g_t})) \mathbf{b}(s_{g_s}, t_{g_t}) J(s_{g_s}, t_{g_t}) J_{\bar{\boldsymbol{\xi}}} \, W_{g_s} W_{g_t} \right] \tag{11.35}$$

where s_{g_s}, t_{g_t} are Gauss point coordinates, $W_{g_s} W_{g_t}$ are weights and G_s, G_t are the number of Gauss points in s, t directions, dependent on the proximity of $\tilde{\boldsymbol{x}}_n$ to the integration region. The numerical integration of the surface term is given by

$$\mathbf{r}_{0n}^{\Gamma} = \sum_{g_s=1}^{G_s} \left[\mathsf{U}(\tilde{\boldsymbol{x}}_n, \hat{\boldsymbol{x}}(s_{g_s}, t_{surf})) \mathbf{b}_n(s_{g_s}, t_{surf}) J(s_{g_s, t_{surf}}) J_{\bar{\xi}} \, W_{g_s} \right] \tag{11.36}$$

where t_{surf} is the value of coordinate t for the surface considered (i.e. $t = 0$ for the bottom and $t = 1$ for the top surface) and $J_{\bar{\xi}} = 0.5$ is the Jacobian of the transformation between s and $\bar{\xi}$ coordinate system.

3-D domains. For 3-D problems we establish a local coordinate system $\boldsymbol{s} = (s, t, r)^{\mathrm{T}} = [0, 1]^3$ as shown in Fig. 11.4.

The global coordinates of a point $\boldsymbol{x}$ with the local coordinates $\boldsymbol{s}$ are given by

$$\boldsymbol{x}(s, t, r) = (1 - r) \, \boldsymbol{x}^I(s, t) + r \, \boldsymbol{x}^{II}(s, t) \tag{11.37}$$

where

$$\boldsymbol{x}^I(s, t) = \sum_{k=1}^{K^I} R_k^I(s, t) \, \boldsymbol{x}_k^I \quad \text{and} \quad \boldsymbol{x}^{II}(s, t) = \sum_{k=1}^{K^{II}} R_k^{II}(s, t) \, \boldsymbol{x}_k^{II}. \tag{11.38}$$

The superscript I relates to the bottom (red) surface and II to the top (green) surface and $\boldsymbol{x}_k^I$, $\boldsymbol{x}_k^{II}$ are control point coordinates. K^I and K^{II} represent the number of control points, $R_k^I(s, t)$ and $R_k^{II}(s, t)$ are NURBS basis functions. Note that there is a one to one mapping between the local coordinates of the surfaces ξ, η and the local coordinates s, t.

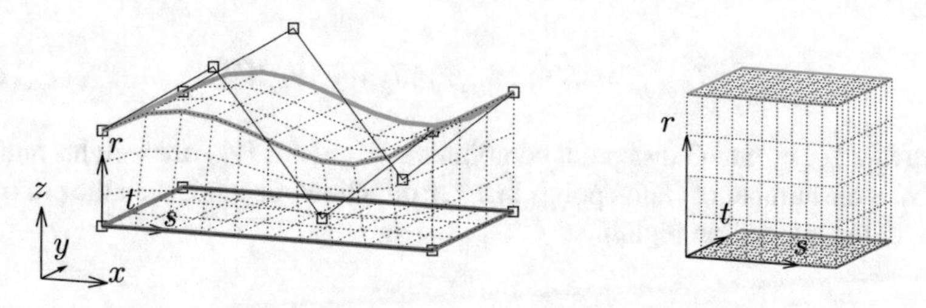

Fig. 11.4 Mapping of 3-D inclusion showing the bottom and top NURBS surfaces and the associated control points defining the inclusion: left in global $\boldsymbol{x}$, right in local $\boldsymbol{s}$ space

The derivatives are given by

$$\begin{aligned}\frac{\partial \boldsymbol{x}(s,t,r)}{\partial s} &= (1-r)\,\frac{\partial \boldsymbol{x}^{I}(s,t)}{\partial s} &+\; r\,\frac{\partial \boldsymbol{x}^{II}(s,t)}{\partial s}\\ \frac{\partial \boldsymbol{x}(s,t,r)}{\partial t} &= (1-r)\,\frac{\partial \boldsymbol{x}^{I}(s,t)}{\partial t} &+\; r\,\frac{\partial \boldsymbol{x}^{II}(s,t)}{\partial t}\\ \frac{\partial \boldsymbol{x}(s,t,r)}{\partial r} &= -\boldsymbol{x}^{I}(s,t) &+\; \boldsymbol{x}^{II}(s,t)\end{aligned} \tag{11.39}$$

where for example:

$$\frac{\partial \boldsymbol{x}^{I}(s,t)}{\partial s} = \sum_{k=1}^{K^{I}} \frac{\partial R_k^{I}(s,t)}{\partial s}\boldsymbol{x}_k^{I} \quad \text{and} \quad \frac{\partial \boldsymbol{x}^{II}(s,t)}{\partial s} = \sum_{k=1}^{K^{II}} \frac{\partial R_k^{II}(s,t)}{\partial s}\boldsymbol{x}_k^{II}. \tag{11.40}$$

The Jacobi matrix of this mapping is

$$\mathbf{J} = \begin{pmatrix} \frac{\partial \boldsymbol{x}}{\partial s} \\ \\ \frac{\partial \boldsymbol{x}}{\partial t} \\ \\ \frac{\partial \boldsymbol{x}}{\partial r} \end{pmatrix} \tag{11.41}$$

and the Jacobian is $J = |\mathbf{J}|$. For the Gauss integration we have to transform form $\boldsymbol{s} = (s,t,r)^{\mathrm{T}} = [0,1]^3$ to $\bar{\boldsymbol{\xi}} = (\bar{\xi},\bar{\eta},\bar{\zeta})^{\mathrm{T}} = [-1,1]^3$

$$s = \frac{1}{2}(1+\bar{\xi}),\;\; t = \frac{1}{2}(1+\bar{\eta}),\;\; r = \frac{1}{2}(1+\bar{\zeta}). \tag{11.42}$$

The Jacobian of this transformation is $J_{\bar{\xi}} = 0.125$.

The numerical integration of the volume term is given by

$$\mathbf{r}_{0n}^{\Omega} = \sum_{g_s=1}^{G_s}\sum_{g_t=1}^{G_t}\sum_{g_r=1}^{G_r} [\mathsf{U}(\tilde{\boldsymbol{x}}_n, \hat{\boldsymbol{x}}(s_{g_s}, t_{g_t}, r_{g_r}))\mathbf{b}(s_{g_s}, t_{g_t}, r_{g_r}) J_{g_s g_t g_r}] \tag{11.43}$$

with

$$J_{g_s g_t g_r} = J(s_{g_s}, t_{g_t}, r_{g_r}) J_{\bar{\xi}}\, W_{g_s} W_{g_t} W_{g_r} \tag{11.44}$$

where $s_{g_s}, t_{g_t}, r_{g_r}$ are Gauss point coordinates, $W_{g_s}, W_{g_t}, W_{g_r}$ are weights and G_s, G_t, G_r is the number of Gauss points in s, t, r directions, dependent on the proximity of $\tilde{\boldsymbol{x}}_n$ to the integration region.

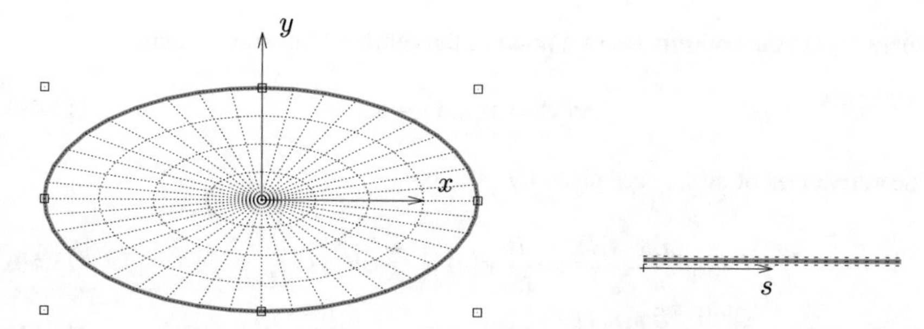

Fig. 11.5 Mapping of a plane inclusion showing the NURBS curve defining the boundary of the inclusion and the associated control points: left in global $\boldsymbol{x}$, right in local s space

The numerical integration of the surface term is given by

$$\mathbf{r}_{0n}^{\Gamma} = \sum_{g_s=1}^{G_s} \sum_{g_t=1}^{G_t} \left[\mathsf{U}(\tilde{\boldsymbol{x}}_n, \hat{\boldsymbol{x}}(s_{g_s}, t_{g_t}, r_{surf})) \mathbf{b}_n(s_{g_s}, t_{g_t}, r_{surf}) J_{g_s g_t r_{surf}} \right] \quad (11.45)$$

with

$$J_{g_s g_t r_{surf}} = J(s_{g_s}, t_{g_t}, r_{surf}) J_{\bar{\xi}} \, W_{g_s} W_{g_t} \quad (11.46)$$

where r_{surf} is the value of coordinate r for the surface considered (i.e. $r = 0$ for the bottom and $r = 1$ for the top surface) and $J_{\bar{\xi}} = 0.25$ is the Jacobian of the transformation between s, t and $\bar{\xi}, \bar{\eta}$ coordinate system.

11.3.3.3 Mapping Method 3

This mapping is for general shapes. Although this can be extended to 3-D, we only show the plane application here.

This mapping uses a cylindrical mapping from a local s, t coordinate system with $t = 0$ at the centre of the domain. First, we define the boundary of the domain by a suitable closed one-dimensional NURBS curve. The coordinates of a point $\boldsymbol{x}(s, 1)$ on the curve ($t = 1$) is given by (see Fig. 11.5)

$$\boldsymbol{x}(s, 1) = \sum_{i=1}^{I} R_i(s) \boldsymbol{x}_i \quad (11.47)$$

where $R_i(s)$ are NURBS basis functions, $\boldsymbol{x}_i$ are control point coordinates and I is the number of control points. Note that there is again a one to one mapping of the local coordinate ξ of the curve and the local coordinate s. The coordinates of a point with the local coordinates s, t are given by

$$\boldsymbol{x}(s, t) = \boldsymbol{x}_0 + \mathbf{v}(s) \, t \quad (11.48)$$

where $\boldsymbol{x}_0$ are the coordinates of a point in the centre of the domain and

$$\mathbf{v}(s) = \boldsymbol{x}(s,1) - \boldsymbol{x}_0. \tag{11.49}$$

The derivatives of $\boldsymbol{x}(s,t)$ are given by

$$\frac{\partial \boldsymbol{x}(s,t)}{\partial s} = \frac{\partial}{\partial s}\mathbf{v}(s) = \frac{\partial}{\partial s}\boldsymbol{x}(s,1)\,t, \tag{11.50}$$

$$\frac{\partial \boldsymbol{x}(s,t)}{\partial t} = \mathbf{v}(s). \tag{11.51}$$

The Jacobian matrix of this mapping is

$$\mathbf{J} = \begin{pmatrix} \frac{\partial x}{\partial s} \\ \\ \frac{\partial x}{\partial t} \end{pmatrix} \tag{11.52}$$

and the Jacobian is $J = |\mathbf{J}|$.

The numerical integration is the same as for the mapping option 2.

11.4 Summary and Conclusions

In this chapter we have introduced procedures that can be used to numerically evaluate the volume integral that arises when effects, that occur inside the domain (resulting in body forces), have to be considered.

Instead of using the traditional cell-based methods we explore here various mapping methods that, using isogeometric concepts and bounding curves or surfaces, map the domain from the global to a local coordinate system, where the numerical integration or differentiation is carried out. The advantage of the proposed approach is that no generation of cell meshes is necessary and that the geometrical information for the bounding curves/surfaces can be taken directly from CAD data.

Different mapping methods are presented for linear domains with a small width, for domains where the extension in one direction is small and for general domains. We will use the mapping methods in the following chapters dealing with inclusions, elasto-plasticity and viscous flow.

Chapter 12
Treatment of Inhomogeneities/Inclusions

12.1 Introduction

The BEM relies on the availability of fundamental solutions of the governing differential equations, which are only available for homogeneous domains. As this would restrict the practical application of the method, we explore here the possibility of considering inhomogeneities. Instead of dealing with a general inhomogeneous domain we concentrate on piecewise heterogeneous domains, i.e. the case where properties are different in certain parts of the domain. There are basically 2 approaches that can be taken. One is where we assemble two or more Boundary Element (BE) regions with different properties in a similar way as in the FEM. The other approach is to solve first for the linear homogeneous problem and to then modify the solution using body force effects.

12.2 Multi-region Approach

In this approach BE regions with different material properties are combined. How this is done is shown in Fig. 12.1 for two regions.

At the interface between the regions, we enforce the compatibility condition, i.e. the primary variable (potential or displacement) has to be the same at the interface

$$\mathbf{u}_1 = \mathbf{u}_2 \tag{12.1}$$

where the subscript denotes the region number. To ensure compatibility at the interface the following conditions must be satisfied for each connecting BE region:

- The basis functions describing the boundary variable must be the same.
- The locations of the collocation points must match.
- The connected parameter spaces must match.

G. Beer et al., *The Isogeometric Boundary Element Method*, Lecture Notes in Applied and Computational Mechanics 90,
https://doi.org/10.1007/978-3-030-23339-6_12

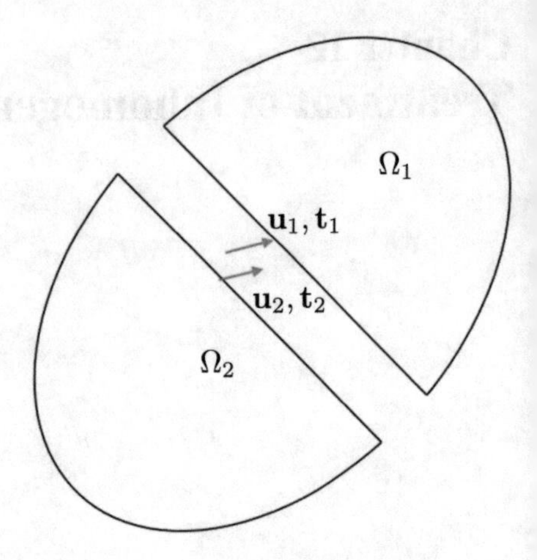

Fig. 12.1 Assembly of 2 regions

If this is the case we can express condition (12.1) in terms of parameter values, i.e. the conditions are implemented by setting the parameters at the collocation points to be the same at the interface.

In addition we have to enforce the continuity or equilibrium condition. Considering there is no applied flux or traction at the interface we have the condition

$$\mathbf{t}_1 + \mathbf{t}_2 = 0 \tag{12.2}$$

For each region we can write

$$\begin{aligned} [\mathsf{U}]_1 \{\mathbf{t}\}_1 &= [\mathsf{T}]_1 \{\mathbf{u}\}_2 \\ [\mathsf{U}]_2 \{\mathbf{t}\}_2 &= [\mathsf{T}]_2 \{\mathbf{u}\}_2 \end{aligned} \tag{12.3}$$

where $[\mathsf{U}]$ and $[\mathsf{T}]$ are assembled system matrices and the subscript refers to the region number. Vectors $\{\mathbf{u}\}_i$ and $\{\mathbf{t}\}_i$ contain values of parameters at collocation points of region i. The square brackets indicate that these are assembled matrices. The curly bracket indicates that the vector contains all parameter values of a region.

We can now solve for $\mathbf{t}$

$$\begin{aligned} \{\mathbf{t}\}_1 = [\mathsf{U}]_1^{-1} [\mathsf{T}]_1 \{\mathbf{u}\}_1 &= [\mathsf{A}]_1 \{\mathbf{u}\}_1 \\ \{\mathbf{t}\}_2 = [\mathsf{U}]_2^{-1} [\mathsf{T}]_2 \{\mathbf{u}\}_2 &= [\mathsf{A}]_2 \{\mathbf{u}\}_2. \end{aligned} \tag{12.4}$$

Next, we partition matrices $[\mathsf{A}]_i$ into a part that multiplies with primary values at the interface ($\{\mathbf{u}\}_{12} = \{\mathbf{u}\}_{21}$) and ones that are not coupled ($\{\mathbf{u}\}_{11}, \{\mathbf{u}\}_{22}$). For region 1 we have

$$\begin{pmatrix} \{\mathbf{t}\}_{11} \\ \{\mathbf{t}\}_{12} \end{pmatrix} = \begin{pmatrix} [\mathsf{A}]_{1,11} & [\mathsf{A}]_{1,12} \\ [\mathsf{A}]_{1,21} & [\mathsf{A}]_{1,22} \end{pmatrix} \begin{pmatrix} \{\mathbf{u}\}_{11} \\ \{\mathbf{u}\}_{12} \end{pmatrix} \tag{12.5}$$

where $[\mathsf{A}]_{1,11}$ etc. are sub-matrices of $[\mathsf{A}]_1$. Similarly for region 2

$$\begin{pmatrix} \{\mathbf{t}\}_{22} \\ \{\mathbf{t}\}_{21} \end{pmatrix} = \begin{pmatrix} [\mathsf{A}]_{2,11} & [\mathsf{A}]_{2,12} \\ [\mathsf{A}]_{2,21} & [\mathsf{A}]_{2,22} \end{pmatrix} \begin{pmatrix} \{\mathbf{u}\}_{22} \\ \{\mathbf{u}\}_{21} \end{pmatrix}. \tag{12.6}$$

Applying the compatibility condition (12.1) and the continuity condition (12.3) we obtain the combined system

$$\begin{pmatrix} \{\mathbf{t}\}_{11} \\ 0 \\ \{\mathbf{t}\}_{22} \end{pmatrix} = \begin{pmatrix} [\mathsf{A}]_{1,11} & [\mathsf{A}]_{1,12} & 0 \\ [\mathsf{A}]_{1,21} & [\mathsf{A}]_{1,11} + [\mathsf{A}]_{1,22} & [\mathsf{A}]_{1,12} \\ 0 & [\mathsf{A}]_{1,21} & [\mathsf{A}]_{1,22} \end{pmatrix} \begin{pmatrix} \{\mathbf{u}\}_{11} \\ \{\mathbf{u}\}_{12} \\ \{\mathbf{u}\}_{22} \end{pmatrix}. \tag{12.7}$$

The assembly process is essentially the same as the one used in the FEM.

The disadvantage of this coupling approach is that it involves the inversion of the system matrices for each region. This is not very efficient, especially if not all nodes of a region are coupled. An alternative has been proposed in [14] and involves the computation of a "stiffness matrix" $\mathbf{K}_n$ for region n, similar to the FEM. $\mathbf{K}_n$ is basically a matrix which provides values of the secondary variable (traction or flow) due to unit values of the primary variable (potential or displacement). This involves the solution of N Dirichlet problems (where N is the number of degrees of freedom) with unit values of the primary variable. The first column of the stiffness matrix is obtained by solving

$$[\mathsf{U}]\,\{\mathbf{t}\}_1 = [\mathsf{T}]\,\{\mathbf{u}\}_1 \tag{12.8}$$

where

$$\{\mathbf{u}\}_1 = \begin{Bmatrix} 1 \\ 0 \\ \vdots \end{Bmatrix}. \tag{12.9}$$

By multiplying $[\mathsf{T}]\,\{\mathbf{u}\}_1$ we can see that the right-hand side of Eq. (12.8) is the first column of matrix $[\mathsf{T}]$.

The computation of $\mathbf{K}_n$ involves the solution of Eq. (12.8) with multiple right-hand sides

$$[\mathsf{U}]\,\{\mathbf{t}\}_i = \{\mathbf{r}\}_i \tag{12.10}$$

where $\{\mathbf{r}\}_i$ is the ith column of $[\mathsf{T}]$. Each solution $\{\mathbf{t}\}_i$ represents a column in $\mathbf{K}_n$

$$\mathbf{K}_n = \left(\{\mathbf{t}\}_1\, \{\mathbf{t}\}_2 \cdots \right). \tag{12.11}$$

The region matrices can be assembled now in the same way as with the FEM. However, it should be noted that in contrast to the FEM $\mathbf{K}_n$ is not symmetric. In the case of partially coupled regions, i.e. where only a portion of the boundary is coupled, one can first solve the system with zero values of the primary variable at the interface(s) to obtain a right-hand side vector and then solve for the interface unknowns. Details can be found in [14].

12.3 Inclusions

The disadvantage of the multi-region approach presented in the previous section is that additional unknowns are introduced at the interface. Also the geometry description is more complicated, because the basis functions and the collocation points have to match at the interfaces, to ensure compatibility.

An alternative is to treat inhomogeneities as inclusions and to solve the system in an iterative way. The advantage is that no additional degrees of freedom are introduced and non-linear behaviour of the inclusions can also be considered. The procedure is explained for elasticity problems in Algorithm 6, but can be also used for potential problems. The basic approach is to solve the problem in an iterative way. First the problem is solved considering an homogeneous domain. Then the solution is modified to account for the presence of inclusions, by considering the arising body forces. In the following we define an increment with an overdot.

Algorithm 6 Algorithm for simulation of inclusions

1: Solve the elastic, homogeneous problem $\mathbf{L}\mathbf{x}_0 = \mathbf{r}$
2: **for** i **to** Number iterations **do**
3: Determine the increment of strain $\{\dot{\epsilon}\}$ inside the inclusion.
4: Determine an increment in initial stress $\{\dot{\sigma}\}^0$ using Eq. (12.13).
5: Convert $\{\dot{\sigma}\}^0$ to body force and traction increments $\dot{\mathbf{b}}_0$, $\dot{\mathbf{t}}_0$.
6: Compute $\mathbf{r}_0$
7: Solve for the new right-hand side ($\mathbf{L}\dot{\mathbf{x}}_i = \mathbf{r}_0$).
8: Accumulate solution: $\mathbf{x}_i = \mathbf{x}_{i-1} + \dot{\mathbf{x}}_i$
9: Compute Residual
10: **if** Residual < Tolerance **then**
11: Exit
12: **end if**
13: **end for**

To compute the initial stress increment for the case where the inclusions have elastic properties that are different from the ones used for the fundamental solutions we use the relation between increments of stress $\dot{\sigma}$ and strain $\dot{\epsilon}$ in Voigt notation

$$\{\dot{\sigma}\} = \mathbf{D}\,\{\dot{\epsilon}\} \tag{12.12}$$

where $\mathbf{D}$ is the constitutive matrix for the domain that has been used for the computation of the fundamental solutions. The difference in stress between the inclusion and the domain and therefore the initial stress increment can be computed by

$$\{\dot{\sigma}\}^0 = (\mathbf{D}_{inc} - \mathbf{D})\,\{\dot{\epsilon}\} \tag{12.13}$$

where $\mathbf{D}_{inc}$ is the constitutive matrix for the inclusion. We can also account for initial stresses due to non-linear material behaviour and this will be discussed in Chap. 13.

12.3.1 Evaluation of the Volume Integrals

The volume integrals to be evaluated (11.3) and (11.4) are re-written in incremental form

$$\mathbf{r}_{0n}^{\Omega_0} = \int_{\Omega_0} \mathsf{U}(\tilde{\boldsymbol{x}}_n, \boldsymbol{x})\dot{\mathbf{b}}_0(\boldsymbol{x})d\Omega_0(\boldsymbol{x}) \tag{12.14}$$

$$\mathbf{r}_{0n}^{\Gamma_0} = \int_{\Gamma_0} \mathsf{U}(\tilde{\boldsymbol{x}}_n, \boldsymbol{x})\dot{\mathbf{t}}_0(\boldsymbol{x})d\Gamma_0(\boldsymbol{x}) \tag{12.15}$$

where $\dot{\mathbf{b}}_0$ is the increment in body force due to $\{\dot{\sigma}\}^0$ and $\dot{\mathbf{t}}_0$ is the associated initial traction. We use an overdot to indicate increment.

For the numerical evaluation of the integrals, we require the values of the body force at the Gauss points, whose locations depend on the proximity of the source point. As the number of Gauss points can be quite large, it would be inefficient to compute the values there. Instead, we define a grid of points inside the inclusion (as shown for example in Fig. 12.2) where the strains and subsequently the initial stresses are computed. The value at a Gauss point with the local coordinates $\boldsymbol{s} = (s, t)^{\mathrm{T}} = [0, 1]^2$ in 2-D or $\boldsymbol{s} = (s, t, r)^{\mathrm{T}} = [0, 1]^3$ in 3-D is then obtained by

$$\dot{\mathbf{b}}_0(\boldsymbol{s}) = \sum_{k=1}^{K} M_k(\boldsymbol{s})\dot{\mathbf{b}}_{0k} \tag{12.16}$$

where $\dot{\mathbf{b}}_{0k}$ is the body force at grid point k with the local coordinate $\boldsymbol{s}_k$. $M_k(\boldsymbol{s})$ are linear or quadratic interpolation functions with local support (i.e. they are only non-zero near $\boldsymbol{s}$).

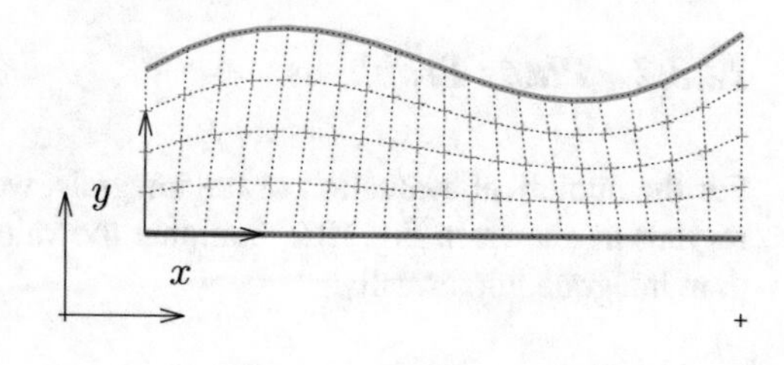

Fig. 12.2 Definition of the grid of internal points and subdivision into integration regions

Since the right-hand side needs to be computed at each iteration, we can optimise the computations by establishing a matrix $[\mathbf{R}_0]$ that multiplies with the body forces. The right-hand side can then be computed by

$$\mathbf{r}_0 = [\mathbf{R}_0]\{\dot{\mathbf{b}}_0\} \tag{12.17}$$

where $\{\dot{\mathbf{b}}_0\}$ is a vector that contains body force values at grid points.

The values of the strain increment $\{\dot{\epsilon}\}$ at the grid points can be computed for the first iteration by

$$\{\dot{\epsilon}\}_0 = \hat{\mathbf{E}}\dot{\mathbf{x}}_0 + \{\bar{\epsilon}\} \tag{12.18}$$

where $\dot{\mathbf{x}}_0$ is the vector containing the solution, $\hat{\mathbf{E}}$ is an assembled matrix and $\{\bar{\epsilon}\}$ is the value of ϵ due to known boundary values (see Chap. 6). For the subsequent iterations the formula is:

$$\{\dot{\epsilon}\}_i = \hat{\mathbf{E}}\dot{\mathbf{x}}_i + \left[\hat{\mathbf{R}}_0\right]\{\dot{\mathbf{b}}_0\}_i \tag{12.19}$$

where $\left[\hat{\mathbf{R}}_0\right]$ is computed in the same way as $[\mathbf{R}_0]$ except that instead of collocation point coordinates the coordinates of the internal points are used.

In the following, we discuss the computation of the body force and the numerical evaluation of the volume integrals first for plane and then 3-D problems using the mapping option 2 of Chap. 11 for the geometrical description of Ω_0. The computation of matrix $[\mathbf{R}_0]$ involves the integration over the boundary Γ_0 and over the domain Ω_0, i.e.

$$[\mathbf{R}_0] = [\mathbf{R}_0]^{\Gamma_0} + [\mathbf{R}_0]^{\Omega_0}. \tag{12.20}$$

The boundary integration and the volume integration are presented separately next.

Remark It should be noted that if an inclusion with different elastic material constants intersects the boundary Γ, then we have to make sure that, because of the expected strain discontinuity, the displacement field can only have a C^0-continuity there. Therefore we have to adjust the basis functions for the approximation of the unknown displacement by inserting multiple knots at this location. As a consequence, there will always be a collocation point located at the intersection.

12.3.2 Plane Problems

For the numerical evaluation of the integrals, we divide the volume into integration regions as shown in Fig. 12.2, compute the values of body force at grid points and then integrate numerically.

12.3.2.1 Computation of Values $\dot{\mathbf{b}}_0$ and $\dot{\mathbf{t}}_0$ at Grid Points

The body force increment $\dot{\mathbf{b}}_0$ is computed from the stresses

$$\dot{\mathbf{b}}_0 = -\begin{pmatrix} \dfrac{\partial \dot{\sigma}_x^0}{\partial x} + \dfrac{\partial \dot{\tau}_{xy}^0}{\partial y} \\ \\ \dfrac{\partial \dot{\tau}_{xy}^0}{\partial x} + \dfrac{\partial \dot{\sigma}_x^0}{\partial y} \end{pmatrix}. \tag{12.21}$$

The initial tractions are

$$\dot{\mathbf{t}}_0 = \begin{pmatrix} \dot{\sigma}_x^0 & \dot{\tau}_{xy}^0 \\ \\ \dot{\tau}_{xy}^0 & \dot{\sigma}_y^0 \end{pmatrix} \mathbf{n}. \tag{12.22}$$

where $\mathbf{n}$ is the unit outward normal vector to the surface Γ_0 which is defined by the two bounding curves (i.e. the top and bottom curves) and the two sides.

It is convenient to compute the derivatives with respect to local coordinates $\boldsymbol{s}$ first and then transform them to global coordinates. For example, the global derivatives of σ_x in terms of local derivatives are given by the transformation

$$\sigma_{x,\boldsymbol{x}} = \mathbf{J}^{-1}\, \sigma_{x,\boldsymbol{s}} \tag{12.23}$$

where

$$\sigma_{x,\boldsymbol{x}} = \begin{pmatrix} \dfrac{\partial \sigma_x}{\partial x} \\ \\ \dfrac{\partial \sigma_x}{\partial y} \end{pmatrix} \quad \text{and} \quad \sigma_{x,\boldsymbol{s}} = \begin{pmatrix} \dfrac{\partial \sigma_x}{\partial s} \\ \\ \dfrac{\partial \sigma_x}{\partial t} \end{pmatrix}. \tag{12.24}$$

and $\mathbf{J}$ is the Jacobi matrix, Eq. (11.33).

The derivatives are numerically computed using finite differences. For grid points inside the inclusion that have other points left and right (or top and bottom) of them, we use a central finite difference, whereas for points that only have one point on a side, we use forward or backward finite differences. Referring to Fig. 12.3 the local derivatives, computed using a forward finite difference scheme, are given as

$$\frac{\partial \sigma}{\partial s} = \frac{\sigma_{n+1,m} - \sigma_{n,m}}{ds} \quad \text{and} \quad \frac{\partial \sigma}{\partial t} = \frac{\sigma_{n,m+1} - \sigma_{n,m}}{dt}. \tag{12.25}$$

12.3.2.2 Computation of the Line Integral over Γ_0

Here we distinguish between the case when the collocation point $\tilde{\boldsymbol{x}}_n$ is outside the integration region (regular integration) or inside (singular integration).

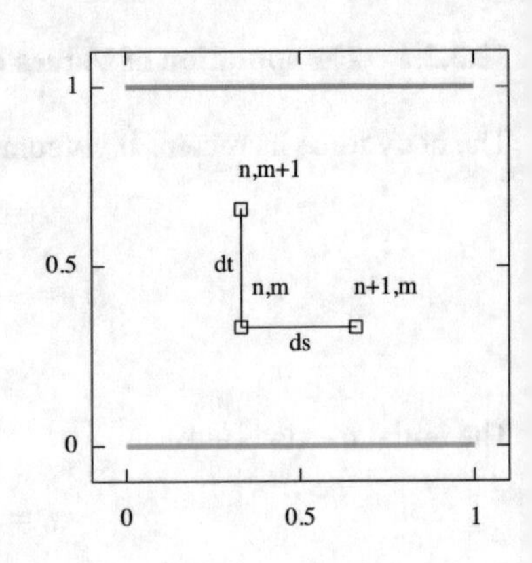

Fig. 12.3 Template for the computation of first derivatives inside an inclusion by forward difference in the local s, t coordinate system for grid point n, m

Regular integration. For the computation of the integral over Γ_0, we first integrate along the two curves defining the inclusion (i.e. for $s = [0, 1]$) and then along the edges (i.e. for $t = [0, 1]$).

For the integration along the bounding curves (*I* and *II*) the global locations of Gauss points are computed by

$$\boldsymbol{x}^I(\xi) = \sum_{k=1}^{K^I} R_k^I(\xi)\, \boldsymbol{x}_k^I \qquad \text{and} \qquad \boldsymbol{x}^{II}(\xi) = \sum_{k=1}^{K^{II}} R_k^{II}(\xi)\, \boldsymbol{x}_k^{II}. \tag{12.26}$$

The Jacobian of this transformation is $J^i = \sqrt{(\frac{dx_1^i}{d\xi})^2 + (\frac{dx_2^i}{d\xi})^2}$, where $i = \{I, II\}$.

Coordinates ξ are expressed in terms of Gauss coordinates $\bar{\xi}$:

$$\xi = \frac{\Delta\xi_{ns}}{2}(1 + \bar{\xi}) + \xi_{ns} \tag{12.27}$$

where $\Delta\xi_{ns}$ is the size of the integration region ns and ξ_{ns} is the start coordinate. The Jacobian of this transformation is $J_{\xi}^{ns} = \frac{\Delta\xi_{ns}}{2}$.

For the integration along the left edge e_1 we have

$$\bar{\boldsymbol{x}}(t) = (1 - t)\, \boldsymbol{x}(\xi = 0)^I + t\, \boldsymbol{x}(\xi = 0)^{II}. \tag{12.28}$$

Assuming for simplicity[1] that there is no subdivision into integration regions in the t-direction we can express the coordinate t in terms of the Gauss coordinate $\bar{\xi}$ by:

$$t = \frac{1}{2}(1 + \bar{\xi}). \tag{12.29}$$

The Jacobian of transformation (12.28) is $J_{e_1} = \frac{1}{2}[\boldsymbol{x}(\xi = 0)^{II} - \boldsymbol{x}(\xi = 0)^I]$.

[1] This will be the case for the examples presented later which involve relatively thin inclusions, but is not a restriction of the method.

For the right edge e_2 we have

$$\bar{\boldsymbol{x}}(t) = (1-t)\,\boldsymbol{x}(\xi=1)^{I} + t\,\boldsymbol{x}(\xi=1)^{II}. \tag{12.30}$$

The Jacobian of this transformation is $J_{e_2} = \frac{1}{2}[\boldsymbol{x}(\xi=1)^{II} - \boldsymbol{x}(\xi=1)^{I}]$. We can now write the sub-vector of $\mathbf{r}_0^{\Gamma_0}$ related to the collocation point n as

$$\begin{aligned}\mathbf{r}_{0n}^{\Gamma_0} &= \sum_{i=1}^{2}\sum_{ns=1}^{N_s}\int_{-1}^{1} \mathsf{U}(\tilde{\boldsymbol{x}}_n,\bar{\boldsymbol{x}})\dot{\boldsymbol{t}}_0(\bar{\boldsymbol{x}})\, J^i\, J_\xi^{ns}\, d\bar{\xi}\\ &+\sum_{j=1}^{2}\int_{-1}^{1} \mathsf{U}(\tilde{\boldsymbol{x}}_n,\bar{\boldsymbol{x}})\dot{\boldsymbol{t}}_0(\bar{\boldsymbol{x}})\, J_{e_j}\, d\bar{\xi}\end{aligned} \tag{12.31}$$

where N_s is the number of subregions. Applying Gauss integration we have

$$\begin{aligned}\mathbf{r}_{0n}^{\Gamma_0} &\approx \sum_{i=1}^{2}\sum_{ns=1}^{Ns}\sum_{g=1}^{G} \mathsf{U}\left(\tilde{\boldsymbol{x}}_n,\bar{\boldsymbol{x}}(\bar{\xi}_g)\right)\dot{\boldsymbol{t}}_0\left(\bar{\boldsymbol{x}}(\bar{\xi}_g)\right) J^i(\bar{\xi}_g)\, J_\xi^{ns}(\bar{\xi}_g)\, W_g\\ &+\sum_{i=1}^{2}\sum_{g=1}^{G} \mathsf{U}\left(\tilde{\boldsymbol{x}}_n,\bar{\boldsymbol{x}}(\bar{\xi}_g)\right)\dot{\boldsymbol{t}}_0\left(\bar{\boldsymbol{x}}(\bar{\xi}_g)\right) J_{e_i}(\bar{\xi}_g)\, W_g.\end{aligned} \tag{12.32}$$

where G is the number of integration points depending on the proximity of the source point, $\bar{\xi}_g$ are the local coordinates of Gauss points and W_g are quadrature weights.

Substituting Eqs. (12.16) and (12.22) we obtain for the sub-matrices of $[\mathbf{R}_0]$ related to the surface integration

$$\begin{aligned}\mathbf{R}_{0nk}^{\Gamma_0} &\approx \sum_{i=1}^{2}\sum_{ns=1}^{Ns}\sum_{g=1}^{G} \mathsf{U}\left(\tilde{\boldsymbol{x}}_n,\bar{\boldsymbol{x}}(\bar{\xi}_g)\right)\mathbf{n}\left(\bar{\boldsymbol{x}}(\bar{\xi}_g)\right) M_k(s(\bar{\xi}_g)) J^i(\bar{\xi}_g) J_\xi^{ns}(\bar{\xi}_g) W_g\\ &+\sum_{i=1}^{2}\sum_{g=1}^{G} \mathsf{U}\left(\tilde{\boldsymbol{x}}_n,\bar{\boldsymbol{x}}(\bar{\xi}_g)\right)\mathbf{n}\left(\bar{\boldsymbol{x}}(\bar{\xi}_g)\right) M_k(s(\bar{\xi}_g))\, J_{e_i}(\bar{\xi}_g)\, W_g.\end{aligned} \tag{12.33}$$

Singular integration. If the collocation point is located inside Γ_0 the integrand tends to infinity with $\mathcal{O}(\ln\frac{1}{r})$. We invoke the procedure used for integrating a $\ln(\frac{1}{r})$ function as explained in Chap. 7.

12.3.2.3 Computation of the Volume Integral over Ω_0

We distinguish between the case when the point $\tilde{\boldsymbol{x}}_n$ is outside the integration region (regular integration) or inside (singular integration).

Regular integration. The transformation from $\boldsymbol{s}$ coordinates to Gauss point coordinates $\bar{\boldsymbol{\xi}} = (\bar{\xi},\bar{\eta})^{\mathrm{T}} = [-1,1]^2$ is given for integration region ns by

$$
\begin{aligned}
s &= \frac{\triangle s_{ns}}{2}(1+\bar{\xi}) + s_{ns} \\
t &= \frac{\triangle t_{ns}}{2}(1+\bar{\eta}) + t_{ns}
\end{aligned} \tag{12.34}
$$

where $\triangle s_{ns} \times \triangle t_{ns}$ denotes the size of the integration region and s_{ns}, t_{ns} are the starting coordinates. The Jacobian of this transformation for the integration over Ω_0 is $J_{\xi}^{ns} = \frac{\triangle s_{ns}\,\triangle t_{ns}}{4}$.

The sub-vector of $\mathbf{r}_0^{\Omega_0}$ related to collocation point n can be written as

$$
\mathbf{r}_{0n}^{\Omega_0} = \sum_{ns=1}^{Ns} \int_{-1}^{1}\int_{-1}^{1} \mathsf{U}\left(\tilde{\boldsymbol{x}}_n, \boldsymbol{x}(\xi,\eta)\right) \dot{\mathbf{b}}_0\left(\bar{\boldsymbol{x}}(\bar{\xi},\bar{\eta})\right) J\, J_{\xi}^{ns}\; d\bar{\xi} d\bar{\eta}. \tag{12.35}
$$

The numerical integration leads to

$$
\mathbf{r}_{0n}^{\Omega_0} \approx \sum_{ns=1}^{Ns}\sum_{g_s=1}^{G_s}\sum_{g_t=1}^{G_t} \mathsf{U}\left(\tilde{\boldsymbol{x}}_n, \boldsymbol{x}(\bar{\xi}_{g_s},\bar{\eta}_{g_t})\right)\; \dot{\mathbf{b}}_0\left(\bar{\boldsymbol{x}}(\bar{\xi}_{g_s},\bar{\eta}_{g_t})\right) J_{g_s g_t}^{ns} \tag{12.36}
$$

with

$$
J_{g_s g_t}^{ns} = J\, J_{\xi}^{ns}\; W_{g_s} W_{g_t}. \tag{12.37}
$$

The number of Gauss points G_s, G_t has to be adjusted according to the proximity of the collocation point. Substitution of Eq. (12.16) leads to the components of matrix $[\mathbf{R}_0]$ related to the volume integration

$$
\mathbf{R}_{0nk}^{\Omega_0} \approx \sum_{ns=1}^{Ns}\sum_{g_s=1}^{G_s}\sum_{g_t=1}^{G_t} \mathsf{U}\left(\tilde{\boldsymbol{x}}_n, \boldsymbol{x}(\bar{\xi}_{g_s},\bar{\eta}_{g_t})\right)\; M_k\left(\boldsymbol{s}(\bar{\xi}_{g_s},\bar{\eta}_{g_t})\right) J_{g_s g_t}^{ns}. \tag{12.38}
$$

Singular integration. If the point $\tilde{\boldsymbol{x}}_n$ is part of domain Ω_0 the integrand tends to infinity as the point is approached. A procedure is invoked that has been used to deal with weakly singular integrals in the three-dimensional BEM, involving triangular subregions as explained in Sect. 7.4.2. The procedure leads to the following expression:

$$
\begin{aligned}
\mathbf{r}_{0n}^{\Omega_0} &= \sum_{nt=1}^{N_t} \int_{-1}^{+1}\int_{-1}^{+1} \mathsf{U}\left(\tilde{\boldsymbol{x}}_n, \boldsymbol{x}\left(\bar{\boldsymbol{x}}(\xi,\eta)\right)\right) \dot{\mathbf{b}}_0\left(\bar{\boldsymbol{x}}(\xi,\eta)\right) J\, J_{\triangle}^{nt}\; d\bar{\xi} d\bar{\eta} \\
&\approx \sum_{nt=1}^{N_t}\sum_{g_s=1}^{G_s}\sum_{g_t=1}^{G_t} \mathsf{U}\left(\tilde{\boldsymbol{x}}_n, \boldsymbol{x}\left(\bar{\boldsymbol{x}}(\bar{\xi}_{g_s},\bar{\eta}_{g_t})\right)\right) \dot{\mathbf{b}}_0\left(\bar{\boldsymbol{x}}(\bar{\xi}_{g_s},\bar{\eta}_{g_t})\right) J J_{\triangle}^{nt} W_{g_s} W_{g_t}
\end{aligned} \tag{12.39}
$$

where N_t is the number of triangles and $J_{\triangle}^{nt}$ is the Jacobian of the transformation from the square to the triangular subregion.

12.3.2.4 Test Example

The example tests the algorithm for the case of a single elastic inclusion. It consists of a square with the dimension 1×1 composed of two different materials. A two-dimensional analysis is carried out using plane stress assumptions and the discretization is shown in Fig. 12.4. The cube is defined by four NURBS curves with knot vectors and control point coordinates as shown in Table 12.1

The inclusion is defined by two NURBS curves as shown in Table 12.2 and assigned a Young's modulus E of half the one used for computing the fundamental solution and no change in the Poisson's ratio ν. The cube is loaded with a moment as shown left in Fig. 12.4 and is fixed at the bottom.

For the analysis, the concept of a geometry independent field approximation was used and the basis functions for approximating the displacements were defined using the knot vectors shown in Table 12.3 with all weights equal to one. This approximation results in a quadratic variation of the displacements in the vertical direction. The insertion of double knots into the knot vector ensures a C^0-continuity (as required)

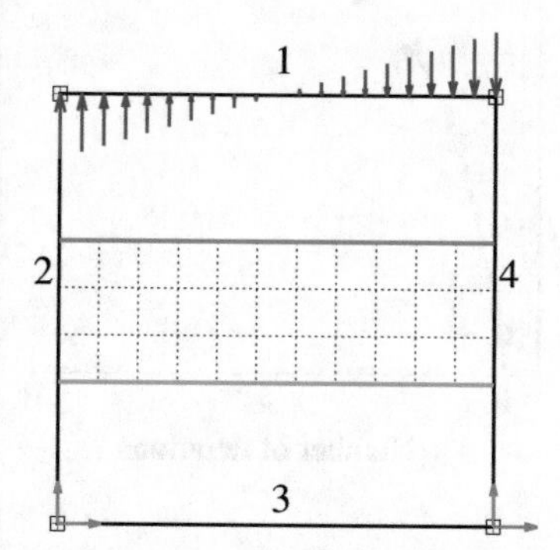

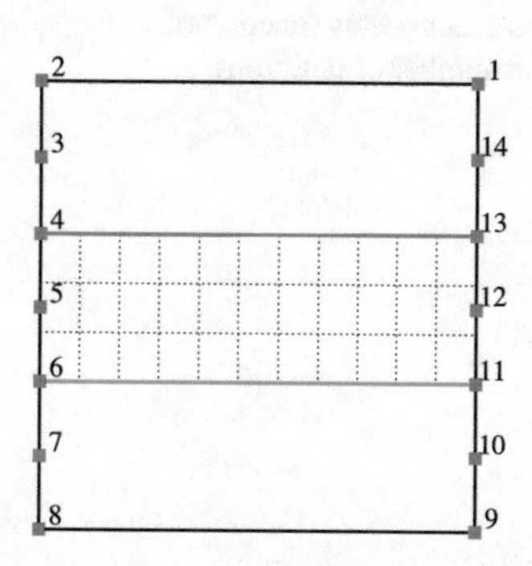

Fig. 12.4 Test example 1: left: geometry definition with 4 numbered NURBS curves showing control points, loading and boundary conditions. The inclusion is defined with 2 colour coded spline curves; right: location of collocation points

Table 12.1 Geometrical definition of cube

Curve	Knot vector	Control point	x	y	z	Weight
1	0, 0, 1, 1	1	1	1	0	1
.	...	2	0	1	0	1
2	0, 0, 1, 1	1	0	1	0	1
.	...	2	0	0	0	1
3	0, 0, 1, 1	1	0	0	0	1
.	...	2	1	0	0	1
4	0, 0, 1, 1	1	1	0	0	1
.	...	2	1	1	0	1

Table 12.2 Geometrical definition of inclusion

Curve	Knot vector	Control point	x	y	z	Weight
1	0, 0, 1, 1	1	0	0.66	0	1
.	...	2	1	0.66	0	1
2	0, 0, 1, 1	1	0	0.33	0	1
.	...	2	1	0.33	0	1

Table 12.3 Knot vectors for the approximation of the displacements

Curve	Knot vector
1	0, 0, 1, 1
2	0, 0, 0, 0.33, 0.33, 0.66, 0.66, 1, 1, 1
3	0, 0, 1, 1
4	0, 0, 0, 0.33, 0.33, 0.66, 0.66, 1, 1, 1

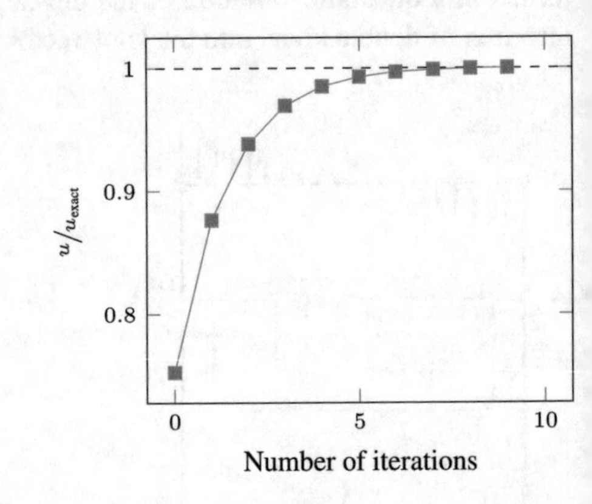

Fig. 12.5 Test example 1: plot of ratio of maximum computed displacement to the exact one as function of the number of iterations

where the lines defining the inclusions intersect with the boundary. The resulting location of the collocation points are shown in Fig. 12.4 on the right and this will give the exact solution for the applied loading. The results are shown in Figs. 12.5 and 12.6. It can be seen that convergence to the exact solution is achieved after about 7 iterations.

12.3.3 3-D Problems

For the numerical evaluation of the integrals in 3-D, we establish a number of internal points and divide the volume into integration regions as shown in Fig. 12.7.

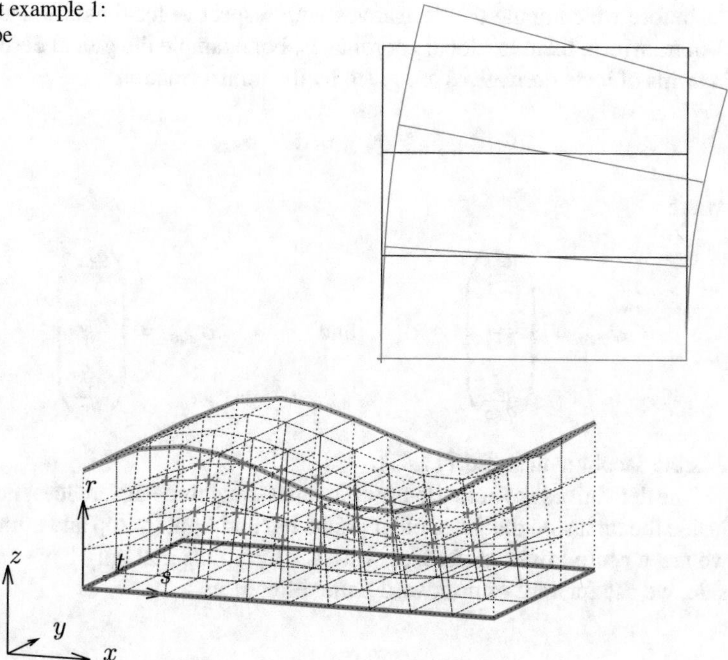

Fig. 12.6 Test example 1: deformed shape

Fig. 12.7 Definition of the grid of internal points and subdivision into integration regions

12.3.3.1 Computation of $\dot{\mathbf{t}}_0$ and $\dot{\mathbf{b}}_0$ at Grid Points

After computing the initial stress increment $\dot{\boldsymbol{\sigma}}_0$ the initial traction increments $\dot{\mathbf{t}}_0$ are computed by

$$\dot{\mathbf{t}}_0 = \begin{pmatrix} \dot{\sigma}_{0x} & \dot{\tau}_{0xy} & \dot{\tau}_{0xz} \\ \dot{\tau}_{0xy} & \dot{\sigma}_{0y} & \dot{\tau}_{0yz} \\ \dot{\tau}_{0zx} & \dot{\tau}_{0zy} & \dot{\sigma}_{0z} \end{pmatrix} \mathbf{n} \tag{12.40}$$

where $\mathbf{n}$ is the unit outward normal vector to the surface Γ_0.

The body force increment $\dot{\mathbf{b}}_0$ can be computed by

$$\dot{\mathbf{b}}_0 = - \begin{pmatrix} \frac{\partial \dot{\sigma}_{0x}}{\partial x_1} + \frac{\partial \dot{\tau}_{0xy}}{\partial x_2} + \frac{\partial \dot{\tau}_{0xz}}{\partial x_3} \\ \\ \frac{\partial \dot{\tau}_{0yx}}{\partial x_1} + \frac{\partial \dot{\sigma}_{0y}}{\partial x_2} + \frac{\partial \dot{\tau}_{0yz}}{\partial x_3} \\ \\ \frac{\partial \dot{\tau}_{0zx}}{\partial x_1} + \frac{\partial \dot{\tau}_{0zy}}{\partial x_2} + \frac{\partial \dot{\sigma}_{0z}}{\partial x_3} \end{pmatrix}. \tag{12.41}$$

As before we compute the derivatives with respect to local coordinates $\boldsymbol{s}$ first and then transform them to global coordinates. For example the global derivatives of σ_x in terms of local derivatives are given by the transformation

$$\boldsymbol{\sigma}_{x,x} = \mathbf{J}^{-1}\,\boldsymbol{\sigma}_{x,s} \tag{12.42}$$

where

$$\boldsymbol{\sigma}_{x,x} = \begin{pmatrix} \frac{\partial \sigma_x}{\partial x_1} \\ \frac{\partial \sigma_x}{\partial x_2} \\ \frac{\partial \sigma_x}{\partial x_3} \end{pmatrix} \quad \text{and} \quad \boldsymbol{\sigma}_{x,s} = \begin{pmatrix} \frac{\partial \sigma_x}{\partial s} \\ \frac{\partial \sigma_x}{\partial t} \\ \frac{\partial \sigma_x}{\partial r} \end{pmatrix} \tag{12.43}$$

$\mathbf{J}$ is the Jacobi matrix, Eq. (11.41).

The derivatives are numerically computed using finite differences. For grid points inside the inclusion that have other points left and right (or top and bottom) of them, we use a central finite difference, whereas for points that only have one point on a side, we use forward or backward finite differences.

12.3.3.2 Computation of the Surface Integral over Γ_0

We distinguish between singular integration when point $\boldsymbol{x}$ is inside the integration region, nearly singular integration when it is close to the region and regular integration when it is further away.

Regular integration. If $\tilde{x}$ is not inside the integration region then the number of integration points depends on the proximity of the point to the region and the size of the region. In the implementation, a Quadtree method introduced in Sect. 7.4.1.1, is used. We first integrate over the two surfaces defining the inclusion (i.e. over $s,t = [0,1]^2$ for $r = 0,1$) and then over the edges. For the integration along the bounding surfaces *I* and *II* the global locations of Gauss points are computed by

$$\boldsymbol{x}^I(\mathrm{s},\mathrm{t}) = \sum_{k=1}^{K^I} R_k^I(s,t)\,\boldsymbol{x}_k^I \quad \text{and} \quad \boldsymbol{x}^{II}(s,t) = \sum_{k=1}^{K^{II}} R_k^{II}(s,t)\,\boldsymbol{x}_k^{II}. \tag{12.44}$$

The Jacobian of this transformation is J^i. The local coordinates s,t are related to the Gauss coordinates $\bar{\boldsymbol{\xi}} = [-1,1]^2$ by

$$s = \frac{\Delta s_n}{2}(1+\bar{\xi}) + s_n \quad \text{and} \quad t = \frac{\Delta t_n}{2}(1+\bar{\eta}) + t_n \tag{12.45}$$

where $\Delta s_n, \Delta t_n$ denotes the size of the integration region n_s and s_n, t_n are the local coordinates of the edge of the subregion. The Jacobian of this transformation is J_s^n.

For the integration along the edges we have for example for edge e_1 at $t = 0$

$$\bar{\boldsymbol{x}}(s,r) = (1-r)\,\boldsymbol{x}(s,t=0)^I + r\,\boldsymbol{x}(s,t=0)^{II}. \tag{12.46}$$

The Jacobian of this transformation is J^{e_1}. The transformation to the $\bar{\xi}$ coordinate system is given by

$$s = \frac{\Delta s_n}{2}(1+\bar{\xi}) + s_n \qquad \text{and} \qquad r = \frac{\Delta r_n}{2}(1+\bar{\eta}) + r_n \tag{12.47}$$

where $\Delta s_n, \Delta r_n$ denotes the size of the integration region n_s and s_n, r_n are the local coordinates of the edge of the subregion. The Jacobian of this transformation is $J^n_{e_1}$.

We can now write the sub-vector of $\mathbf{r}_0^{\Gamma_0}$ related to the collocation point $\boldsymbol{x}_n$ as

$$\begin{aligned}\mathbf{r}_{0n}^{\Gamma_0} &= \sum_{i=1}^{2}\sum_{n_s=1}^{N_s}\int_{-1}^{1}\int_{-1}^{1} \mathsf{U}(\tilde{\boldsymbol{x}}_n,\bar{\boldsymbol{x}})\boldsymbol{t}_0(\bar{\boldsymbol{x}})\,J^i\,J_s^{n_s}\,d\bar{\xi}d\bar{\eta}\\ &+\sum_{j=1}^{4}\sum_{n_s=1}^{N_s}\int_{-1}^{1}\int_{-1}^{1} \mathsf{U}(\tilde{\boldsymbol{x}}_n,\bar{\boldsymbol{x}})\boldsymbol{t}_0(\bar{\boldsymbol{x}})\,J^{e_j}\,J_{e_j}^{n_s}\,d\bar{\xi}d\bar{\eta}\end{aligned} \tag{12.48}$$

where N_s is the number of subregions.

Singular integration. If the collocation point is located on Γ_0 the integrand approaches infinity as the point is approached. Here we apply the method that has been used for dealing with weakly singular integrals over surface Γ. It involves the transformation to a local coordinate system where the Jacobian tends to zero as the collocation point is approached.

12.3.3.3 Computation of the Volume Integral over Ω_0

Regular integration. For integration region n_s the transformation from $\mathbf{s}$ coordinates to $\bar{\boldsymbol{\xi}} = (\bar{\xi},\bar{\eta},\bar{\zeta})^{\mathrm{T}} = [-1,1]^3$ is given by

$$\begin{aligned} s &= \tfrac{\Delta s_n}{2}(1+\bar{\xi}) + s_{n_s}\\ t &= \tfrac{\Delta t_n}{2}(1+\bar{\eta}) + t_{n_s}\\ r &= \tfrac{\Delta r_n}{2}(1+\bar{\zeta}) + r_{n_s}\end{aligned} \tag{12.49}$$

where $\Delta s_n \times \Delta t_n \times \Delta r_n$ denotes the size of the integration region and s_n, t_n, r_n are the edge coordinates. The Jacobian of this transformation is $J^n_{\bar{\xi}} = \frac{1}{8}\,\Delta s_n\,\Delta t_n\,\Delta r_n$.

The sub-vector of $\mathbf{r}_0^{\Omega_0}$ related to collocation point n can be written as:

$$\mathbf{r}_{0n}^{\Omega_0} = \sum_{n_s=1}^{N_s} \int_{-1}^{1}\int_{-1}^{1}\int_{-1}^{1} \mathsf{U}\left(\tilde{\boldsymbol{x}}_n, \bar{\boldsymbol{x}}(\bar{\xi},\bar{\eta},\bar{\zeta})\right) \dot{\mathbf{b}}_0\left(\bar{\boldsymbol{x}}(\bar{\xi},\bar{\eta},\bar{\zeta})\right) J(\boldsymbol{s})\, J_{\xi}^{n_s}\, d\bar{\xi}d\bar{\eta}d\bar{\zeta} \tag{12.50}$$

where $J(\mathrm{s})$ is the Jacobian of the mapping between $\boldsymbol{s}$ and $\boldsymbol{x}$ coordinate systems.

Applying Gauss integration we have:

$$\mathbf{r}_{0n}^{\Omega_0} \approx \sum_{n_s=1}^{N_s}\sum_{g_s=1}^{G_s}\sum_{g_t=1}^{G_t}\sum_{g_r=1}^{G_r} \mathsf{U}\left(\tilde{\boldsymbol{x}}_n, \bar{\boldsymbol{x}}(\bar{\xi}_{g_s},\bar{\eta}_{g_t},\bar{\zeta}_{g_r})\right) \dot{\mathbf{b}}_0\left(\bar{\boldsymbol{x}}(\bar{\xi}_{g_s},\bar{\eta}_{g_t},\bar{\zeta}_{g_r})\right) J_{g_s g_t g_r}^{n_s} \tag{12.51}$$

with

$$J_{g_s g_t g_r}^{n_s} = J(\boldsymbol{s})\, J_{\xi}^{n_s}\, W_{g_s}\, W_{g_t}\, W_{g_r} \tag{12.52}$$

where N_s is the number of integration regions and G_s, G_t and G_r are the number of integration points in s, t and r directions, respectively. To determine the number of Gauss points necessary for an accurate integration we consider that, whereas there is usually a moderate variation of body force, the Kernel U is $\mathcal{O}(r^{-1})$ so the number of integration points has to be increased if $\boldsymbol{x}_n$ is close to Ω_0.

Singular integration. If the integration region includes the collocation point $\boldsymbol{x}_n$, then the integrand tends to infinity as the point is approached. To deal with the integration involving the weakly singular Kernel we perform the integration in a local coordinate system, where the Jacobian tends to zero as the singularity point is approached. For this we divide the integration region into tetrahedral sub-regions. The transformation from the local $\bar{\boldsymbol{\xi}}$ coordinate system, in which the Gauss coordinates are defined, to global coordinates involves the following transformation steps:

1. from $\bar{\boldsymbol{\xi}}$ to a local system $(\sigma, \tau, \rho)^{\mathrm{T}} = [0, 1]^3$
2. from (σ, τ, ρ) to $\boldsymbol{s}$
3. from $\boldsymbol{s}$ to $\boldsymbol{x}$

Steps 1 and 3 have already been discussed, so we concentrate on explaining the second step. Referring to Fig. 12.8 we assume that the singular point is an edge point of the integration region.

For this case the transformation is as follows: First we determine the local coordinates $\boldsymbol{s}_1$ to $\boldsymbol{s}_5$ of the edge points of the tetrahedron, with 5 being the singularity point. Next we define a linear plane NURBS surface with points 1–4 and map the coordinates of the point (σ, τ) onto this surface:

$$\boldsymbol{s}_0(\sigma, \tau) = \sum_{i=1}^{4} R_i(\sigma, \tau)\; \boldsymbol{s}_i \tag{12.53}$$

where $R_i(\sigma, \tau)$ are linear basis functions. The final map is obtained by interpolation in the ρ-direction:

$$\boldsymbol{s}(\sigma, \tau, \rho) = (1 - \rho)\; \mathbf{s}_0(\sigma, \tau) + \rho\; \boldsymbol{s}_5 \tag{12.54}$$

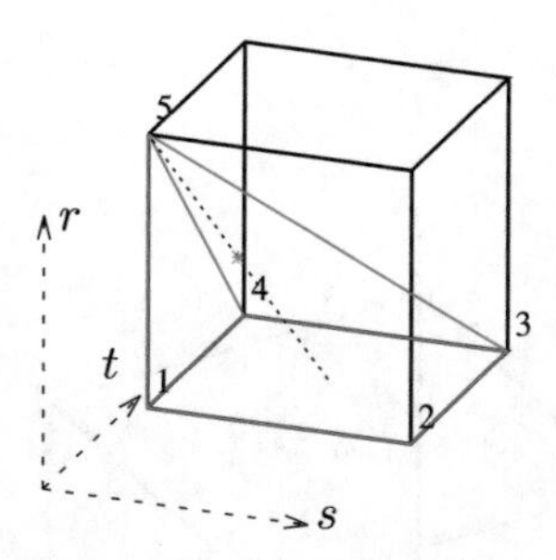

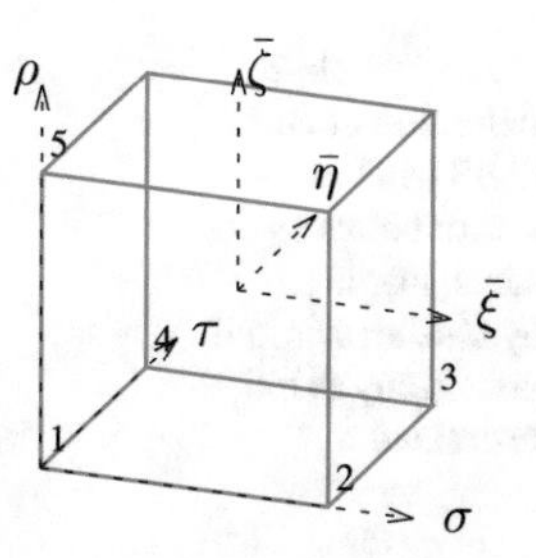

Fig. 12.8 Singular volume integration, showing a tetrahedral subregion of an integration region and the mapping from the s to the σ, τ, ρ coordinate system. A point with the local coordinates $\sigma = \tau = \rho = 0.5$ (i.e. $\bar{\xi} = \bar{\eta} = \bar{\zeta} = 0$) is shown as a red star

The Jacobi matrix of this transformation is given by:

$$\mathbf{J} = \begin{pmatrix} (1-\rho)\frac{\partial \mathbf{s}_0}{\partial \sigma} \\ (1-\rho)\frac{\partial \mathbf{s}_0}{\partial \tau} \\ \mathbf{s}_5 - \mathbf{s}_0 \end{pmatrix} \tag{12.55}$$

The Jacobian of this transformation tends to zero as the singular point ($\rho = 1$) is approached.

12.3.3.4 Test Example

The example tests the algorithm for the case of a single elastic inclusion. It consists of a unit cube $[0, 1]^3$, composed of two different materials. Its geometry is defined by six NURBS surfaces (patches) with basis functions of order 1. The cube is loaded with a moment and fixed at the bottom as shown in Fig. 12.9. The inclusion is defined by two linear NURBS surfaces and assigned a Young's modulus E of half the one used for computing the fundamental solution and no change in the Poisson's ratio ν. For the analysis, the concept of a geometry independent field approximation was used and the basis functions for approximating the displacements were defined using the knot vectors as shown in Table 12.4.

This approximation results in a quadratic variation of the displacements in the vertical direction with a C^0-continuity at the interface between materials. The resulting location of the collocation points are shown in Fig. 12.10 and this will give the exact solution for the applied loading. The results are shown in Figs. 12.11 and 12.12. It can be seen that convergence to the exact solution is achieved after about 7 iterations.

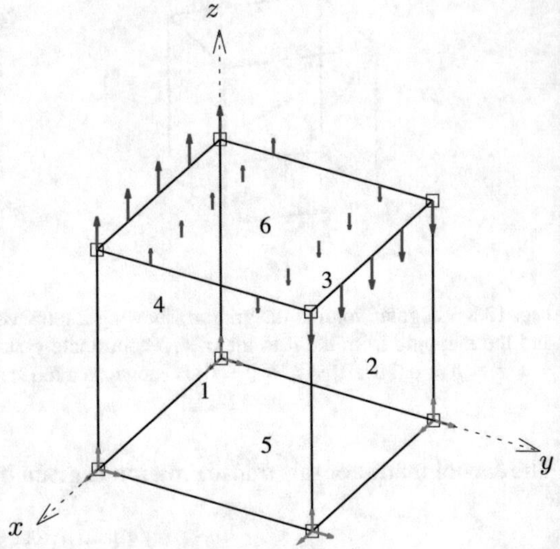

Fig. 12.9 Test example 2: geometry definition of cube with 6 NURBS patches showing control points as hollow squares, loading indicated by blue arrows and boundary conditions as red arrows in restrained directions

Table 12.4 Knot vectors for the approximation of the displacements

Patch	Ξ_1	Ξ_2
1	0, 0, 1, 1	0, 0, 0, 0.5, 0.5, 1, 1, 1
2	0, 0, 1, 1	0, 0, 0, 0.5, 0.5, 1, 1, 1
3	0, 0, 1, 1	0, 0, 0, 0.5, 0.5, 1, 1, 1
4	0, 0, 1, 1	0, 0, 0, 0.5, 0.5, 1, 1, 1
5	0, 0, 1, 1	0, 0, 1, 1
6	0, 0, 1, 1	0, 0, 1, 1

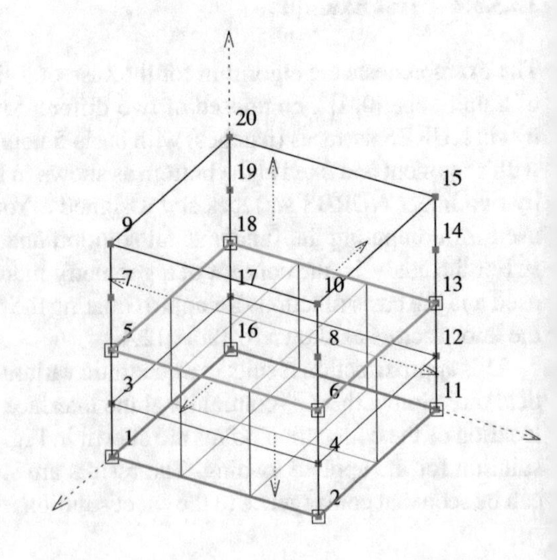

Fig. 12.10 Test example 1: definition of inclusion with 2 linear NURBS surfaces shown in green and red with the associated control points marked with hollow squares. Collocation points are numbered and shown as filled squares

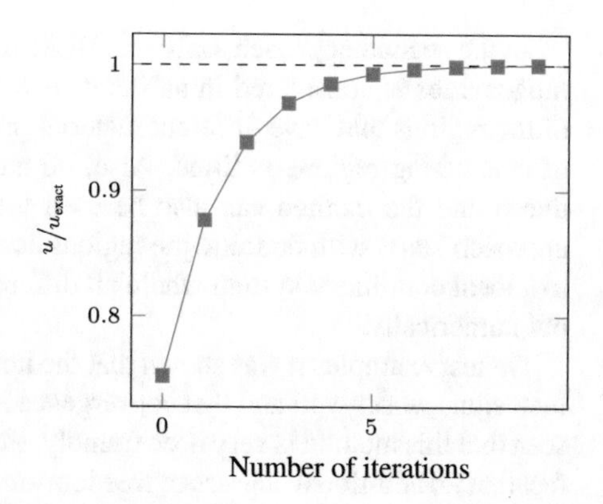

Fig. 12.11 Test example 1: plot of ratio of the maximum computed displacement to the exact one as function of the number of iterations

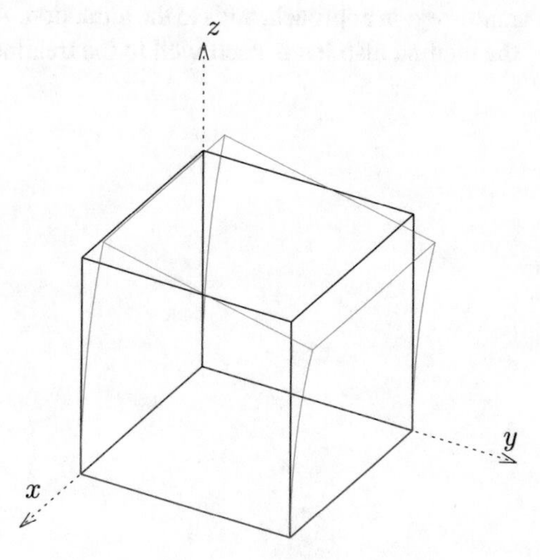

Fig. 12.12 Test example 1: deformed shape

12.4 Summary and Conclusions

In this chapter, we have extended the capabilities of the BEM to allow simulations with heterogeneous domains. While it is theoretically possible to solve for a general heterogeneous material, we concentrated here on solving piecewise heterogeneous problems, i.e. where parts of the domain have material properties different from the ones used for computing the Kernels.

Two methods were presented. In the multi-region approach two or more boundary element regions with different material properties were assembled much in the same way as in the FEM, by enforcing compatibility and continuity/equilibrium conditions at the region interfaces. The disadvantage of this method is that a greater effort has to be spent in mesh generation (especially in 3-D) when there are a great number of regions, since regions have to be properly connected. Also the number of unknowns increases with the number of BEM regions.

In the second approach, only one BEM region is used and the influence of inhomogeneities is considered in an iterative way. The advantage is that the definition of the regions that have different material properties is easier since the requirement of connecting regions is lifted. Also, no additional degrees of freedom are introduced and the method can also be used for modelling non-linear behaviour. The approach starts with defining the regions as outlined in Chap. 11 and mapping them to a local coordinate system where all differentiations or integrations can be carried out numerically.

On test examples it was shown that the iterative method of solving problems with inclusions works well and that convergence is achieved with few iterations. We have seen that this method is very user-friendly, since no mesh generation is involved. The fact that no additional degrees of freedom are introduced, as would be the case for the multi-region approach, adds to the attraction. As we will see in the following chapter, the method also lends itself well to the treatment of non-linear material behaviour.

Chapter 13
Material Non-linear Behaviour

13.1 Introduction

The boundary integral equations presented so far can only deal with linear material behaviour, which means that there must be a linear relationship between potential and flow or stress and strain. If this is not the case, we can obtain a solution but we must consider the effect of body forces and consider volume integrals. Therefore the first step in dealing with non-linear material behaviour is to define the domain, where non-linear effects take place, Ω_0, as an inclusion, as outlined in Chap. 11.

The general approach is then to solve the problem iteratively using a modified Newton–Raphson method. The implementation is shown in Algorithm 7. In this method, the left-hand side is unchanged during the iteration. While it is possible to implement a full Newton–Raphson method, where the left-hand side is changed during the iteration, this is rather complicated. More details about the implementation of the full Newton–Raphson method can be found in [65].

Algorithm 7 Algorithm for modified Newton–Raphson method

1: Linear solution: $\mathbf{L}\dot{\mathbf{x}}_0 = \mathbf{r}$
2: **for** i **to** Max. number iterations **do**
3: Determine initial flow $\mathbf{q}^0$ or initial stress $\{\sigma\}^0$
4: Compute new right-hand side $\mathbf{r}_0^i$
5: Solve $\mathbf{L}\dot{\mathbf{x}}_i = \mathbf{r}_0^i$
6: Accumulate $\mathbf{x}_i = \mathbf{x}_{i-1} + \dot{\mathbf{x}}_i$
7: Compute Residual
8: **if** Residual < Tolerance **then**
9: Exit
10: **end if**
11: **end for**
12: **return** $\mathbf{x}$

G. Beer et al., *The Isogeometric Boundary Element Method*, Lecture Notes in Applied and Computational Mechanics 90,
https://doi.org/10.1007/978-3-030-23339-6_13

13.2 Computation of Initial Flow/Stress

The initial flow/stress is computed from a non-linear relationship

$$\mathbf{q}^0 = \Phi(\mathbf{q}, a, b \ldots) \tag{13.1}$$

$$\{\boldsymbol{\sigma}\}^0 = \Phi(\{\boldsymbol{\sigma}\}, a, b \ldots) \tag{13.2}$$

where Φ is a function of $\mathbf{q}$ or $\{\boldsymbol{\sigma}\}$ and a, b are parameters. Here, we concentrate on solid mechanics problems, i.e. explain how the initial stress is computed. The procedures described here, however, can also be used for non-linear potential problems. For more details see also [17, 19].

13.2.1 Solid Mechanics

13.2.1.1 Elasto-Plasticity

We define a yield function F that determines when the stress state ceases to be elastic. There are two possibilities:

$$F(\{\boldsymbol{\sigma}\}) < 0 \quad \text{Elastic} \tag{13.3}$$

$$F(\{\boldsymbol{\sigma}\}) = 0 \quad \text{Elasto-plastic.} \tag{13.4}$$

States with $F(\{\boldsymbol{\sigma}\}) > 0$ are not allowed. When the elasto-plastic state is reached the total strain is composed of an elastic strain $\{\boldsymbol{\epsilon}\}^e$ and a plastic strain $\{\boldsymbol{\epsilon}\}^p$. To compute the increment in plastic strain, we require a flow-law $Q(\{\boldsymbol{\sigma}\})$. The increment in plastic strain is given by:

$$d\{\boldsymbol{\epsilon}\}^p = d\epsilon^p \, \frac{\partial Q}{\partial \{\boldsymbol{\sigma}\}} \tag{13.5}$$

where $d\epsilon^p$ is an equivalent plastic strain. In the elastic state, the linear relationship between increments of stress and strain is

$$d\{\boldsymbol{\sigma}\} = \mathbf{D}\, d\{\boldsymbol{\epsilon}\} \tag{13.6}$$

$$d\{\boldsymbol{\epsilon}\} = \mathbf{D}^{-1}\, d\{\boldsymbol{\sigma}\} \tag{13.7}$$

where $\mathbf{D}$ is the constitutive matrix. For perfect plasticity the equivalent plastic strain can be computed from the condition that once the elastic limit has been reached no increase in $F(\{\boldsymbol{\sigma}\})$ is possible. The increment in total strain at yield ($F(\{\boldsymbol{\sigma}\}) = 0$) can be divided into an elastic and plastic part

$$d\{\boldsymbol{\epsilon}\} = d\{\boldsymbol{\epsilon}\}^e + d\{\boldsymbol{\epsilon}\}^p = \mathbf{D}^{-1} d\{\boldsymbol{\sigma}\} + \frac{\partial Q}{\partial \{\boldsymbol{\sigma}\}}\, d\epsilon^p. \tag{13.8}$$

Using a first order Taylor expansion we have the condition

$$dF = \left(\frac{\partial F}{\partial\{\boldsymbol{\sigma}\}}\right)^{\mathrm{T}} d\{\boldsymbol{\sigma}\} = 0. \tag{13.9}$$

Combining Eqs. (13.8) and (13.9), we can write the following system of equations

$$\begin{pmatrix} d\{\epsilon\} \\ 0 \end{pmatrix} = \begin{pmatrix} \mathbf{D}^{-1} & \frac{\partial Q}{\partial\{\boldsymbol{\sigma}\}} \\ \left(\frac{\partial F}{\partial\{\boldsymbol{\sigma}\}}\right)^{\mathrm{T}} & 0 \end{pmatrix} \begin{pmatrix} d\{\boldsymbol{\sigma}\} \\ d\epsilon^{p} \end{pmatrix}. \tag{13.10}$$

Multiplying the first equation with $\mathbf{D}\left(\frac{\partial F}{\partial\{\boldsymbol{\sigma}\}}\right)^{\mathrm{T}}$ and subtracting it form the second equation we can solve for $d\epsilon^{p}$

$$d\epsilon^{p} = \frac{1}{C}\mathbf{D}\left(\frac{\partial F}{\partial\{\boldsymbol{\sigma}\}}\right)^{\mathrm{T}} d\{\epsilon\} \tag{13.11}$$

where $C = \left(\frac{\partial F}{\partial\{\boldsymbol{\sigma}\}}\right)^{\mathrm{T}} \mathbf{D}\frac{\partial Q}{\partial\{\boldsymbol{\sigma}\}}$.

Substitution of this result into the first equation of (13.10) yields

$$d\{\boldsymbol{\sigma}\} = \mathbf{D}_{ep} d\{\epsilon\} \tag{13.12}$$

where

$$\mathbf{D}_{ep} = \mathbf{D} - \frac{1}{C}\mathbf{D}\frac{\partial Q}{\partial\{\boldsymbol{\sigma}\}}\left(\frac{\partial F}{\partial\{\boldsymbol{\sigma}\}}\right)^{\mathrm{T}}\mathbf{D}. \tag{13.13}$$

The initial stress is the difference between the elastic stress and the non-linear stress just computed

$$\{\boldsymbol{\sigma}\}^{0} = (\mathbf{D} - \mathbf{D}_{ep})d\{\epsilon\}. \tag{13.14}$$

In practice the situation where the yield condition will be satisfied everywhere will not be achieved after an iteration, so to avoid drift and enforce the condition $F(\{\boldsymbol{\sigma}\}) = 0$, special procedures termed "return algorithms" have to be implemented. It is beyond the aim of this book to elaborate on this. Readers are referred to [171] for a detailed discussion.

13.2.1.2 Visco-Plasticity

In visco-plasticity [130], we allow (temporarily) a condition that $F(\{\boldsymbol{\sigma}\}) > 0$ and specify a visco-plastic strain rate as

$$\frac{\partial\{\epsilon\}^{vp}}{\partial t} = \frac{1}{\eta}\Phi(F)\frac{\partial Q}{\partial\{\sigma\}} \tag{13.15}$$

where η is a viscosity parameter, F is the yield function, Q the plastic potential and

$$\Phi(F) = 0 \quad \text{for } F < 0 \tag{13.16}$$
$$\Phi(F) = F \quad \text{for } F > 0. \tag{13.17}$$

The visco-plastic strain increment during a time increment Δt can be computed by an explicit time integration scheme

$$\{\dot{\epsilon}\}^{vp} = \frac{\partial\{\epsilon\}^{vp}}{\partial t} \cdot \Delta t \tag{13.18}$$

where the time increment Δt cannot be chosen freely and if chosen too large, oscillatory behaviour will occur in the solution (for a suitable time step size see [45]). The initial stress increment is given by

$$\{\dot{\sigma}\}^0 = \mathbf{D} \cdot \{\dot{\epsilon}\}^{vp}. \tag{13.19}$$

As the iteration converges the state $F(\{\sigma\}) = 0$ will be approached.

13.2.1.3 Yield Conditions and Flow Laws

Here we mention only three yield conditions that are used in this book.

Mohr–Coulomb. This yield function is for soil and rock and is defined as [48]

$$F(\{\sigma\}) = \frac{\sigma_1 + \sigma_2}{2}\sin\phi - \frac{\sigma_1 - \sigma_3}{2} - c\,\cos\phi \tag{13.20}$$

where $\sigma_1, \sigma_2, \sigma_3$ are principal stresses, ϕ is the friction angle and c the cohesion. The flow-law is

$$Q(\{\sigma\}) = \frac{\sigma_1 + \sigma_2}{2}\sin\psi \tag{13.21}$$

where ψ is the dilation angle.

Von Mises. This yield function is used for metals and is defined as

$$F(\{\sigma\}) = \sigma_{eq} - \sigma_Y \tag{13.22}$$

where σ_Y is the yield stress and

$$\sigma_{eq} = \sqrt{\sigma_1\,\sigma_1 + \sigma_2\,\sigma_3 + \sigma_3\,\sigma_1}. \tag{13.23}$$

Limited stress. This yield function limits the maximum stress and is given by

$$F(\{\sigma\}) = \sigma_{max} - \sigma_c \tag{13.24}$$

where σ_{max} is the maximum stress and σ_c is the allowed stress.

13.3 Evaluation of the Volume Integrals

To be able to evaluate the volume integrals, we need to define the domain Ω_0 where initial stresses are generated. This is done by defining inclusions as outlined in Chap. 11. Of course, the extent of a plastic zone is not known beforehand. For finite domain problems it is possible to define the whole domain as an inclusion but this may not be the most efficient way. For infinite domains this is not possible. However, plasticity usually starts from the boundary and if not, from an area of weaker material. So the obvious choice is to define Ω_0 in such a way that it extends a certain way from the boundary. If during the iterations it is determined that the plasticity zone would extend beyond the specified domain, it is possible to extend the zone by redefining the bounding curves or surfaces. If the plasticity starts at a zone of weaker material there will be an inclusion defined there in any case.

The solution proceeds exactly as explained in Chap. 12, except that the initial stresses are now computed using Eq. (13.14) or (13.19).

13.4 Plane Problems

13.4.1 Test Example 1: Cube with Inelastic Inclusion

The test geometry is the same as the one with an elastic inclusion in the previous chapter and shown in Fig. 13.1.

A limited stress material law was implemented, that restricts the maximum normal stress to 80% of the maximum elastic stress. At the end of the iteration, it is checked that the limited stress yield condition is satisfied everywhere. An indication of convergence is that the internal moment is equal to the external moment. Figure 13.2 shows the convergence of the iteration based on this criterion after 8 iterations. Figure 13.3 shows the history of stress changes.

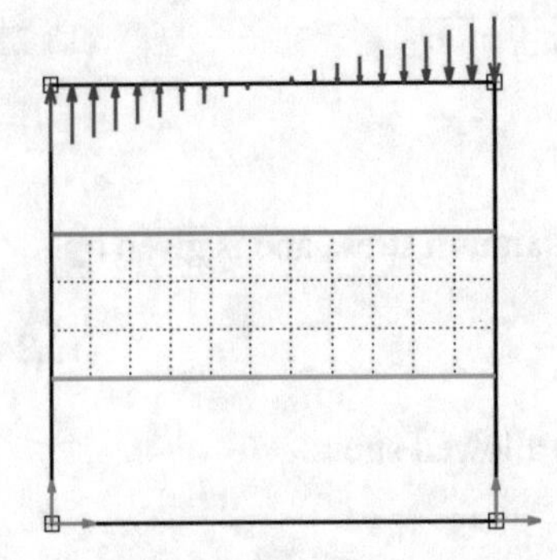

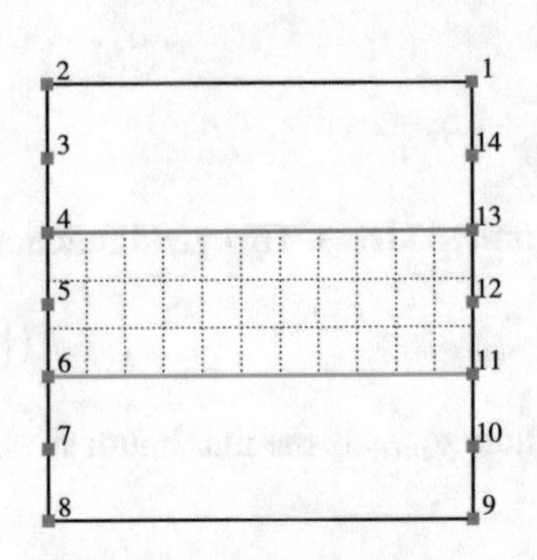

Fig. 13.1 Test example 1: left: geometry definition with 4 NURBS patches and control points, loading and boundary conditions. The inclusion is defined with 2 colour coded NURBS; right: collocation points

Fig. 13.2 Test example 1: plot of the ratio internal and external moment versus number of iterations

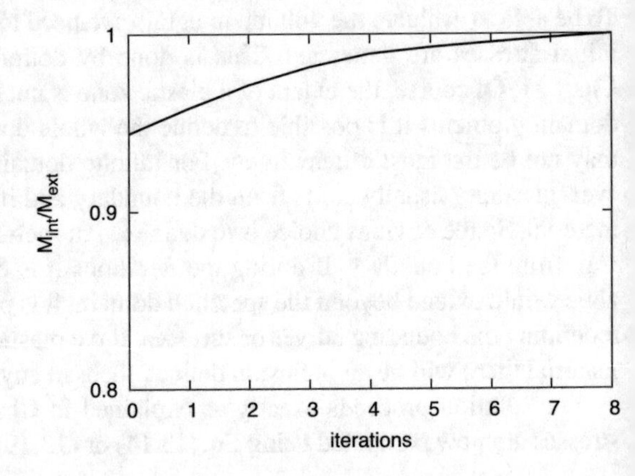

Fig. 13.3 Test example 1: change in the distribution of normal stress on the left half of the cube during iterations. Initial (elastic) stress state in grey, final stress state in black

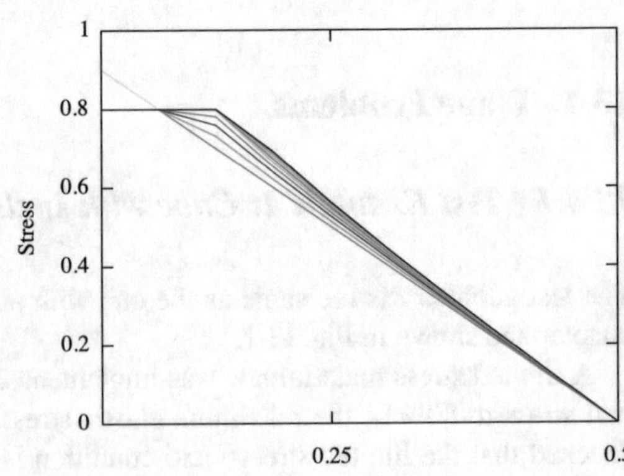

13.4.2 Test Example 2: Hole in an Infinite Domain

For the previous example $\dot{\mathbf{b}}_0$ was zero everywhere because the derivatives of the initial stress were zero and therefore the term associated with the integration over the domain was zero. Test example 2 is designed to test the volume integration. It is a hole in an infinite domain with an inelastic inclusion on top of it.

The geometry of the problem is shown in Fig. 13.4. The hole is defined by two NURBS curves as shown in Table 13.1. This exactly describes the geometry of a circle. The inclusion was defined by two NURBS curves as shown in Table 13.2.

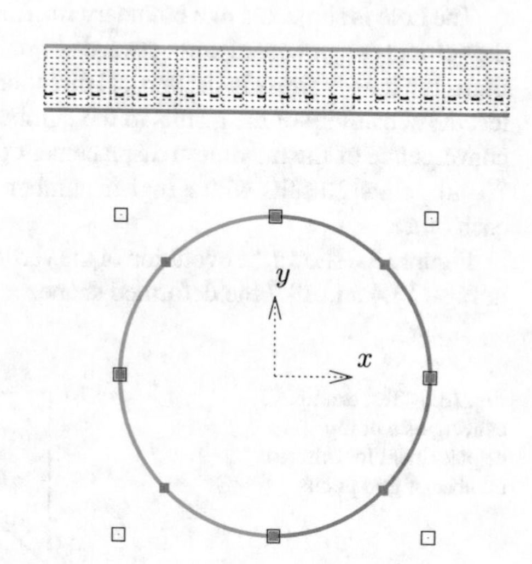

Fig. 13.4 Test example 2: the geometry of the circle is defined with two NURBS patches. Control points are shown as hollow squares, collocation points as filled red squares. The inclusion above the hole is defined by two linear NURBS curves. Also shown is the line (dashed) along which the stresses are plotted

Table 13.1 Geometrical definition of circular excavation

Patch	Knot vector	Control point	x	y	z	Weight
1	0, 0, 0.5, 0.5, 1, 1	1	0	0.5	0	1
.	...	2	0.5	0.5	0	0.707
.	...	3	0.5	0	0	1
.	...	4	0.5	–0.5	0	0.707
.	...	5	0	–0.5	0	1
2	0, 0, 0.5, 0.5, 1, 1	1	0	–0.5	0	1
.	...	2	–0.5	–0.5	0	0.707
.	...	3	–0.5	0	0	1
.	...	4	–0.5	0.5	0	0.707
.	...	5	0	0.5	0	1

Table 13.2 Geometrical definition of the nonlinear inclusion

Patch	Knot vector	Control point	x	y	z	Weight
1	0, 0, 1, 1	1	–0.25	0.75	0	1
.	...	2	1.25	0.75	0	1
2	0, 0, 1, 1	1	–0.25	0.95	0	1
.	...	2	1.25	0.95	0	1

The hole is subjected to a boundary traction along Γ of $\mathbf{t} = (0, n_2)^{\mathrm{T}}$, where n_2 is the vertical component of unit outward normal to Γ. The limited stress yield condition limits the tensile stress in the vertical direction to 0.5. This example was also used to test the sensitivity of the results to the number of grid points. Figure 13.5 shows the convergence of the maximum displacement (at the top of the circle) for the case of 20–40 points. Results with a higher number of points were indistinguishable from each other.

Figure 13.6 shows the evolution of the vertical stress along the dashed line depicted in Figs. 13.4 and 13.7 the deformed shape.

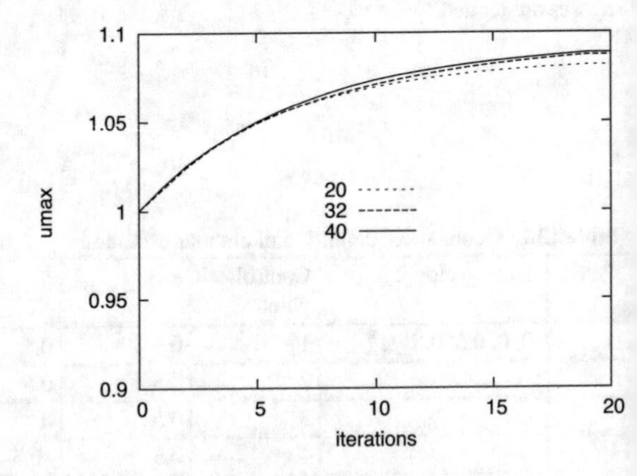

Fig. 13.5 Test example 2: convergence of top displacement for different number of grid points

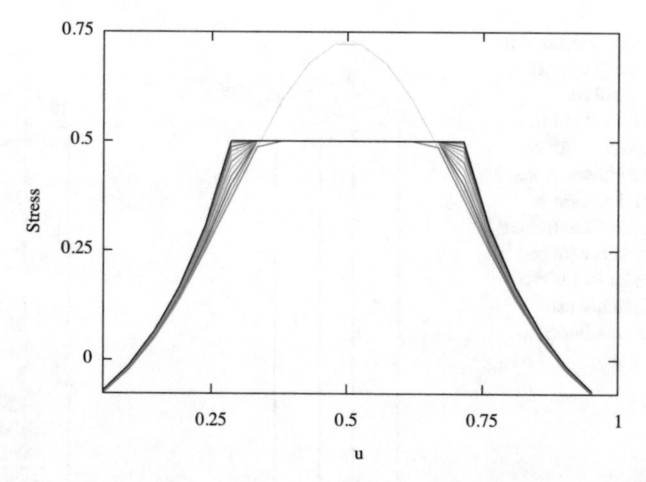

Fig. 13.6 Test example 2: evolution of vertical stress during iteration. Grey line shows the initial elastic stress and black line the final stress

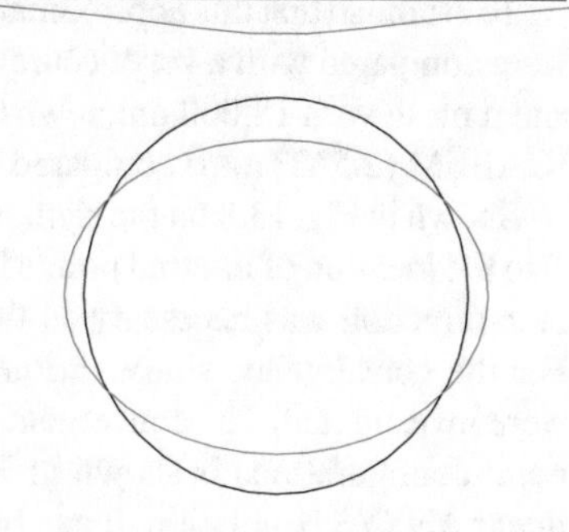

Fig. 13.7 Test example 2: deformed shape

13.5 3-D Problems

13.5.1 Test Example: Cantilever Beam with an Inelastic Inclusion

The example tests the capability of the method to simulate inelastic material behaviour in 3-D. It relates to the analysis of a cantilever beam of dimension $1 \times 1 \times 5\,\mathrm{m}$ with an equally distributed load of $0.049\,\mathrm{kN/m^2}$ at the end. The elastic modulus was assumed to be $10\,\mathrm{kPa}$ with a zero Poisson's ratio. A *von Mises* material law was used with a yield stress of $1\,\mathrm{kPa}$. The geometry is defined by 8 control points and 6 patches (Fig. 13.8 left). Figure 13.8 right shows the collocation points obtained after order elevating two times (from linear to cubic) in the z-direction and one time (from linear to quadratic) in the y-direction. The simulation has $24 \times 3 = 72$ unknowns.

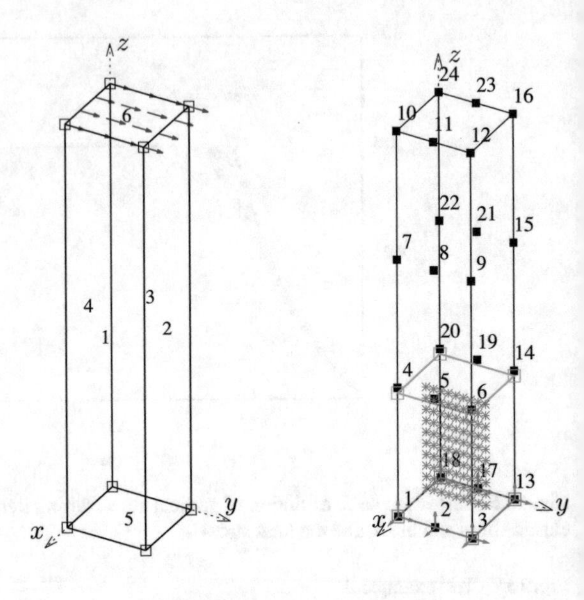

Fig. 13.8 Test example: left: definition of the geometry with 6 linear NURBS patches (associated control points are shown as hollow squares). Also shown is the loading; right: location of collocation points (restrained points are marked with red arrows). Also shown is the definition of the inelastic inclusion with the internal points

To establish that this approximation of the unknowns is adequate, the elastic results were compared with a very accurate FEM result using the commercial code ANSYS and a mesh with 18,000 unknowns. The elastic maximum displacement obtained by IGABEM (2.5223 mm) compared well with the ANSYS solution (2.5235 mm).

Shown in Fig. 13.8 on the right is the definition of the inelastic inclusion, depicting also the location of internal points for one of the test runs. For this case only one point in x-direction was necessary as the variation of stress is constant in this direction. For the convergence study various combinations of the number of internal points were investigated. The convergence of the displacement with three different internal point configurations is shown in Fig. 13.9 and compared with the results of a non-linear ANSYS simulation. It can be seen that the run with $1 \times 9 \times 10$ internal points (i.e. one in x-direction 9 in y-direction and 10 in z-direction) gives the best agreement with the ANSYS results.

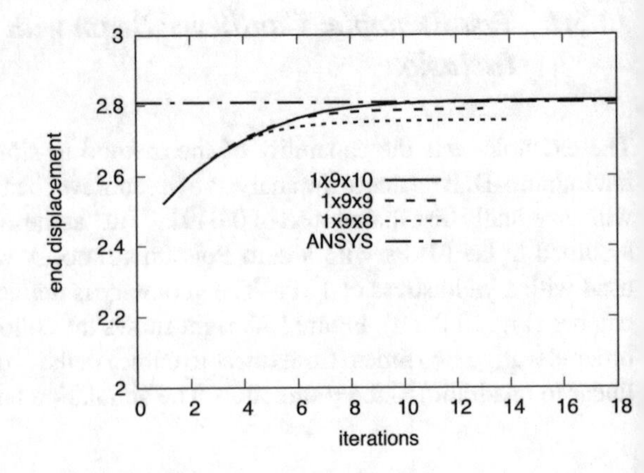

Fig. 13.9 Test example: convergence of the maximum displacement as a function of the number of iterations for three configurations of internal points. The results are compared with an ANSYS finite element simulation

13.6 Summary and Conclusions

Here we have shown how the capability of the BEM can be extended to deal with material non-linear problems. This has been done by considering body forces, i.e. the initial stresses inside the domain. This required the definition of the volume where these non-linear effects occur. Of course one does not know this beforehand but one can make either an educated guess or start with a small domain near the boundary, which is then extended during the iterations. The definition of the non-linear domain is quite different from established methods, because inclusions are not defined using a cell mesh but by 2 bounding curves or surfaces.

Here we have discussed only a limited range of material non-linear behaviour, namely elasto-plasticity with only a few yield conditions. The method of solution presented here, however is generally applicable to a wide range of material non-linear problems.

The method for dealing with material non-linear problems can be combined with the one presented in the previous chapter, dealing with inclusions that have different linear properties. Indeed, the only difference is that two types of initial stresses are added, one stemming from the difference in elastic properties and one from the non-linear behaviour. Care has to be taken in the iterative procedure as the change in linear properties influences the computation of the stresses and therefore the yield function. If the linear material properties of the inclusion are very different from the properties of the domain the iteration may not converge to the right solution. One remedy to this would be to apply the loading in small increments.

Having expanded the range of applications of the IGABEM, we now turn our attention to solving real problems. This will be in our area of expertise, geomechanics.

Chapter 14
Applications in Geomechanics

14.1 Introduction

The BEM is ideally suited for applications in geomechanics because it can handle infinite domain problems without truncation. Furthermore, the IGABEM is able to describe smooth excavation geometries with few parameters. Design geometries for tunnels, for example, are usually specified using arcs which can be exactly represented with NURBS. Geological features are often described in a CAD database. Since we propose to use NURBS to specify them as inclusions there is a possibility to take geological information directly from CAD data. In the following, we will first explain how to simulate excavation processes, how to define geometries of underground excavations and then proceed to the definition of geological inclusions. Finally, some examples are presented. Due to the expertise of the authors, we focus on underground excavations (caverns and tunnels). For more details the reader is referred to [13, 15, 115].

14.2 Simulation of the Excavation Process

In the examples, we will deal with the simulation of excavations. This means that we simulate the removal of material in a ground that is prestressed with a *virgin stress* and we are seeking the response of the ground. To get the final result, we must consider two stages:

1. Virgin stage (before the excavation)
2. Change in ground conditions due to excavation.

The virgin ground conditions can be simulated in two ways. One can calculate the virgin ground conditions by switching on gravity or specify the virgin stresses directly. The first approach, also known as "gravity switch on", requires an additional simulation and suffers from the fact that it does not consider tectonic effects that may

G. Beer et al., *The Isogeometric Boundary Element Method*, Lecture Notes in Applied and Computational Mechanics 90,
https://doi.org/10.1007/978-3-030-23339-6_14

change the gravity-induced stresses considerably. In one of the examples presented here (based on a real project) the horizontal stress components were considerably higher than would have been determined by a "gravity switch on" analysis. In the second approach one usually specifies a gravitational vertical virgin stress and a k_0 factor that determines the other stress components, based on on-site measurements.

As material is removed to create the excavation, the tractions on the excavated boundary are gradually reduced to zero. In the examples presented here, however, we assume that the reduction to zero is instantaneous. To achieve a traction-free excavation surface we have to apply the tractions, that existed before excavation, in the opposite direction. The tractions due to the virgin stresses σ_{ij}^{v} can be computed by

$$t_i = \sigma_{ij}^{\mathrm{v}}\, n_j. \tag{14.1}$$

This is then applied in the negative sense as a loading and an increment in displacement and stress $\dot{\boldsymbol{\sigma}}$ is obtained. The final stress distribution in the ground is an addition of the two stages:

$$\boldsymbol{\sigma} = \dot{\boldsymbol{\sigma}} + \boldsymbol{\sigma}^{\mathrm{v}} \tag{14.2}$$

Usually, we are only interested in the incremental displacement due to the removal of material. It should be noted that in most simulations the gradual removal of material is approximated and in the simplest case the excavation is assumed to be made in one stage as it is done for the examples presented here. However, it is possible to simulate the excavation in stages, with ground support being introduced at each stage. Simulation of staged excavation may have a profound effect on the results, if the material behaviour is non-linear and if ground support is introduced after each stage. For more details on modelling sequential excavation with the BEM see [11, 12].

14.3 Simulation of Ground Support

Excavations, especially tunnels, require support. This can be either a shell of sprayed concrete (shotcrete) or rock bolts or jet-grouted columns. Ground support can be modelled as an inclusion. In the case of rock bolts they are treated as linear inclusions with a circular cross-section as explained in Chap. 11. For more details on modelling rock bolts, the reader is referred to [14, 145, 146]. In the case of shotcrete, there is also a possibility to couple with shell finite elements [57, 58, 137, 144].

14.4 Geometry Definition

In order to avoid stress concentrations in underground excavations, it is very important to design a smooth excavation geometry without kinks. This can be best achieved by specifying arcs in such a way that there is a C^1-continuity between them.

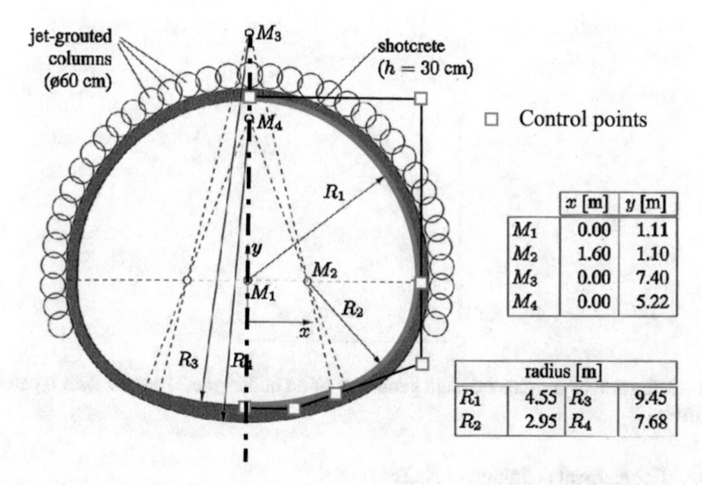

	x [m]	y [m]
M_1	0.00	1.11
M_2	1.60	1.10
M_3	0.00	7.40
M_4	0.00	5.22

radius [m]			
R_1	4.55	R_3	9.45
R_2	2.95	R_4	7.68

Fig. 14.1 Example of geometry description with NURBS: NATM tunnel definition as designed and description of right half with a NURBS patch of order 2 (in red), showing control points and control polygon

14.4.1 Tunnels

As an example we show the description of the geometry of the cross-section of an NATM[1] tunnel, where the design shape is specified by arcs (centre, radius and extent) as shown in Fig. 14.1. One half of the tunnel can be described with 1 NURBS patch of order 2 and only 7 control points.

The control point coordinates and weights can be computed from arc centres, radii and start/end angles using a simple formula (see Fig. 4.2 and [9]) and are shown in Table 14.1.

This describes exactly the design geometry. Note that, even though double knots appear in the knot vector, C^1-continuity is guaranteed due to the location of the control points as discussed in Sect. 4.1.2.

14.4.2 Caverns

The accurate description of the excavation geometry of caverns can also be easily achieved with NURBS as shown in Fig. 14.2 where the design geometry of an underground cavern is described by 6 patches and 9 control points.

[1] NATM stands for New Austrian Tunnelling Method, which is actually quite old. In this excavation method the installation of ground support is delayed to allow deformation with the aim that a ground arch develops. In other sense, the ground surrounding the tunnel is asked to help support it.

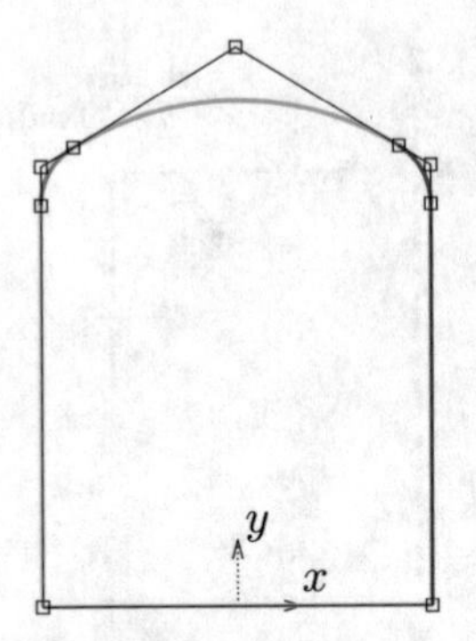

Fig. 14.2 Definition of the exact design geometry of an underground cavern with 6 patches and 9 control points

Table 14.1 Geometrical definition of NATM tunnel

Patch	Knot vector	Control point	x	y	z	Weight
1	0, 0, 0, 0.5, 0.5, 0.8, 0.8, 1, 1, 1	1	0	5.65	0	1
.	...	2	4.55	5.65	0	0.707
.	...	3	4.55	1.1	0	1
.	...	4	4.55	–0.97	0	0.82
.	...	5	2.61	–1.67	0	1
.	...	6	1.33	–2.04	0	0.99
.	...	7	0	–2.04	0	1

14.5 Definition of Geology

Geological features are defined as inclusions as outlined in Chap. 11. Figure 14.3 shows the geological conditions at an actual project where an underground cavern is excavated, the simulation of which is discussed later. Also shown is the final excavation stage. The geology consists mainly of siltstone and mudstone, which is the weaker of the two materials and their consideration is considered important in the simulation. Therefore the zones with mudstone were defined as inclusions. In Fig. 14.4 we show the definition of the inclusions by bounding curves. As explained in Chap. 12, it is important that at the intersection of the excavation boundary with the bounding curves defining the inclusion, the continuity of the displacements there must be changed to C^0.

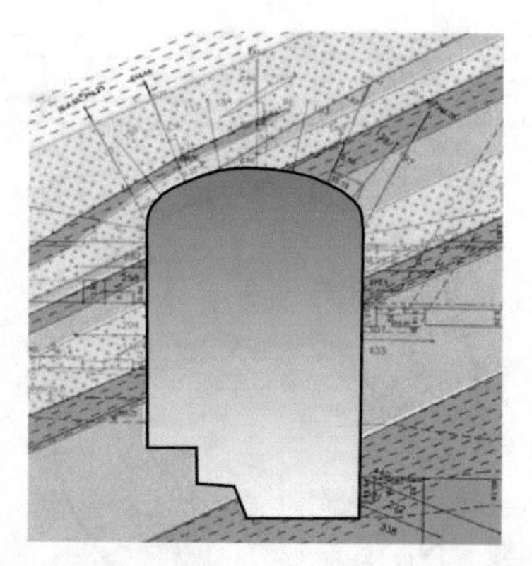

Fig. 14.3 Geological conditions for an underground cavern. The purple areas define mudstone

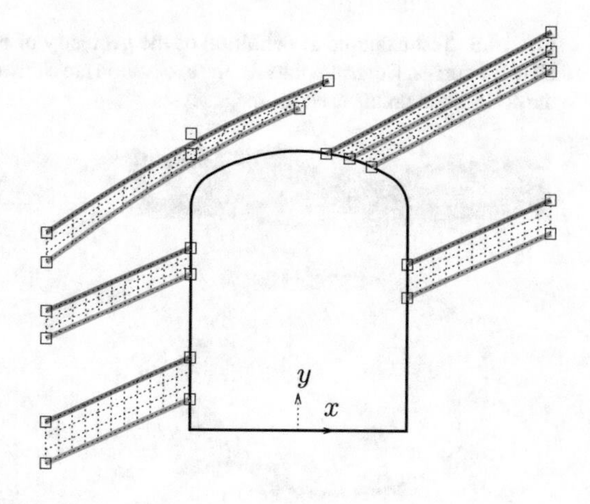

Fig. 14.4 Geological inclusions defined by (colour coded) bounding curves. Associated control points are shown as squares. One geological feature on the top right was defined by 2 inclusions in order to better follow the excavation boundary

14.6 Test Examples

In the following, we show some simple examples that allow the verification of the implementation.

14.6.1 Example 1

Here we test the capability of the method to simulate geological inclusions with non-linear material properties. The example is that of a circular excavation in an infinite domain with the following properties: E = 1 kPa, $\nu = 0$, virgin stress $\sigma_x^{\mathrm{v}} = 0, \sigma_y^{\mathrm{v}} = -1, \tau_{xy}^{\mathrm{v}} = 0$ kPa. The geological inclusion above the excavation is assigned a Mohr–Coulomb yield condition with an angle of friction $\phi = 10°$, cohesion $c = 0$, and dilation angle $\psi = 0$. The description of the geometry is shown in Fig. 14.5. The

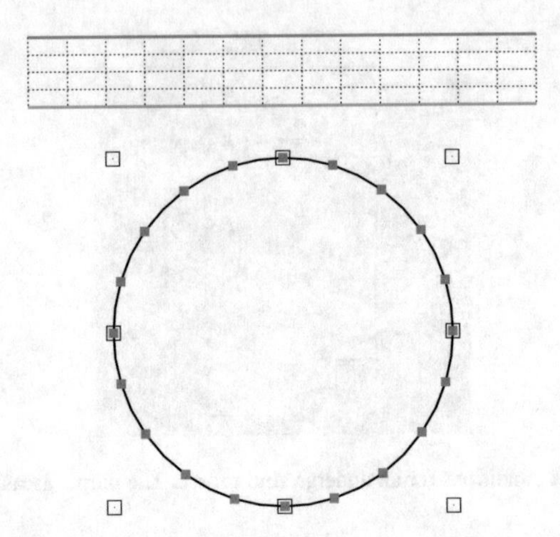

Fig. 14.5 Test example 1: definition of the geometry of the excavation and the inclusion with 2 NURBS curves. Control points for the excavation are shown as hollow squares, red squares indicate the collocation points used for the analysis

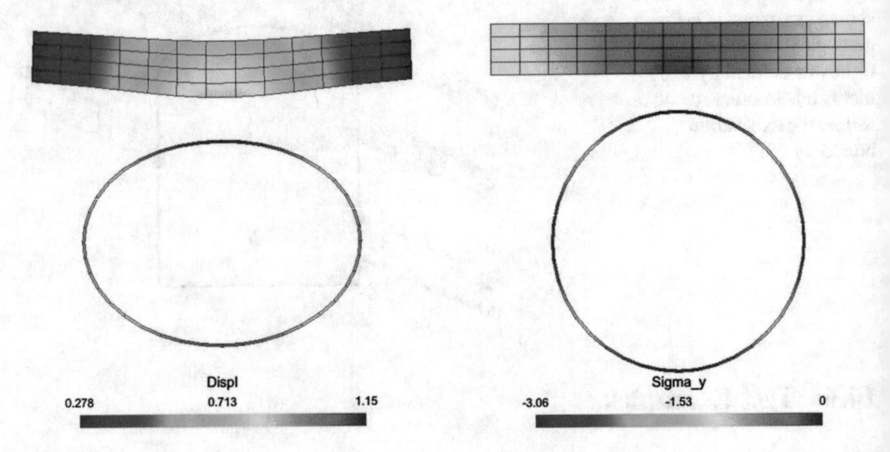

Fig. 14.6 Results of the simulation. Left: deformed shape; Right: vertical stress

solution was achieved by order elevating the basis functions used for the description of the geometry three times resulting in the collocation points in Fig. 14.5. The results of the simulation are shown in Fig. 14.6. Figure 14.7 shows that the simulation converges after 8 iterations and that the end result compares well with a coupled Boundary/Finite element analysis using the software BEFE [14].

14.6.2 Test Example 2

This example is designed to test the 3-D implementation with geological inclusions that intersect the boundary. The test setup is shown in Fig. 14.8.

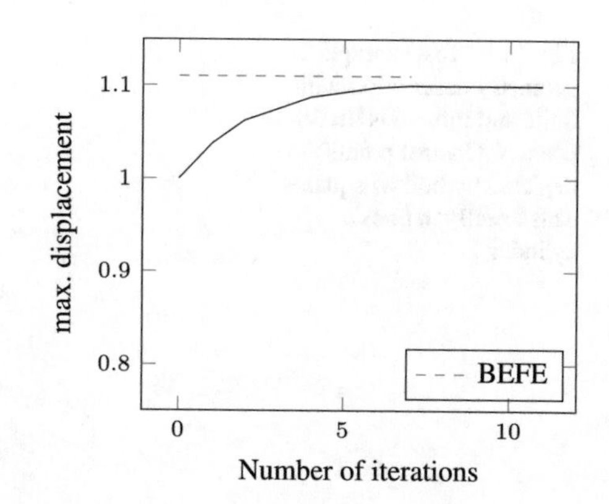

Fig. 14.7 Example 1: convergence of maximum displacement as a function of the number of iterations

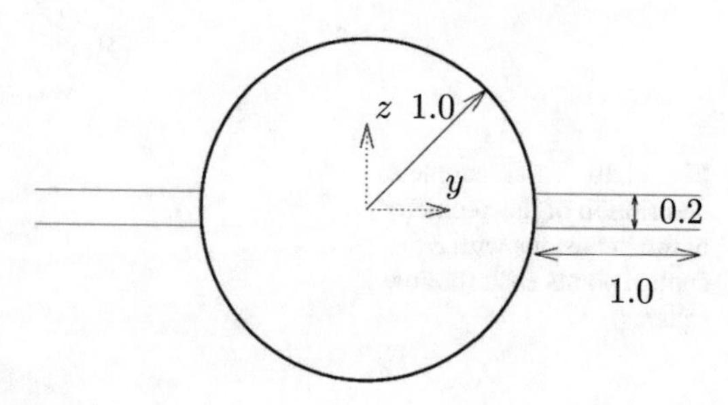

Fig. 14.8 Test example 2: cross-section of the 3-D geometry of the excavation with an inclusion depicted by red lines

It relates to an infinite cylindrical excavation in an infinite domain excavated under a virgin stress field of $\sigma_x^{\mathrm{v}} = \sigma_y^{\mathrm{v}} = -0.5, \sigma_z^{\mathrm{v}} = -0.5, \tau_{xz}^{\mathrm{v}} = -1$ MPa, all other shear stresses being zero. It is assumed that the material of the inclusion obeys the Mohr–Coulomb yield condition with an angle of friction $\phi = 30°$ and a cohesion of $c = 0$. A non-associate flow-law is used with a dilation angle $\psi = 0°$. The diameter of the excavation is 2 m and the modulus of elasticity of the domain is $E = 10$ MPa. The horizontal inclusion has the same modulus of elasticity and extends to 1 m each side of the excavation. The description of the geometry of this problem using NURBS is shown in Fig. 14.9 and consists of 4 finite patches with order 2 in the finite direction and order 1 in the infinite direction. There are 8 matching infinite plane strain patches that simulate the infinite extent, where the displacements are assumed to be constant in the infinite direction, simulating plane strain conditions. The inclusions are described by linear surfaces as shown in Fig. 14.10. Six internal points were used in the y-direction, assuming a constant variation in x- and z-direction. For the approximation of the unknown, two knots were inserted at the intersections with the inclusion, allowing the continuity to be decreased there from C^1 to C^0. The resulting collocation points are shown in Fig. 14.11.

The convergence of the maximum displacement at the top of the excavation is shown in Fig. 14.12 and compared with a BEFE solution.

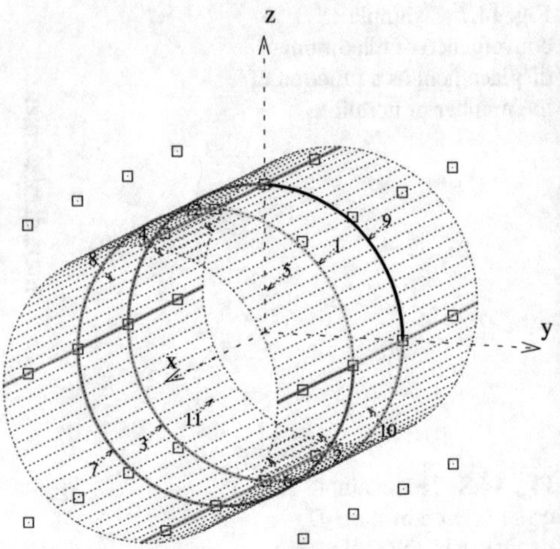

Fig. 14.9 Test example 2: geometry description with finite and infinite NURBS patches. Control points are depicted by hollow squares. This exactly defines a cylinder

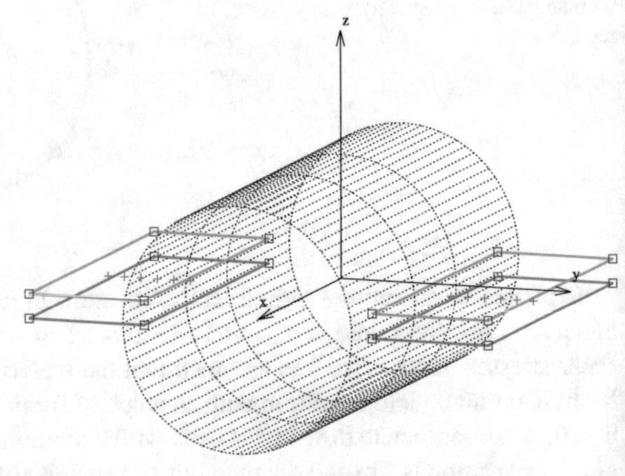

Fig. 14.10 Test example 2: description of the geometry of the inclusions with 8 control points each (hollow squares)

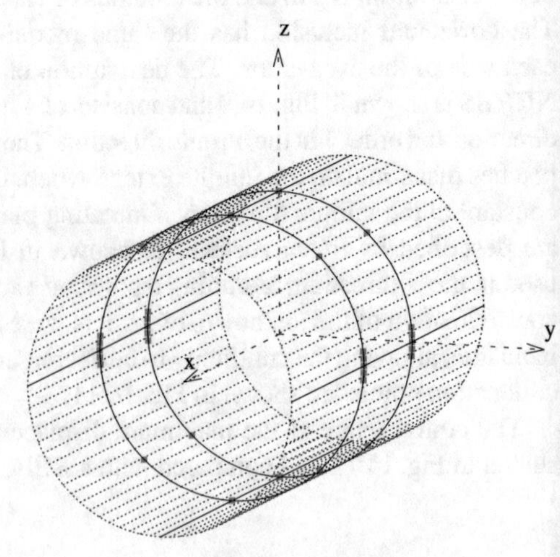

Fig. 14.11 Test example 2: location of collocation points

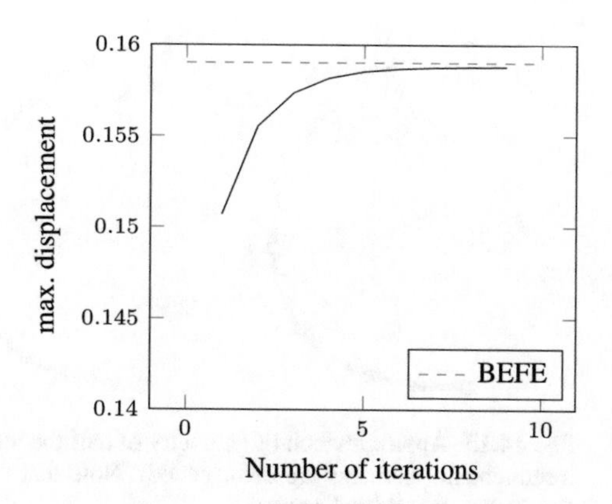

Fig. 14.12 Test example 2: convergence of maximum displacement and comparison with BEFE solution

14.7 Practical Examples

Here we present some practical examples. In the first example, we compare our results with the conventional BEM.

14.7.1 NATM Tunnel

The definition of the geometry of the NATM tunnel has already been shown in Fig. 14.1. The tunnel is assumed to be excavated in a homogeneous elastic domain with the following properties: $E = 10{,}000\,\text{MPa}$, $\nu = 0.3$. The virgin stress field was assumed to be $\sigma_x^{\text{v}} = -0.4, \sigma_y^{\text{v}} = -1.0, \tau_{xy}^{\text{v}} = -0\,\text{MPa}$. We show the salient differences between conventional BEM and IGABEM by comparing the two.

14.7.1.1 Conventional BEM Analysis

In the conventional BEM analysis, Lagrange polynomials are used for the approximation of the geometry and the unknown (=isoparametric boundary elements). Figure 14.13 shows the approximation of the geometry of half of the tunnel and of the unknowns by increasing the number of quadratic elements. We see that for this refinement strategy both a better approximation of the geometry as well of the unknown is achieved.

14.7.1.2 IGABEM

Here half of the tunnel is discretised with 1 NURBS patch of order 2. As explained above, this is an exact representation of the design geometry. The simulation starts

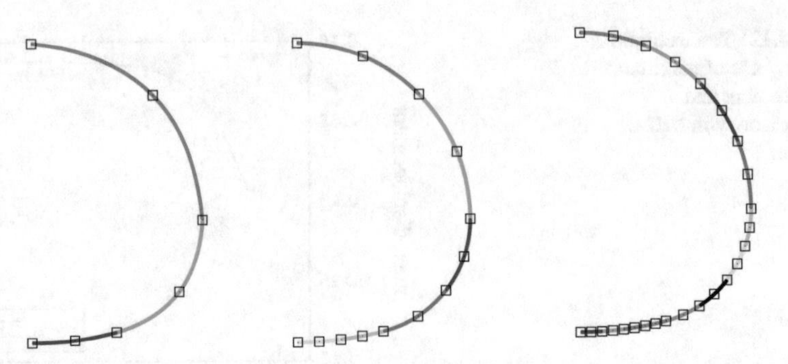

Fig. 14.13 Approximation of geometry of half the tunnel with conventional BEM and a classical h-refinement (elements are color coded). Note that the approximation of the geometry changes (i.e. improves) with refinement

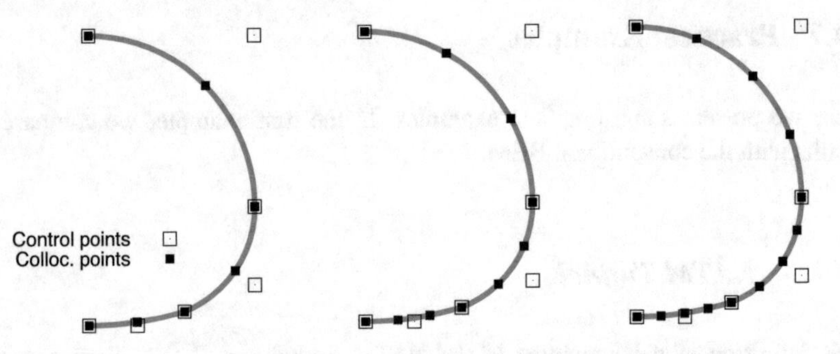

Fig. 14.14 Definition of geometry of half the tunnel using one NURBS patch of order 2. Refinement of solution by order elevation showing collocation points (from left to right order 2 to 4). Note that in contrast to the classical BEM the description of the geometry (already exact) does not change

with assuming the same basis functions for the approximation of the displacements as for the geometry. Refinement of the results are achieved by k-refinement. Figure 14.14 shows the location of collocation points for the original and the first two stages of refinement by order elevation.

14.7.1.3 Results and Comparison

First, we compare the variation of the tangential stress[2] along the tunnel surface, as a function of the number of degrees of freedom in Fig. 14.15.

It can be seen that even without refinement the isogeometric BEM gives good results with a smooth stress distribution. The conventional BEM analysis shows a jump in stresses between elements, which decreases as the number of degrees of freedom is increased. This is because the approximation of the geometry with Lagrange polynomials does not have a unique tangent and a small kink appears. In

[2]The computation of the tangential stress using the stress recovery method is explained in Chap. 6.

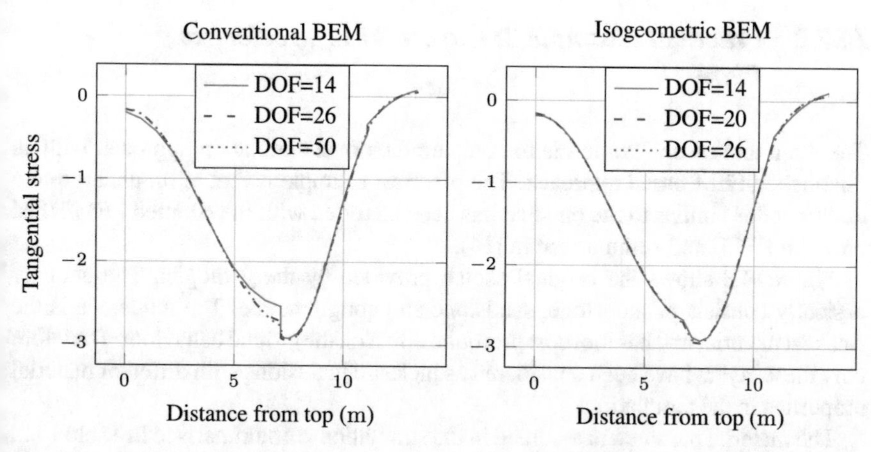

Fig. 14.15 Distribution of tangential stress along tunnel wall for different degrees of freedom (DOF)

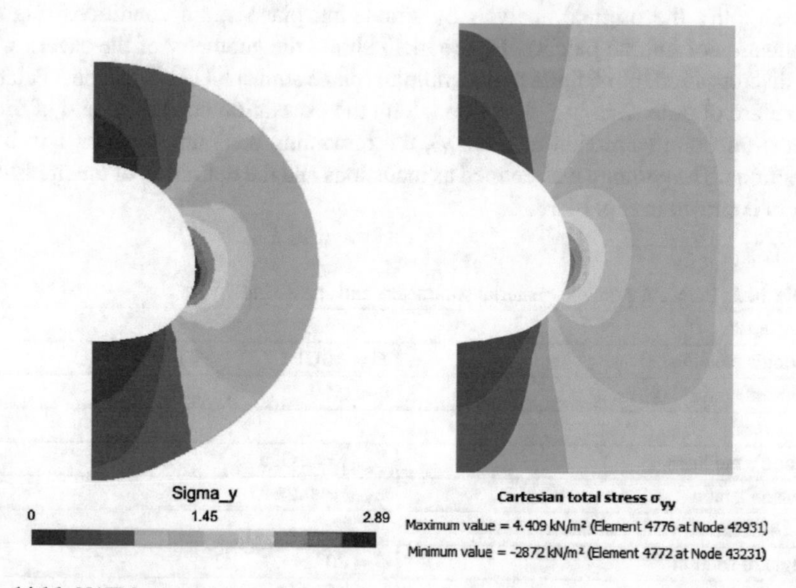

Fig. 14.16 NATM tunnel: comparison of contours of vertical stress (left) for the BEM analysis and (right) with Plaxis. Note that since the contours were plotted by different programs, a slightly different colour coding was used

Fig. 14.16 we compare the results of the unrefined NURBS simulation which only had 14 degrees of freedom with the result of a simulation with the Finite Element program Plaxis. The contours look very similar and the maximum compressive vertical stress of 2.89 $\mathrm{MN/m^2}$ compares well with the value of 2.87 obtained by Plaxis. It can be seen that very good results can be obtained with few degrees of freedom.

14.7.2 Practical Example 2: Cavern of Hydroelectric Plant

The main aim of this section is to compare the novel simulation approach with an established FEM based approach. The practical example is that of an underground cavern and is similar to the one that has been analysed with the coupled FEM/BEM program BEFE and summarised in [14].

Figure 14.3 shows the original sketch provided by the geologist. The geology basically consists of mudstone, sandstone and conglomerate. The mudstone is the weakest material and has the most profound effect on the ground behaviour. Therefore only these layers have been considered as inelastic inclusions with different material properties in the simulation.

The material parameters assumed in the simulation are summarised in Table 14.2. The depth of the top of the cavern is 310 m. For revisiting the BEFE analysis with the proposed isogeometric approach, we only consider the final excavation stage and simplify the original analysis by simulating plane strain conditions, via the application of infinite patches. Figure 14.17 shows the geometry of the cavern with the discretisation into 6 finite and 12 infinite (plane strain) NURBS patches. Patches 1 to 3 are of order 2 in in ξ-direction (along the excavation boundary) and of order 1 in η-direction (direction to infinity), the remaining ones are of order 1 in both directions. The geology was defined as inclusions and the definition of the mudstone layers is shown in Fig. 14.18.

Table 14.2 Practical example: material parameters and stress field

Rock mass	
Young's modulus	$E = 10\,\text{GPa}$
Poisson's ratio	$\nu = 0.20$
Inclusion	
Young's modulus	$E_{incl} = 6\,\text{GPa}$
Poisson's ratio	$\nu_{incl} = 0.25$
Mohr–Coulomb yield condition	
Angle of friction	$\phi = 30°$
Cohesion	$c = 0.73\,\text{MPa}$
Dilation angle	$\psi = 0°$
Virgin stress field	$\sigma_x^{\text{v}} = -0.027\cdot$ depth MPa $\sigma_y^{\text{v}} = \sigma_x^{\text{v}} \cdot 0.5$

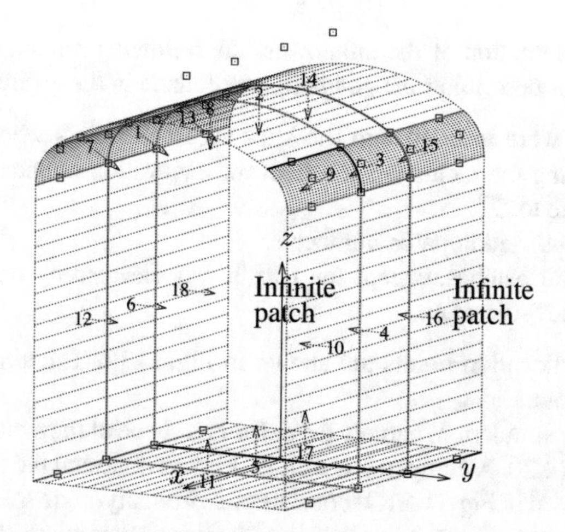

Fig. 14.17 Underground cavern: description of geometry of the cavern with finite and infinite patches, showing control points as hollow squares and patch numbering. This describes the design geometry exactly

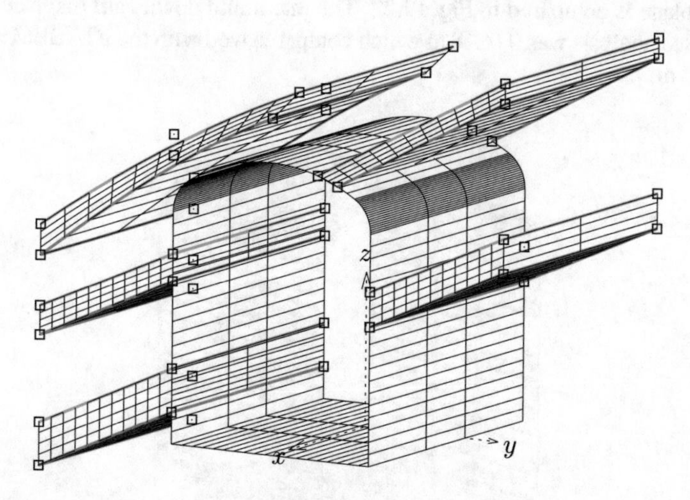

Fig. 14.18 Underground cavern: definition of the mudstone layers as inclusions, showing the colour coded bounding surfaces and the associated control points as hollow squares

For the approximation of the unknowns, the following refinements to the basis functions used for describing the geometry were made in the ξ-direction only:

- Double knots were inserted into the knot vector at points where the boundary surfaces defining the inclusions intersect the excavation boundary, changing the continuity there to C^0.
- 4 knots were inserted in patch number 2.
- The order of all patches, except for 1 to 3, was elevated by one from linear to quadratic in the ξ-direction.

The resulting collocation points are shown in Fig. 14.19. The simulation has 192 degrees of freedom.

A result of the simulation, namely the deformed shape of the excavation boundary is shown in Fig. 14.20. A comparison with BEFE was done and the mesh used for the simulation is shown in Fig. 14.21. It consists of 239 twenty-node solid finite elements (where the displacement was restrained in x-direction to simulate plane strain conditions) and 96 eight-node iso-parametric boundary elements including infinite (plane strain) boundary elements that simulate the infinite domain. A vertical plane of symmetry is assumed. The mesh has 4036 degrees of freedom. The deformed shape in the y–z plane is compared in Fig. 14.22. The maximum downward displacement of the coupled analysis was 0.0248 m which compares well with the IGABEM analysis of 0.0245 m.

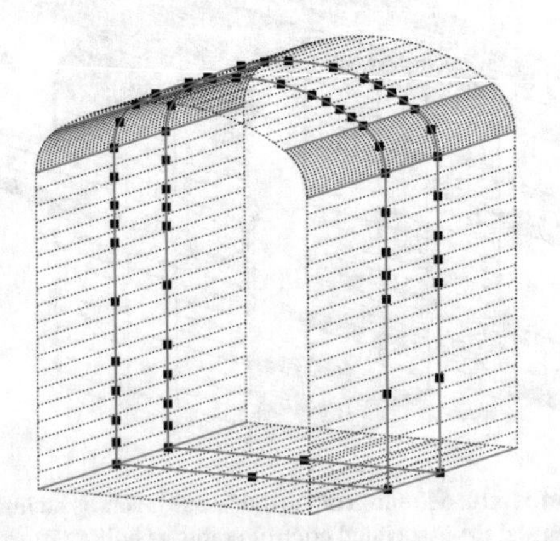

Fig. 14.19 Underground cavern: location of collocation points

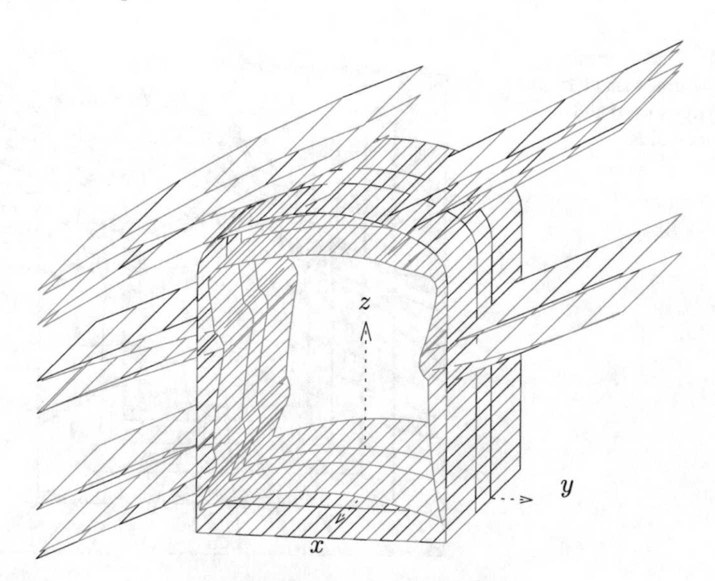

Fig. 14.20 Underground cavern: deformed shape

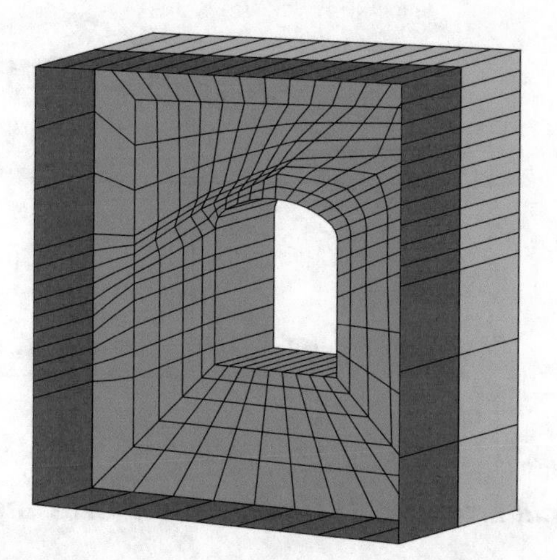

Fig. 14.21 Underground cavern: view of BEFE mesh with vertical symmetry plane. Finite elements are dark cyan, finite boundary elements light blue and infinite boundary elements red

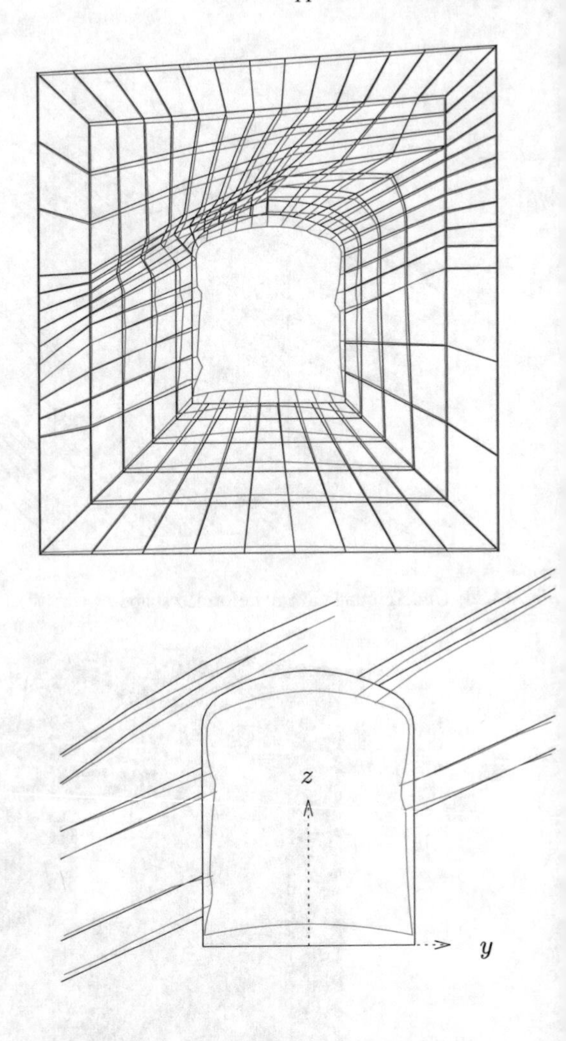

Fig. 14.22 Underground cavern: deformed shape from BEFE analysis (top) and current simulation

14.7.3 Practical Example 3: Hudson River Crossing Project

This is an actual project related to the construction of a rail crossing under the Hudson River between New York and New Jersey. The authors performed a preliminary analysis, but the project did not go ahead. The planned crossing consisted of two tunnels at different elevations that were connected by cross passages at certain locations. The CAD model of the tunnel system was provided by the consultant. Figure 14.23 shows an axonometric view of part of the tunnel system. The simulation presented here differs from the previous one in that the geometry was taken directly from the CAD model and since the model contained trimmed patches, the procedures as outlined in Chap. 10 were applied. No approximation of the geometry by a mesh is involved.

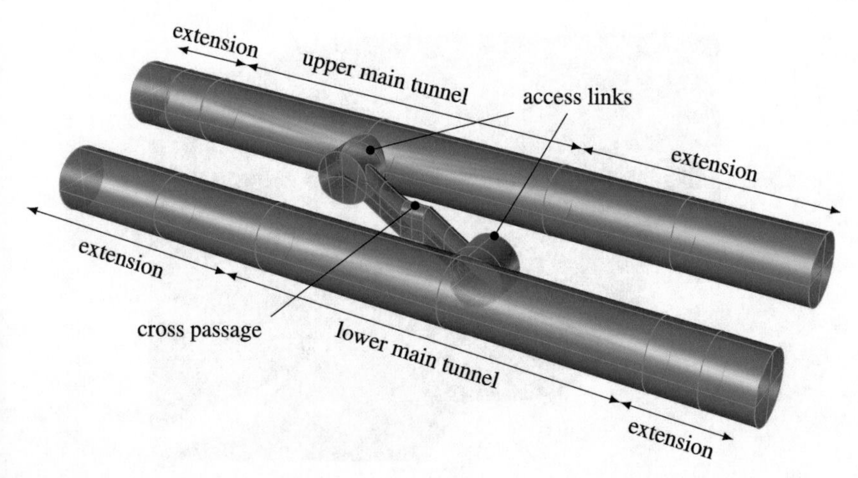

Fig. 14.23 CAD model of the tunnel cross passage

Each main tube consists of three sections: the main tunnel and extensions at both of its ends. Even though the tunnels continue at both ends they were closed after the extensions, sufficiently far from the region of interest. The connection of the main tubes is also divided into three parts: the cross passage and two access links. The intersections of these parts result in various complex trimming cases. In total, the geometry is represented by 9 trimmed patches and 6 regular ones (i.e., the extensions of the main tube and the closed cross-sections of the access links).

A geometry independent field approximation was used and the refinement was achieved by knot insertion, with the order of the linear patches being elevated to $p = 2$. In particular, the main tunnels and all parts of their connection were refined. Extended THB-splines using $c_e = p/2$ for the stabilisation of the trimmed basis and a discontinuous collocation was used. Figure 14.24 shows the computed integration elements at the connection of the upper tunnel tube with the cross passage. The total number of degrees of freedom is 12993.

The excavation process is simulated by subjecting the boundary Γ to excavation forces that are determined by the virgin stress field given by $\sigma_x^{\mathrm{v}} = \sigma_y^{\mathrm{v}} = 1.375\,\mathrm{MPa}$ and $\sigma_z^{\mathrm{v}} = 2.75\,\mathrm{MPa}$ (all compression). The material properties are: Poisson ratio $\nu = 0.2$, Young's modulus $E = 313\,\mathrm{MPa}$. One result of the simulation, namely the displacements due to the excavation process is shown in Fig. 14.25.

Since a discontinuous collocation was used, the jumps of the displacement field across intersections could be seen as a kind of error indicator. Note that we obtain a smooth displacement field at the intersections of the access links with the main tunnels as illustrated in the contour-plots in Fig. 14.25(b, c). Despite the fact that the displacements for the trimmed patches are evaluated separately for each patch, the displacements along intersections match very well. This indicates that a good quality result is obtained.

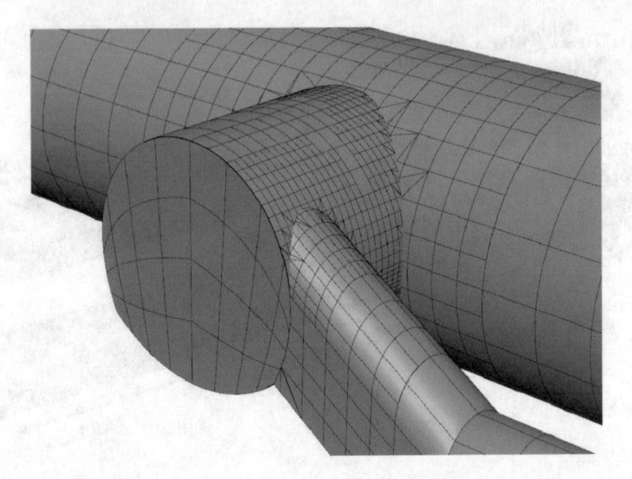

Fig. 14.24 The resulting integration elements for $c_e = p/2$ along the intersection of the cross passage and the upper main tunnel

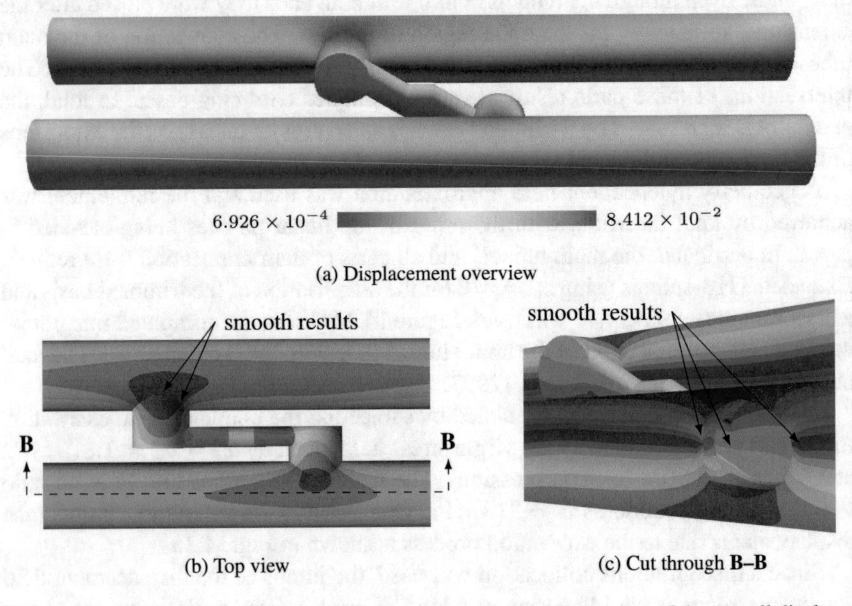

Fig. 14.25 Displacement of the tunnel excavation example: (a) overview of the overall displacement and (b, c) corresponding contour-plots. The data range of all plots is the same. In (b, c), arrows point to the crucial areas of the example where trimmed surfaces intersect

14.8 Summary and Conclusions

Here we have discussed applications in geomechanics, especially related to underground excavation problems. Since this involves an infinite domain, it is where the BEM is at its best. We have shown how smooth boundaries, essential for efficient and safe excavation, can be defined accurately with only a few parameters. Geological conditions can be defined as inclusions by bounding curves or surfaces and the data can be taken directly from CAD programs.

After demonstrating the applicability of the proposed procedures to simulate excavations with geological features on simple test examples, applications in practice are shown. The example of the underground cavern shows how easily smooth geometrical shapes can be defined and how good quality results can be obtained with few degrees of freedom as compared with commonly used software. The second example of an underground passage shows how, using the procedures outlined in this book, CAD data can be used directly for the simulation, without the requirement of mesh generation. This example also shows that discontinuous collocation can be used safely.

This approach to modelling is in stark contrast to the sometimes tedious mesh generation required in the FEM. As has been demonstrated in the second practical example, the use of CAD data directly without a mesh generation can be achieved. In addition the number of degrees of freedom required for accurate IGABEM simulations is significantly reduced, when compared to the FEM or coupled FEM/BEM.

The most computation time spent is in the numerical evaluation of integrals, especially the domain integrals and is significantly greater than for the FEM. However, as has been shown in Chap. 12 matrices may be recomputed, that depend only on the geometry and that multiply with the body forces. This means that the computational overhead is mainly in the pre-computation of these matrices. Once this is done, the actual simulation, with different material properties or virgin stress fields can be very fast.

Chapter 15
Viscous Flow Problems

15.1 Introduction

In this chapter, we discuss the implementation of the isogeometric BEM for steady state viscous incompressible flows [18, 22]. The integral equations for viscous flow has been derived in Chap. 2 as

$$\begin{aligned} c_{ij}(\tilde{\boldsymbol{x}})\,\dot{u}_i(\tilde{\boldsymbol{x}}) = & \int_{\Gamma} [\mathsf{U}_{ij}(\tilde{\boldsymbol{x}},\hat{\boldsymbol{x}})t_i(\hat{\boldsymbol{x}}) - \mathsf{T}_{ij}(\tilde{\boldsymbol{x}},\hat{\boldsymbol{x}})\dot{u}_i(\hat{\boldsymbol{x}})]\,d\Gamma(\hat{\boldsymbol{x}}) \\ & - \int_{\Gamma_0} \mathsf{U}_{ij}(\tilde{\boldsymbol{x}},\hat{\boldsymbol{x}})\; t_j^0(\hat{\boldsymbol{x}})d\Gamma_0(\hat{\boldsymbol{x}}) \\ & + \int_{\Omega_0} \mathsf{U}_{ij,k}(\tilde{\boldsymbol{x}},\hat{\boldsymbol{x}})\; b_{jk}^0(\hat{\boldsymbol{x}})d\Omega_0(\hat{\boldsymbol{x}}) \end{aligned} \tag{15.1}$$

where $\dot{u}_i$ is the perturbation velocity, b_{jk}^0 is the body force and t_j^0 the boundary traction due to the body force.

The collocation (incremental) integral equation in matrix notation is

$$\begin{aligned} \mathbf{c}\,(\tilde{\boldsymbol{x}}_n)\;\dot{\boldsymbol{u}}\,(\tilde{\boldsymbol{x}}_n) = & \int_{\Gamma} \mathsf{U}\,(\tilde{\boldsymbol{x}}_n,\hat{\boldsymbol{x}})\,\mathbf{t}\,(\hat{\boldsymbol{x}})\,d\Gamma(\hat{\boldsymbol{x}}) - \int_{\Gamma} \mathsf{T}\,(\tilde{\boldsymbol{x}}_n,\hat{\boldsymbol{x}})\,\dot{\mathbf{u}}\,(\hat{\boldsymbol{x}})\,d\Gamma(\hat{\boldsymbol{x}}) \\ & - \int_{\Gamma_0} \mathsf{U}\,(\tilde{\boldsymbol{x}}_n,\hat{\boldsymbol{x}})\,\dot{\mathbf{t}}_0\,(\hat{\boldsymbol{x}})\,d\Gamma_0(\hat{\boldsymbol{x}}) \\ & + \int_{\Omega_0} \mathsf{U}'\,(\tilde{\boldsymbol{x}}_n,\hat{\boldsymbol{x}})\,\dot{\mathbf{b}}_0\,(\hat{\boldsymbol{x}})\,d\Omega_0(\hat{\boldsymbol{x}}) \end{aligned} \tag{15.2}$$

G. Beer et al., *The Isogeometric Boundary Element Method*, Lecture Notes in Applied and Computational Mechanics 90,
https://doi.org/10.1007/978-3-030-23339-6_15

with $n = \{1, \ldots, N\}$ and where the overdot specifies an increment. The fundamental solutions U and T are listed in Appendix A. U' will be derived later.

After the assembly described in Chap. 6 the following system of equations is obtained:

$$\mathbf{Lx} = \mathbf{r} + \mathbf{r}_0 \tag{15.3}$$

where $\mathbf{L}$ and $\mathbf{r}$ are assembled left-hand and right-hand sides due to loading and $\mathbf{x}$ contains a mixture of $\dot{\mathbf{u}}$ and $\mathbf{t}$ parameters for the initial solution, depending on specified boundary conditions. The sub-vector of $\mathbf{r}_0$ associated with collocation point $\boldsymbol{x}_n$ is given by:

$$\mathbf{r}_{0n} = \int\limits_{\Gamma_0} \mathsf{U}\left(\tilde{\boldsymbol{x}}_n, \hat{\boldsymbol{x}}\right) \dot{\mathbf{t}}_0\left(\hat{\boldsymbol{x}}\right) d\Gamma_0(\hat{\boldsymbol{x}}) - \int\limits_{\Omega_0} \mathsf{U}'\left(\tilde{\boldsymbol{x}}_n, \hat{\boldsymbol{x}}\right) \dot{\mathbf{b}}_0\left(\hat{\boldsymbol{x}}\right) d\Omega_0(\hat{\boldsymbol{x}}). \tag{15.4}$$

For the computation of $\mathbf{r}_0$ we define the region Ω_0 where non-linear behaviour is expected to take place. The domain is then subdivided into a grid of points and into integration elements as explained in Chap. 12. The right-hand side can now be written as

$$\mathbf{r}_0 = [\mathbf{R}_0]\{\dot{\mathbf{b}}_0\} \tag{15.5}$$

where $\{\dot{\mathbf{b}}_0\}$ is a vector of body force values at grid points and $[\mathbf{R}_0]$ is a matrix linking grid point values to values of $\mathbf{r}_0$ as explained in Sect. 12.3.1.

The initial traction vector is given by

$$\dot{\mathbf{t}}_0 = \mathbf{N}\, \dot{\mathbf{b}}_0 \tag{15.6}$$

where $\mathbf{N}$ contains components of the vector normal to the boundary.

The sub-matrices of $[\mathbf{R}_0]$ are given by

$$\mathbf{R}_{0nk} = \int\limits_{\Gamma_0} \mathsf{U}\left(\tilde{\boldsymbol{x}}_n, \hat{\boldsymbol{x}}\right)\ \mathbf{N}\, M_k\, d\Gamma_0 - \int\limits_{\Omega_0} \mathsf{U}'\left(\tilde{\boldsymbol{x}}_n, \hat{\boldsymbol{x}}\right) M_k\, d\Omega_0 \tag{15.7}$$

where M_k are interpolation functions.

The perturbation velocity vector $\dot{\mathbf{v}}$ at any internal point $\boldsymbol{x}$ can be computed by[1]:

$$\begin{aligned} \dot{\mathbf{v}}\left(\boldsymbol{x}\right) = & \int\limits_{\Gamma} \mathsf{U}\left(\boldsymbol{x}, \hat{\boldsymbol{x}}\right) \mathbf{t}\left(\hat{\boldsymbol{x}}\right) d\Gamma(\hat{\boldsymbol{x}}) - \int\limits_{\Gamma} \mathsf{T}\left(\boldsymbol{x}, \hat{\boldsymbol{x}}\right) \dot{\mathbf{u}}\left(\hat{\boldsymbol{x}}\right) d\Gamma(\hat{\boldsymbol{x}}) \\ & - \int\limits_{\Gamma_0} \mathsf{U}\left(\boldsymbol{x}, \hat{\boldsymbol{x}}\right) \dot{\mathbf{t}}_0\left(\hat{\boldsymbol{x}}\right) d\Gamma_0(\hat{\boldsymbol{x}}) + \int\limits_{\Omega_0} \mathsf{U}'\left(\boldsymbol{x}, \hat{\boldsymbol{x}}\right) \dot{\mathbf{b}}_0\left(\hat{\boldsymbol{x}}\right) d\Omega_0(\hat{\boldsymbol{x}}) \end{aligned} \tag{15.8}$$

[1] To distinguish it from the velocity at the boundary we use a different notation here.

The equations be written in matrix notation as:

$$\{\dot{\mathbf{v}}\} = \mathbf{E}\,\mathbf{x} + \{\bar{\mathbf{v}}\} + \left[\hat{\mathbf{R}}_0\right]\{\mathbf{b}_0\} \tag{15.9}$$

where $\{\dot{\mathbf{v}}\}$ contains perturbation velocities at point $\boldsymbol{x}$, matrix $\mathbf{E}$ is an assembled system matrix (see Chap. 6) and $\mathbf{x}$ is the solution vector. $\{\bar{\mathbf{v}}\}$ contains values of $\{\dot{\mathbf{v}}\}$ due to known boundary conditions and matrix $\left[\hat{\mathbf{R}}_0\right]$ is similar to matrix $[\mathbf{R}_0]$ of Eq. (15.7) except that the internal point $\boldsymbol{x}$ is substituted for the source point $\boldsymbol{x}_n$.

15.2 Solution Process

For the solution of the non-linear equations, we use an iterative process which can be a modified Newton–Raphson or full Newton–Raphson. In the first approach, the left-hand side of the system remains unchanged in the second approach the left-hand side changes after each iteration.

15.2.1 Modified Newton–Raphson

The iterative procedure for modified Newton–Raphson is shown in Algorithm 8. Since the problem is highly non-linear we have to use a relaxation scheme with the relaxation parameter β.

Algorithm 8 Algorithm for modified Newton–Raphson

1: Solve for boundary values $\dot{\mathbf{x}}^0$
2: Compute perturbation velocities $\dot{\mathbf{v}}^0$ inside Ω_0
3: Compute body force increments $\dot{\mathbf{b}}_0$ and compute right-hand side $\mathbf{r}_0$
4: **for** i **to** Number of iterations **do**
5: Solve for boundary values $\mathbf{x}^i$
6: Set $\mathbf{x} = \beta\mathbf{x}^i + (1-\beta)\mathbf{x}^{i-1}$
7: Compute velocities $\dot{\mathbf{v}}^i$ at internal points
8: Set $\dot{\mathbf{v}} = \beta\dot{\mathbf{v}}^i + (1-\beta)\dot{\mathbf{v}}^{i-1}$
9: Compute body force increments $\dot{\mathbf{b}}_0$
10: Compute $\mathbf{r}_0^i$
11: Compute Residual
12: **if** Residual < Tolerance **then**
13: Exit
14: **end if**
15: **end for**

15.2.2 Newton–Raphson

For the full Newton–Raphson we re-write Eqs. (15.3) and (15.9):

$$\mathbf{R}_x = -\mathbf{L}\mathbf{x} + \mathbf{r} + [\mathbf{R}_0]\{\mathbf{b}_0\} = 0 \tag{15.10}$$

$$\mathbf{R}_v = -\{\dot{\mathbf{v}}\} + \mathbf{E}\mathbf{x} + \{\bar{\mathbf{v}}\} + \left[\hat{\mathbf{R}}_0\right]\{\mathbf{b}_0\} = 0 \tag{15.11}$$

where $\mathbf{R}_x$ and $\mathbf{R}_v$ are residuals and then use a first order Taylor expansion

$$\frac{\partial \mathbf{R}_x}{\partial \mathbf{x}}\triangle\mathbf{x} + \frac{\partial \mathbf{R}_x}{\partial\{\dot{\mathbf{v}}\}}\triangle\{\dot{\mathbf{v}}\} = 0 \tag{15.12}$$

$$\frac{\partial \mathbf{R}_v}{\partial \mathbf{x}}\triangle\mathbf{x} + \frac{\partial \mathbf{R}_v}{\partial\{\dot{\mathbf{v}}\}}\triangle\{\dot{\mathbf{v}}\} = 0 \tag{15.13}$$

with

$$\frac{\partial \mathbf{R}_x}{\partial \mathbf{x}} = -\mathbf{L} + [\mathbf{R}_0]\frac{\partial\{\mathbf{b}_0\}}{\partial \mathbf{x}} \quad \text{and} \quad \frac{\partial \mathbf{R}_x}{\partial \dot{\mathbf{v}}} = [\mathbf{R}_0]\frac{\partial\{\mathbf{b}_0\}}{\partial\{\dot{\mathbf{v}}\}} \tag{15.14}$$

$$\frac{\partial \mathbf{R}_v}{\partial \mathbf{x}} = \mathbf{E} + \left[\hat{\mathbf{R}}_0\right]\frac{\partial\{\mathbf{b}_0\}}{\partial \mathbf{x}} \quad \text{and} \quad \frac{\partial \mathbf{R}_v}{\partial\{\dot{\mathbf{v}}\}} = -[\mathbf{I}] + \left[\hat{\mathbf{R}}_0\right]\frac{\partial\{\mathbf{b}_0\}}{\partial\{\dot{\mathbf{v}}\}} \tag{15.15}$$

This can be combined into a system of equations

$$\begin{bmatrix} -\mathbf{L} + [\mathbf{R}_0]\left[\frac{\partial\{\mathbf{b}_0\}}{\partial \mathbf{x}}\right], & [\mathbf{R}_0]\left[\frac{\partial\{\mathbf{b}_0\}}{\partial\{\dot{\mathbf{v}}\}}\right] \\ \mathbf{E} + \left[\hat{\mathbf{R}}_0\right]\left[\frac{\partial\{\mathbf{b}_0\}}{\partial \mathbf{x}}\right], & \left[\hat{\mathbf{R}}_0\right]\left[\frac{\partial\{\mathbf{b}_0\}}{\partial\{\dot{\mathbf{v}}\}}\right] - [\mathbf{I}] \end{bmatrix}^i \begin{Bmatrix} \dot{\mathbf{x}} \\ \{\dot{\mathbf{v}}\} \end{Bmatrix}^{i+1} = \begin{pmatrix} \mathbf{0} \\ \mathbf{0} \end{pmatrix} \tag{15.16}$$

or

$$[\mathbf{L}']^i \begin{Bmatrix} \dot{\mathbf{x}} \\ \{\dot{\mathbf{v}}\} \end{Bmatrix}^{i+1} = \begin{pmatrix} \mathbf{0} \\ \mathbf{0} \end{pmatrix} \tag{15.17}$$

This system of equations can be solved for the increments of the unknown $\mathbf{x}$ and for the increments of perturbation velocities at interior points $\{\dot{\mathbf{v}}\}$.

The iterative procedure for full Newton–Raphson is shown in Algorithm 9.

In the following, we discuss the computation of body effects and show examples separately for plane and 3-D problems.

Algorithm 9 Algorithm for full Newton–Raphson

1: Solve for boundary values $\dot{\mathbf{x}}^0$
2: Compute perturbation velocities $\dot{\mathbf{v}}^0$ inside Ω_0
3: Compute body force increments $\dot{\mathbf{b}}_0$ and compute right-hand side $\mathbf{r}_0$
4: **for** i **to** Number of iterations **do**
5: Compute derivatives of $\dot{\mathbf{b}}_0$ and tangent operator $[\mathbf{L}']^i$
6: Solve for boundary value increments $\dot{\mathbf{x}}^{i+1}$ and internal values $\{\dot{\mathbf{v}}\}^{i+1}$
7: Accumulate boundary values and internal values
8: Compute new body force increments $\dot{\mathbf{b}}_0$
9: Compute Residual
10: **if** Residual < Tolerance **then**
11: Exit
12: **end if**
13: **end for**

15.3 Plane Problems

15.3.1 Implementation

To establish the matrix $\mathbf{U}'$ in Eq. (15.2), we use the Einstein summation convention:

$$U_{1j,k}(\tilde{\boldsymbol{x}},\hat{\boldsymbol{x}})\, b^0_{jk}(\hat{\boldsymbol{x}}) = U'_{111}\, b^0_{11} + U'_{112}\, b^0_{12} + U'_{121}\, b^0_{21} + U'_{122}\, b^0_{22} \tag{15.18}$$

$$U_{2j,k}(\tilde{\boldsymbol{x}},\hat{\boldsymbol{x}})\, b^0_{jk}(\hat{\boldsymbol{x}}) = U'_{211}\, b^0_{11} + U'_{212}\, b^0_{12} + U'_{221}\, b^0_{21} + U'_{222}\, b^0_{22} \tag{15.19}$$

This can be converted into a matrix multiplication involving the matrix

$$\mathbf{U}' = \frac{1}{4\,\pi\,\mu\,r}\begin{pmatrix} r_1 - 2\,r_1^3, & -r_2 - 2\,r_1^2\,r_2, & r_2 - 2\,r_1^2\,r_2, & r_1 - 2\,r_2^2\,r_1 \\ r_2 - 2\,r_1^2\,r_2, & r_1 - 2\,r_2^2\,r_1, & -r_1 - 2\,r_2^2\,r_1, & r_2 - 2\,r_2^3 \end{pmatrix} \tag{15.20}$$

where μ is the viscosity. The vector $\dot{\mathbf{b}}_0$ is defined as

$$\dot{\mathbf{b}}_0 = \begin{pmatrix} b^0_{11} \\ b^0_{12} \\ b^0_{21} \\ b^0_{22} \end{pmatrix} = \rho \begin{pmatrix} \mathrm{v}_1\,\dot{\mathrm{v}}_1 \\ \mathrm{v}_1\,\dot{\mathrm{v}}_2 \\ \mathrm{v}_2\,\dot{\mathrm{v}}_1 \\ \mathrm{v}_2\,\dot{\mathrm{v}}_2 \end{pmatrix} \tag{15.21}$$

where ρ is the density, $\mathrm{v}_i = \dot{\mathrm{v}}_i + u^0_i$ and u^0_i is the free stream velocity. The initial traction vector is given by

$$\dot{\mathbf{t}}_0 = \mathbf{N}\,\dot{\mathbf{b}}_0 \tag{15.22}$$

where

$$\mathbf{N} = \begin{pmatrix} n_x & n_y & 0 & 0 \\ 0 & 0 & n_x & n_y \end{pmatrix}. \tag{15.23}$$

The derivatives of the body force are given by

$$\frac{\partial \mathbf{b}_0}{\partial \dot{\mathbf{v}}} = \begin{bmatrix} 2\dot{v}_1 + u_1^0 & 0 \\ \dot{v}_2 + u_2^0 & \dot{v}_1 \\ \dot{v}_2 & \dot{v}_1 + u_1^0 \\ 0 & 2\dot{v}_2 + u_2^0 \end{bmatrix} \tag{15.24}$$

which can be combined into a matrix

$$\left[\frac{\partial\{\mathbf{b}_0\}}{\partial\{\dot{\mathbf{v}}\}}\right] = \begin{bmatrix} \frac{\partial \mathbf{b}_{01}}{\partial\{\dot{\mathbf{v}}\}} & \{0\} & \cdots & \{0\} \\ \{0\} & \frac{\partial \mathbf{b}_{02}}{\partial\{\dot{\mathbf{v}}\}} & \cdots & \{0\} \\ \vdots & \vdots & \ddots & \vdots \\ \{0\} & \cdots & \cdots & \frac{\partial \mathbf{b}_{0K}}{\partial\{\dot{\mathbf{v}}\}} \end{bmatrix}. \tag{15.25}$$

The coefficients of $\left[\frac{\partial\{\mathbf{b}_0\}}{\partial \mathbf{x}}\right]$ are zero if the parameter value in $\mathbf{x}$ relates to a traction. If the parameter value relates to a velocity then Eq. (15.24) is applied.

The implementation of the theory is tested here on two examples. In the first one the results are compared with an available fine grained solution in order to ascertain that good quality of results can be obtained. The second one is used to demonstrate the superiority of the proposed approach for describing more complex practical geometries.

15.3.2 Flow in Cavity

The driven cavity problem has become a standard test problem for fluid dynamics codes. An incompressible fluid of uniform viscosity ($\mu = 1$) is confined within a square region of dimension $H = 1 \times 1$. The fluid velocities on the bottom, left and right are fixed at zero, while a uniform velocity $u_1 = 1$ in horizontal direction is specified at the top, which is tapered off to zero very near the corners. The Reynold-s number is defined as $Re = \rho\ u_1\ H/\mu$. The example is tested for two different Reynolds numbers (=100, 400) by changing the value of ρ.

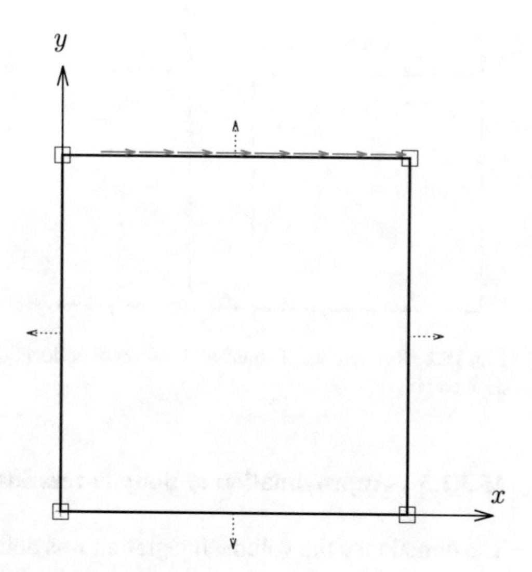

Fig. 15.1 Flow in cavity: definition of geometry with 4 linear patches. Control points are depicted by hollow squares. Also shown is the Dirichlet boundary condition on top

15.3.2.1 Definition of Geometry and Boundary Conditions

The boundary of the problem is defined by 4 linear NURBS patches as shown in Fig. 15.1. Using a geometry independent field approximation, the non-zero Dirichlet boundary condition along the top NURBS patch was defined using the following knot vector for the basis function $\bar{R}_k(\xi)$

$$\Xi = (0, 0, 0.05, 0.95, 1, 1) \tag{15.26}$$

with all weights equal to 1. The parameters were specified as:

$$\dot{\mathbf{u}}^e = \begin{pmatrix} 0 & 1 & 1 & 0 \\ 0 & 0 & 0 & 0 \end{pmatrix} \tag{15.27}$$

This means that the velocity vector at the top is tapered off to zero very near to the corners.

15.3.2.2 Approximation of the Unknown and Refinement

The approximation of the boundary unknown (in this case **t**) was achieved by order elevating the basis functions for describing the geometry from order 1 to 2 and inserting knots. Three different refinements were investigated and the resulting locations of collocation points are shown in Fig. 15.2.

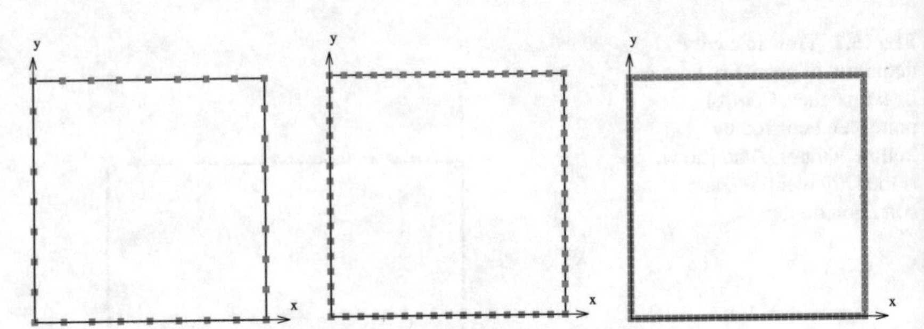

Fig. 15.2 Refinement of solution: location of collocation points for 3, 7 and 15 knot insertions for each patch

15.3.2.3 Approximation of Body Forces Inside Domain

The domain for the volume integration was defined by 2 NURBS curves and covers the whole domain Ω. The refinement of the boundary values was accompanied by an increased number of internal points as shown in Fig. 15.3. Quadratic interpolation between the points was assumed. The number of degrees of freedom and the number of internal points for the different meshes is shown in Table 15.1.

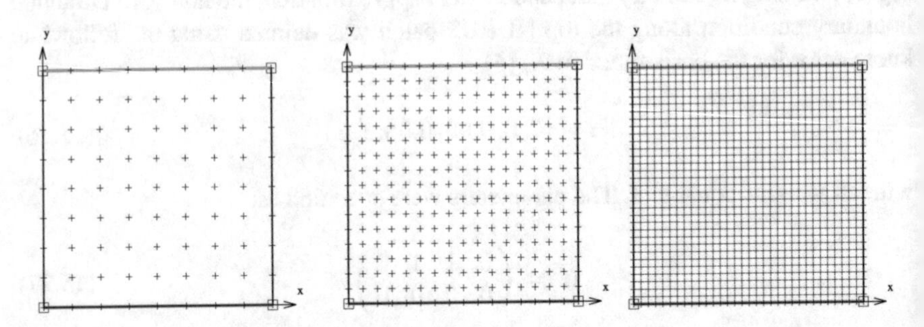

Fig. 15.3 Definition of domain for volume integration with two NURBS curves marked red and green and location of the internal points for the three refinement stages

Table 15.1 Mesh statistics

Mesh	Degrees of freedom	No. of internal points
mesh 1	64	81
mesh 2	128	289
mesh 3	256	1089
Reference [68]	–	16641

15.3.2.4 Results and Comparison

Results were computed for the three refinements (meshes) and two different Reynolds numbers with a modified and full Newton–Raphson method. Figure 15.4 shows the velocity vectors for the two Reynolds numbers. A shift in the vortex centre can be clearly seen. Figures 15.5 and 15.6 show a comparison of the results obtained with the different meshes and iteration methods. The variation of the horizontal component of the velocity vector along a vertical line though the middle agree well

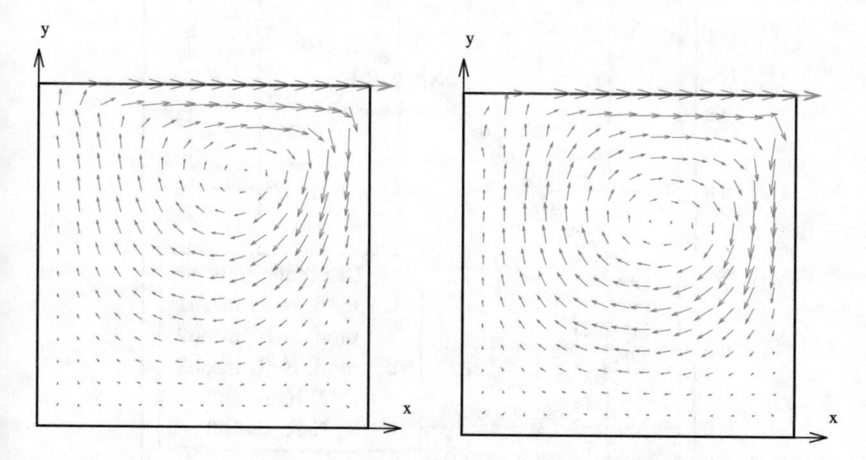

Fig. 15.4 Forced cavity flow: resulting velocity vectors for Re = 100 (left) and (right) Re = 400

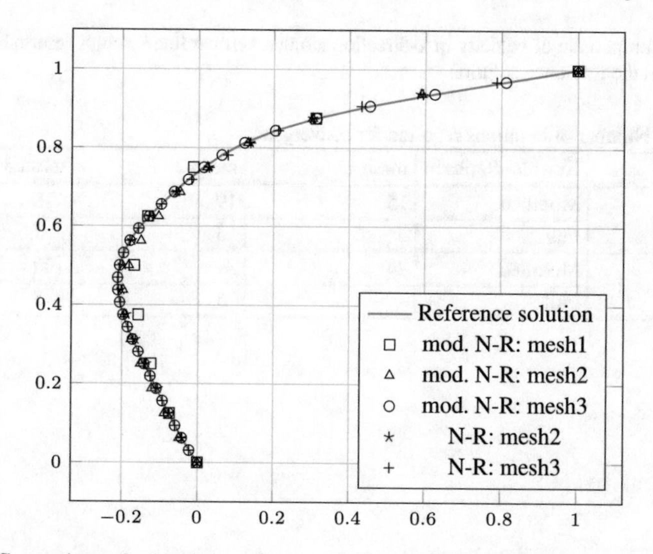

Fig. 15.5 Comparison of velocity in x-direction along a vertical line through centre for Re = 400 together with the reference solution

with the extremely accurate published solution, except for mesh 1 and Re = 400. It seems that for this Reynolds number mesh 1 is not adequate.

There is very little difference between the results obtained with modified and full Newton–Raphson. However, as shown in Table 15.2 there is a large difference with respect to the number of iterations required to achieve convergence to a tolerance of 10^{-4} with the modified Newton–Raphson requiring a significant higher number of iterations.

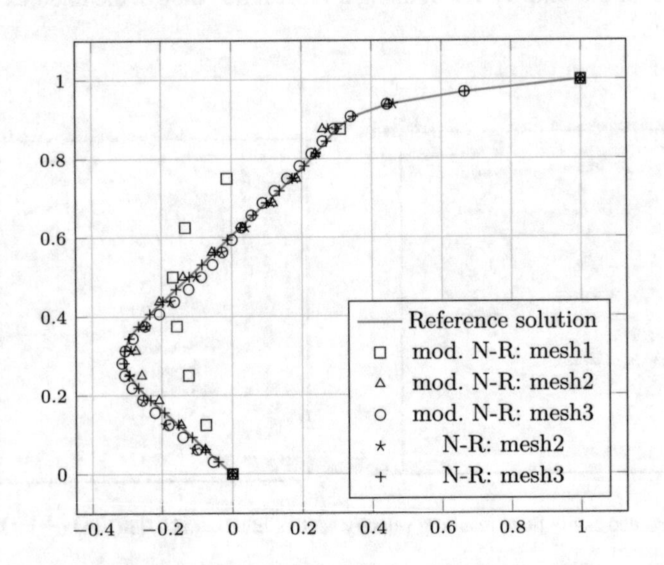

Fig. 15.6 Comparison of velocity in x-direction along a vertical line through centre for Re = 400 together with the reference solution

Table 15.2 Number of iterations required for convergence

Re	Newton–Raphson	mesh 1	mesh 2	mesh 3
100	Modified	15	19	18
	Full	3	3	3
400	Modified	24	39	100
	Full	–	5	5

15.3.3 Airfoil

As a practical example, we show the simulation of the flow past an airfoil. We chose the NACA0018 airfoil, where detailed coordinates are available on www.airfoiltools.com. The airfoil is placed in an infinite domain and subjected to a stream. The free stream velocity is defined by $u_1^0 = 0.984, u_2^0 = 0.173$ which corresponds to a unit velocity vector inclined at 10°.

15.3.3.1 Description of Boundary Geometry

NURBS are ideally suited for describing such shapes with very few parameters. Only 5 control points and basis functions of order 2 are able to describe the shape fairly accurately as is shown in Fig. 15.7. More accurate descriptions can be obtained with more control points and higher basis function order. An important fact is that the approximated shape has a C^1 continuity throughout each patch, which would not be the case if piecewise continuous Lagrange polynomials are used.

15.3.3.2 Approximation of the Boundary Unknown

We use the geometry independent field approximation approach for the variation of the unknown boundary values by inserting 7 knots into the knot vector that is used to describe the boundary, for each patch. The resulting locations of the collocation points are shown in Fig. 15.8.

15.3.3.3 Approximation of the Body Forces

The task left is the description of the domain for the volume integration. Here we use 4 subdomains defined by 2 NURBS curves each. The discretisation and the locations of the internal points are shown in Fig. 15.9. As with the previous example a quadratic interpolation between points is used.

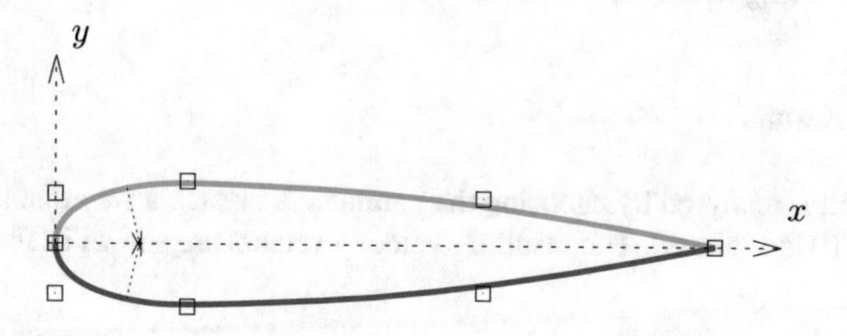

Fig. 15.7 Description of NACA0018 airfoil with two patches and 5 control points and basis functions of order 2 each

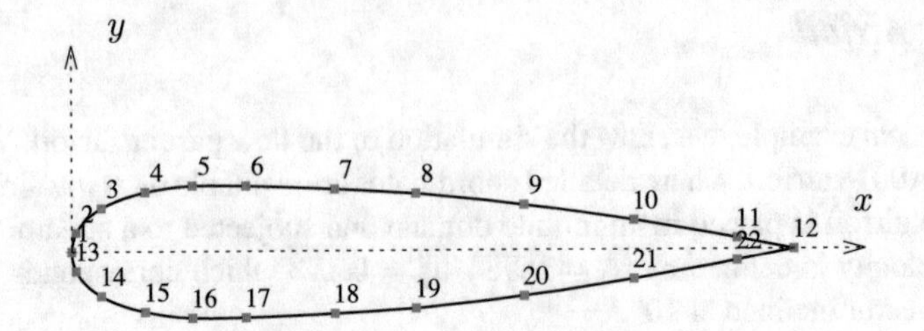

Fig. 15.8 Location of collocation points after refinement by knot insertion

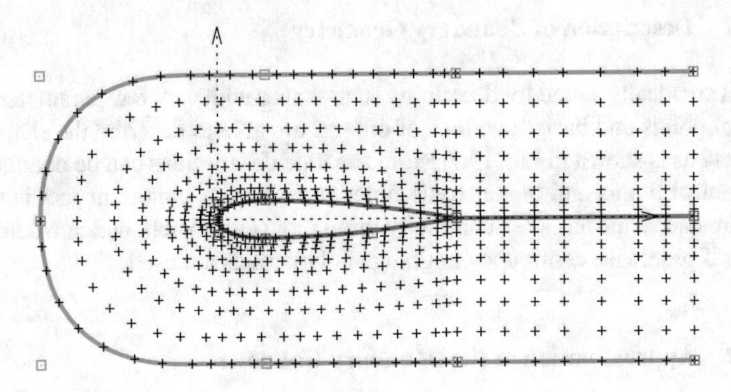

Fig. 15.9 Definition of domain for volume integration

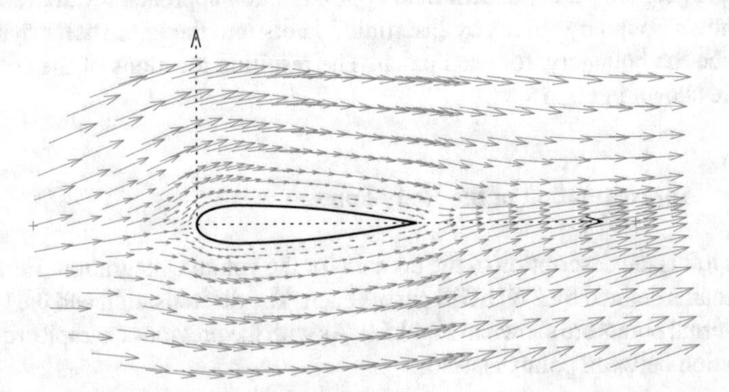

Fig. 15.10 Resulting velocity vectors for Re = 10

15.3.3.4 Results

The problem is analysed by choosing the parameters in such a way that a Reynolds number of 10 is achieved. The resulting velocity vectors are shown in Fig. 15.10.

15.4 3-D Problems

Here we discuss the implementation of viscous flow problems in three dimensions.

15.4.1 Implementation

Using the Einstein summation convention we can write for the last integrand in Eq. (15.2):

$$\begin{aligned} U_{ij,k}(\tilde{\boldsymbol{x}}, \hat{\boldsymbol{x}})\, b^0_{jk}(\hat{\boldsymbol{x}}) = {} & U'_{i11}\, b^0_{11} + U'_{i12}\, b^0_{12} + U'_{i13}\, b^0_{13} \\ & + U'_{i21}\, b^0_{21} + U'_{i22}\, b^0_{22} + U'_{i23}\, b^0_{23} \\ & + U'_{i31}\, b^0_{31} + U'_{i32}\, b^0_{32} + U'_{i33}\, b^0_{33} \end{aligned} \tag{15.28}$$

This can be converted into a matrix multiplication involving the matrix

$$\mathbf{U}' = \begin{pmatrix} U'_{111} & U'_{112} & U'_{113} & U'_{121} & U'_{122} & U'_{123} & U'_{131} & U'_{132} & U'_{133} \\ U'_{211} & U'_{212} & U'_{213} & U'_{221} & U'_{222} & U'_{223} & U'_{231} & U'_{232} & U'_{233} \\ U'_{311} & U'_{312} & U'_{313} & U'_{321} & U'_{322} & U'_{323} & U'_{331} & U'_{332} & U'_{333} \end{pmatrix} \tag{15.29}$$

where

$$U'_{ijk} = \frac{1}{8\pi\mu r^2}\left(-\delta_{ij}\, r_k + \delta_{jk}\, r_i + \delta_{ik}\, r_j - 3\, r_i\, r_j\, r_k\right) \tag{15.30}$$

and the vector

$$\mathbf{b}_0(\hat{\boldsymbol{x}}) = \begin{pmatrix} b^0_{11} & b^0_{12} & b^0_{13} & b^0_{21} & b^0_{22} & b^0_{23} & b^0_{31} & b^0_{32} & b^0_{33} \end{pmatrix}^{\mathrm{T}}. \tag{15.31}$$

The initial traction vector is given by

$$\mathbf{t}_0 = \mathbf{N}\,\mathbf{b}_0 \tag{15.32}$$

where

$$\mathbf{N} = \begin{pmatrix} n_x & n_y & n_z & 0 & 0 & 0 & 0 & 0 & 0 \\ 0 & 0 & 0 & n_x & n_y & n_z & 0 & 0 & 0 \\ 0 & 0 & 0 & 0 & 0 & 0 & n_x & n_y & n_z \end{pmatrix}. \tag{15.33}$$

The derivatives of $\mathbf{b}_0$ are given by

$$\frac{\partial \mathbf{b}_0}{\partial \dot{\mathbf{v}}} = \begin{bmatrix} (2\dot{\mathrm{v}}_1 + u_1^0) & 0 & 0 \\ \mathrm{v}_2 & \dot{\mathrm{v}}_1 & 0 \\ \mathrm{v}_3 & 0 & \dot{\mathrm{v}}_1 \\ \dot{\mathrm{v}}_2 & \mathrm{v}_1 & 0 \\ 0 & 2\dot{\mathrm{v}}_2 + u_2^0 & 0 \\ 0 & \mathrm{v}_3 & \dot{\mathrm{v}}_2 \\ \dot{\mathrm{v}}_3 & 0 & \mathrm{v}_1 \\ 0 & \dot{\mathrm{v}}_3 & \mathrm{v}_2 \\ 0 & 0 & 2\dot{\mathrm{v}}_3 + u_3^0 \end{bmatrix}. \tag{15.34}$$

15.4.2 Numerical Results

We use the same example of the flow in a cavity that has been solved as a plane problem. To be able to compare with the plane solution there are two possibilities for the definition of the 3-D problem. One is to extend the discretisation into the third dimension for a distance and then truncate it. In this case errors will be introduced due to the truncation. The other possibility is to define a closed box of dimension $1 \times 1 \times 0.25$ and to apply slip boundary conditions on the two surfaces in the third direction. This approach has the advantage of not producing any truncation error. However, it results in a significantly higher number of degrees of freedom (DOF) as compared with the first approach. In the following, we compare the two approaches with respect to accuracy and DOF.

15.4.3 Truncated Geometry

15.4.3.1 Description of Geometry

In this approach we extend the geometry in the x-direction and truncate it without producing a closed surface. The geometry is defined by 4 linear patches with 4 control points each as shown in Fig. 15.11. We investigate the error introduced by the truncation by extending the truncation distance.

Using a geometry independent field approximation, the non-zero Dirichlet boundary condition along the top NURBS patch was defined using the following knot vectors for the basis function $\bar{R}_{ij}(\xi, \eta)$

$$\Xi_1 = (0, 0, 0.05, 0.95, 1, 1) \tag{15.35}$$

$$\Xi_2 = (0, 1) \tag{15.36}$$

with all weights equal to 1. The parameters were specified as

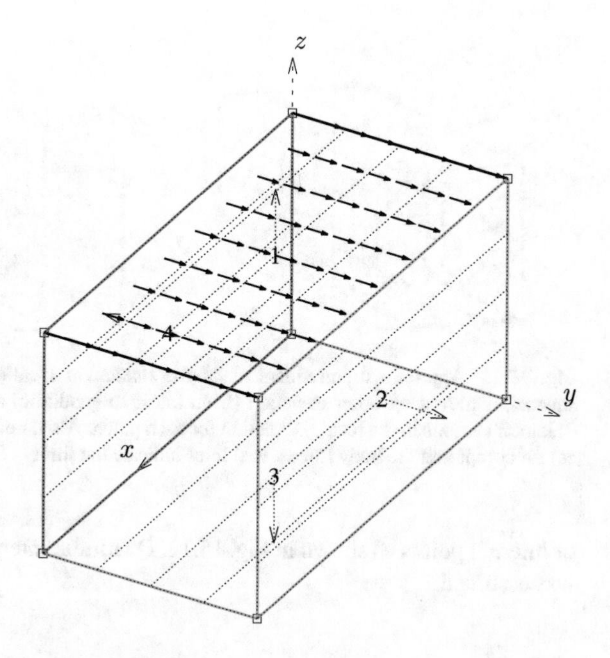

Fig. 15.11 Forced cavity flow: definition of the geometry and boundary conditions for the truncated mesh

$$\dot{\mathbf{u}}^e = \begin{pmatrix} 0 & 0 & 0 & 0 \\ 0 & 1 & 1 & 0 \\ 0 & 0 & 0 & 0 \end{pmatrix} . \tag{15.37}$$

This means that the velocity vector at the top is constant in the y-direction but tapered off to zero very near to the corners.

15.4.3.2 Approximation of the Unknown and Refinement

The approximation of the boundary unknown (in this case $\mathbf{t}$) was achieved by inserting knots and by order elevating the basis functions for describing the geometry (from linear to quadratic) in the local ξ-direction. In the η-direction (in the direction of truncation) the order was reduced (from linear to constant). Two different refinements were investigated and the resulting locations of collocation points are shown in Fig. 15.12.

15.4.3.3 Approximation of Body Forces Inside Domain

The domain for the volume integration was defined by 2 NURBS surfaces which were identical to the top and bottom patches for defining the problem geometry. The refinement of the boundary values was accompanied by an increased number

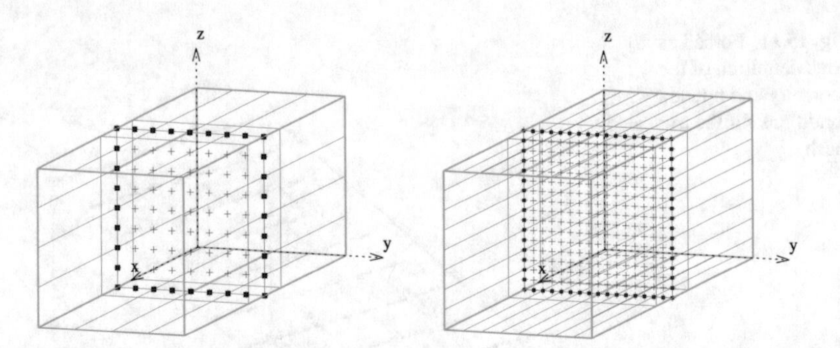

Fig. 15.12 Approximation/refinement of the unknown: location of collocation points (filled squares) achieved by order elevation (from linear to quadratic) and by 3 (mesh 1) and 7 (mesh 2) knot insertions in the local ξ-direction for each patch. Also shown are the internal points (crosses) for computing the body forces. Red lines indicate the limits of the integration regions

of internal points as shown in Fig. 15.12. Quadratic interpolation between the points was assumed.

15.4.3.4 Results for Re = 0

To ascertain the errors introduced by truncation, we compute the results for Re = 0 first and compare with the reference plane solution [68]. The convergence of the solution is shown for mesh 1 in Fig. 15.13 as a function of the distance of the truncation. It can be seen that for truncating the mesh at a distance of 3 a fairly good agreement can be obtained. Next we investigate the influence of the discretisation on the results. It can be seen in Fig. 15.14 that the results for mesh 2 (7 knot insertions) agree well with the reference curve. Mesh 1 has 96 and mesh 2 has 192 DOF.

15.4.3.5 Results for Re = 100

In Fig. 15.15 we show the results for the mesh truncated at a distance of 3 and for mesh 1 and mesh 2. A fairly good agreement can be seen with the reference solution for all meshes and either modified or full Newton–Raphson (referred to as NR in the figure).

15.4.3.6 Results for Re = 400

In Fig. 15.16 we show the results for the mesh truncated at a distance of 3 and for mesh 2. It can be seen that the results do not agree well with the reference solution, with the full Newton–Raphson being closer to it.

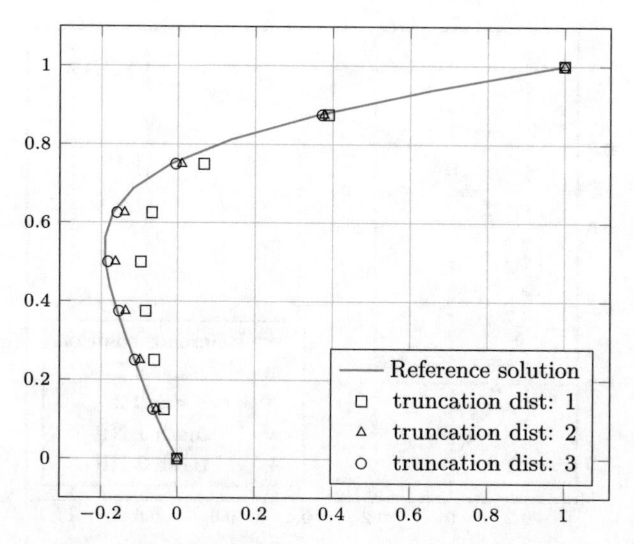

Fig. 15.13 Convergence of the solution for Re = 0 for different distances of the truncated boundary. Plotted is the magnitude of the horizontal velocity component along a vertical line through the middle

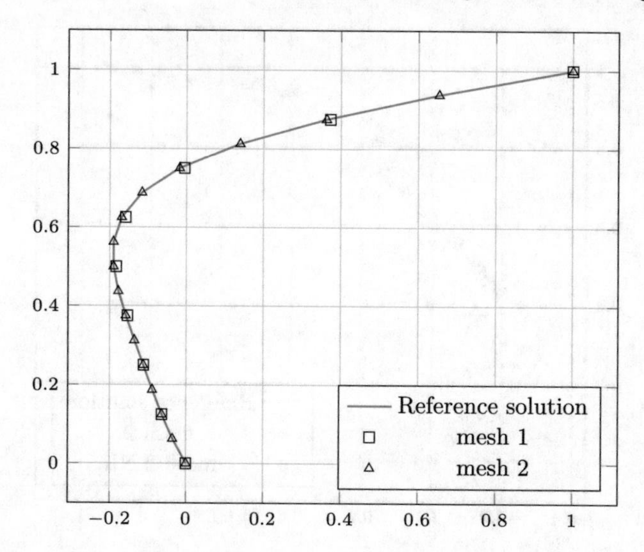

Fig. 15.14 Convergence of the solution for Re = 0 for two different discretisations

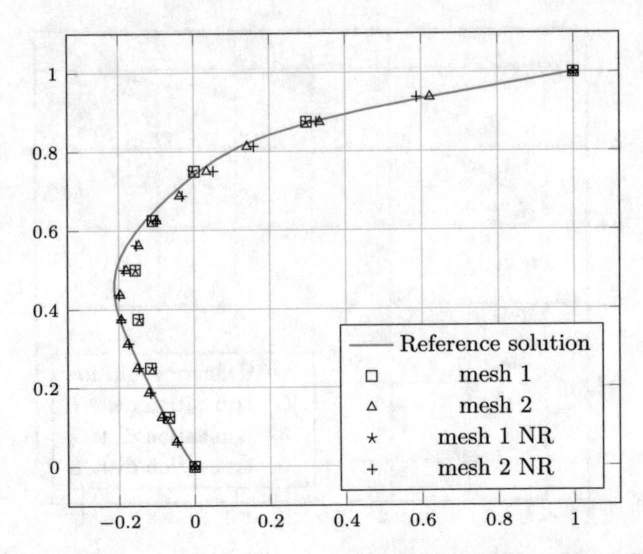

Fig. 15.15 Vertical velocity profile for Re = 100 with modified and full Newton–Raphson (NR) method

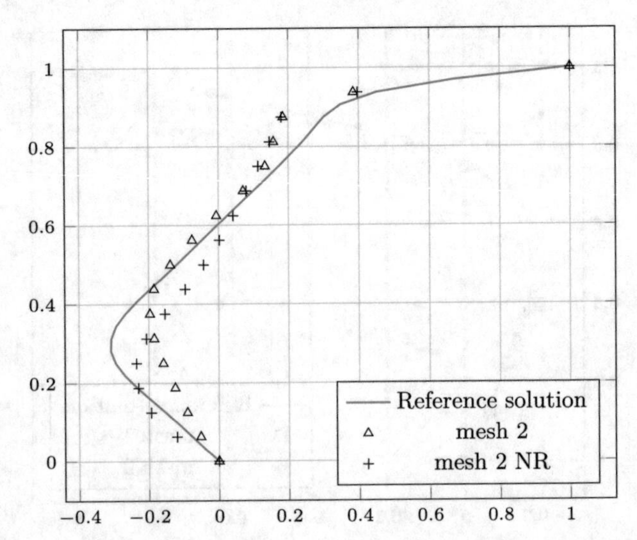

Fig. 15.16 Vertical velocity profile for Re = 400 using modified and full Newton–Raphson (NR)

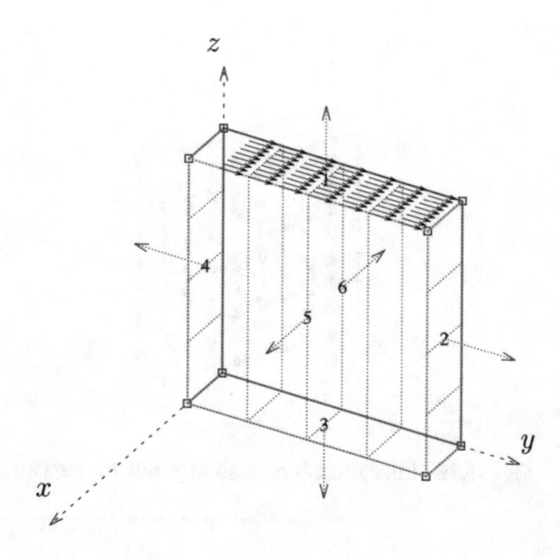

Fig. 15.17 Geometry description for non-truncated problem

15.4.4 *Non-truncated Geometry*

In this approach, the geometry is not truncated and 2 surfaces are included parallel to the y–z plane at a distance of 0.25.

15.4.4.1 Geometry Description and Boundary Conditions

The geometry is described by 6 linear NURBS patches as shown in Fig. 15.17.

The following boundary conditions are applied:

- Patch 1: Dirichlet BC with $u_1 = 0, u_2 = 1, u_3 = 0$
- Patches 2–4: Dirichlet BC with $u_1 = 0, u_2 = 0, u_3 = 0$
- Patches 5 and 6: Mixed BC with $u_1 = 0, t_2 = 0, t_3 = 0$.

For patch 1 the velocities were tapered off towards the corner as shown for the truncated mesh.

15.4.4.2 Approximation of the Unknown and Refinement

The approximation of the boundary unknowns was achieved by inserting knots and by order elevating the basis functions for describing the geometry (from linear to quadratic) in the ξ-direction. In the η-direction, the basis functions for describing the geometry were used for patches 1–4. For patches 5 and 6 the same refinement as in ξ-direction was used for the η-direction. Two different refinements were investigated and the resulting locations of collocation points computed using Greville abscissae are shown in Fig. 15.18. The meshes 1 and 2 have 486 and 1734 DOF respectively. Mesh 3, that has one additional knot inserted, has 2166 DOF.

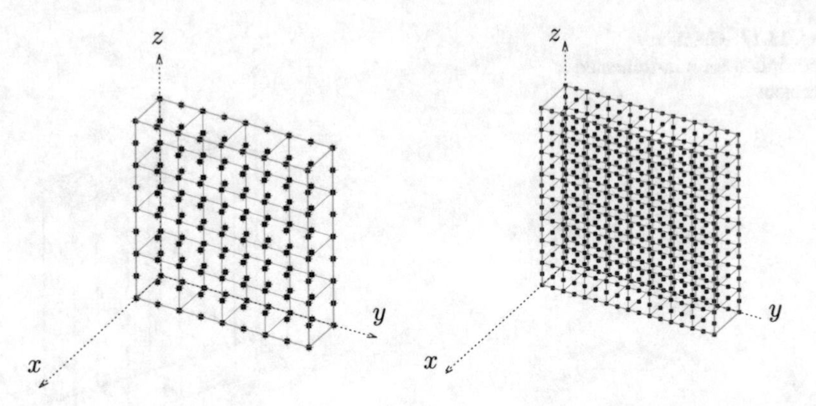

Fig. 15.18 Location of collocation points for the two refinements. Left: mesh 1, right: mesh 2

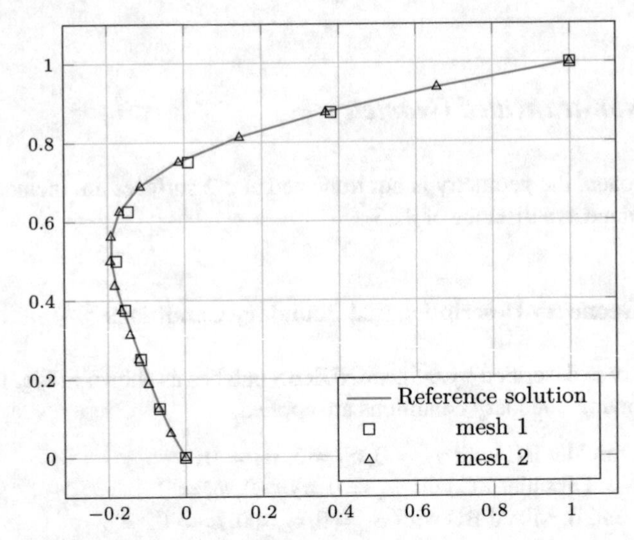

Fig. 15.19 Vertical velocity profile for Re = 0 and comparison with plane reference solution

15.4.4.3 Results for Re = 0

The results for Re = 0 are compared with the plane reference solution in Fig. 15.19 and good agreement can be found.

15.4.4.4 Results for Re = 100

In Fig. 15.20, we show the results for Re = 100 for the modified and full Newton–Raphson method. All results are in good agreement.

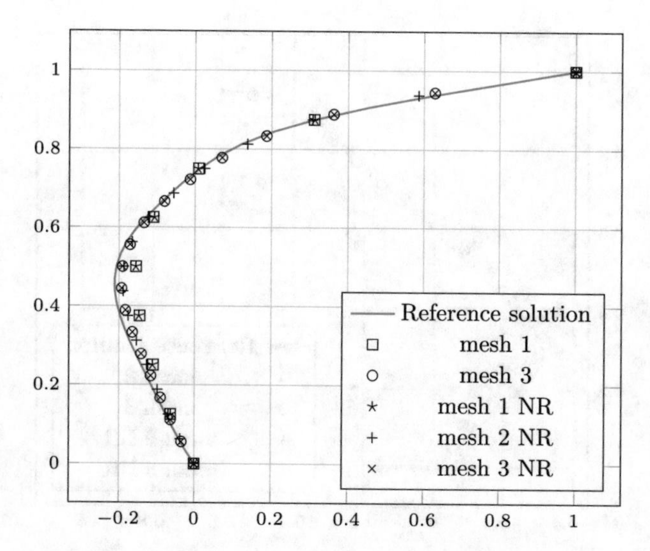

Fig. 15.20 Vertical velocity profile for Re = 100 and comparison with plane reference solution

15.4.4.5 Results for Re = 400

In Fig. 15.21 we show the results for Re = 400. For this Reynolds number the modified Newton–Raphson method requires a high number of iterations with a very low relaxation factor. For meshes 2 and 3 with full Newton–Raphson convergence is achieved with few iterations and the results are improved. It can be seen that as the number of unknowns is increased the reference solution is approached. The resulting velocity vectors are shown in Fig. 15.22.

15.4.5 Practical Example

This relates to the airfoil example that has been solved as a 2-D problem. The airfoil cross-section is lofted in one direction to create a 3-D wing.

15.4.5.1 Description of Geometry

Figure 15.23 shows the description of the boundary with 2 NURBS patches with order 2 in ξ-direction and order 1 in η-direction. The mesh is truncated at a distance of 2, to simulate plane conditions.

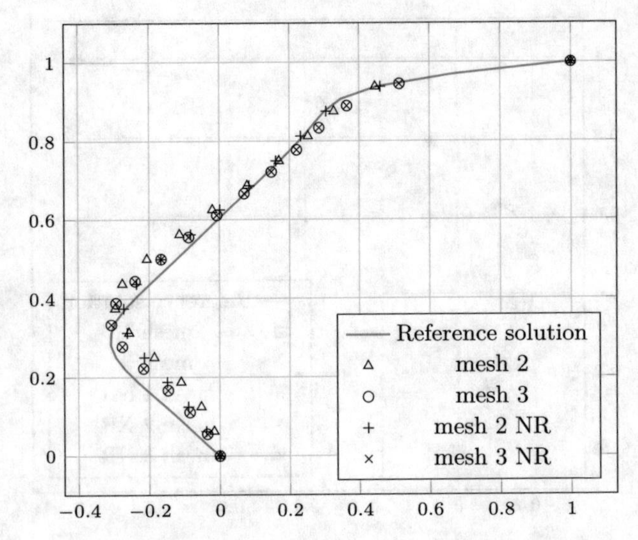

Fig. 15.21 Vertical velocity profile for Re = 400 and comparison with plane reference solution

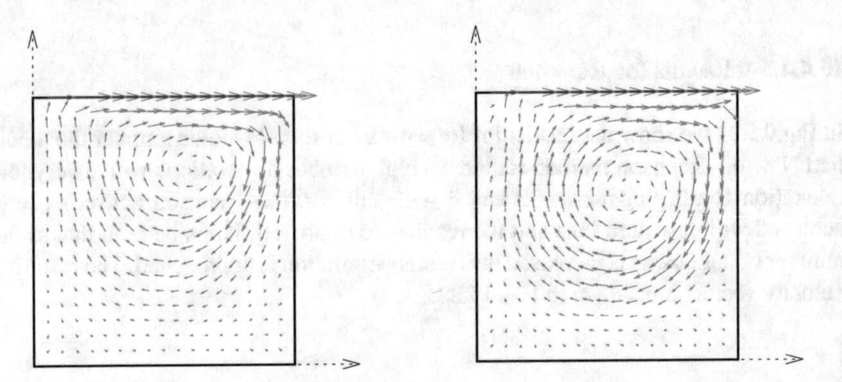

Fig. 15.22 Velocity vectors for (left) Re = 100 and (right) Re = 400

15.4.5.2 Description of the Unknown

Following the geometry independent field approximation philosophy we refine the NURBS basis functions in the ξ-direction by inserting 7 knots at (0.25, 0.375, 0.65, 0.75, 0.85, 0.91, 0.95). In the η-direction we decrease the order from 1 to 0 (constant variation). The resulting collocation points are shown in Fig. 15.24. The model has only 66 DOF.

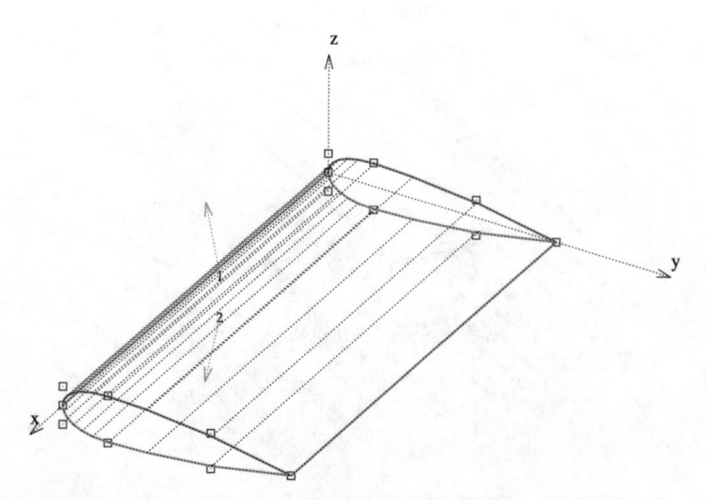

Fig. 15.23 Airfoil: description of the geometry with 2 NURBS patches, showing control points

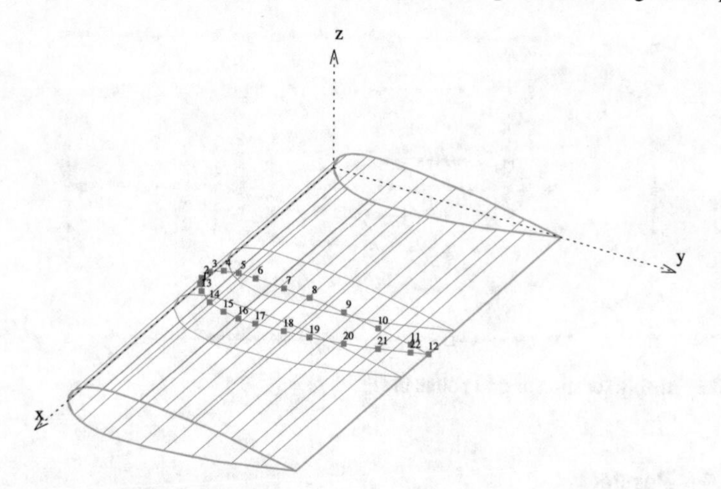

Fig. 15.24 Airfoil: location of collocation points

15.4.5.3 Description of Ω_0

The volume near the airfoil, where we assume non-linear effects are significant, is described by NURBS surfaces as shown in Fig. 15.25. The location of the grid points that were used to determine the Gauss point values with a linear interpolation are shown in Fig. 15.26.

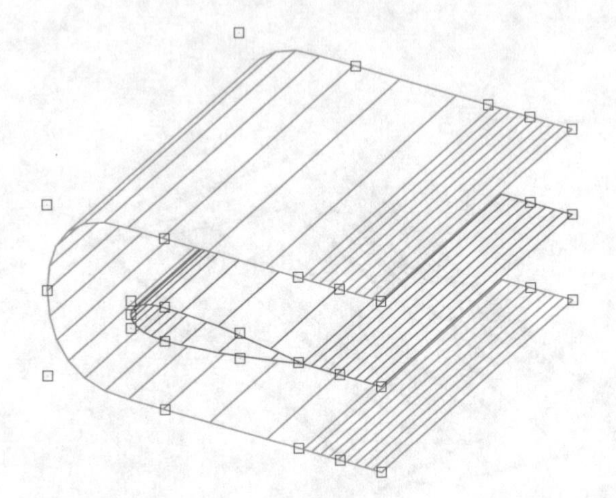

Fig. 15.25 Airfoil: definition of Ω_0 with 2 NURBS surfaces

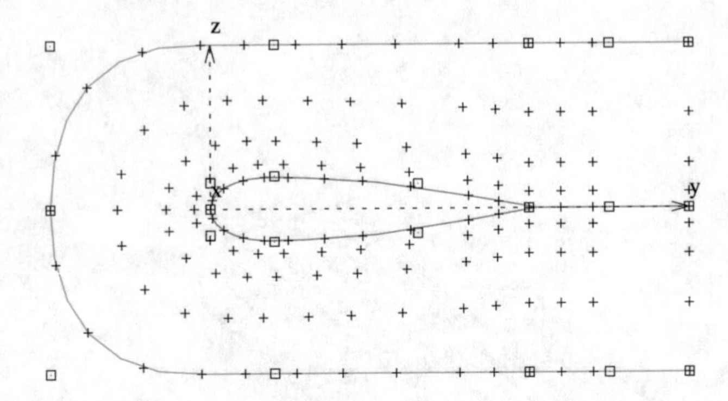

Fig. 15.26 Airfoil: location of grid points in Ω_0

15.4.5.4 Results

For this relatively coarse discretisation of Ω_0, the simulation only converges for a low Reynolds number. The resulting velocity vectors for Re = 10 are shown in Fig. 15.27.

15.5 Summary and Conclusions

Here we extended the BEM to problems of viscous flow. Applications of the BEM in this area are few and the majority of published results use Finite Elements or Finite Differences. However, it can be seen that good results can be obtained even though the problem becomes highly non-linear as the Reynolds number is increased. Even using a full Newton–Raphson method our approach has difficulty dealing with

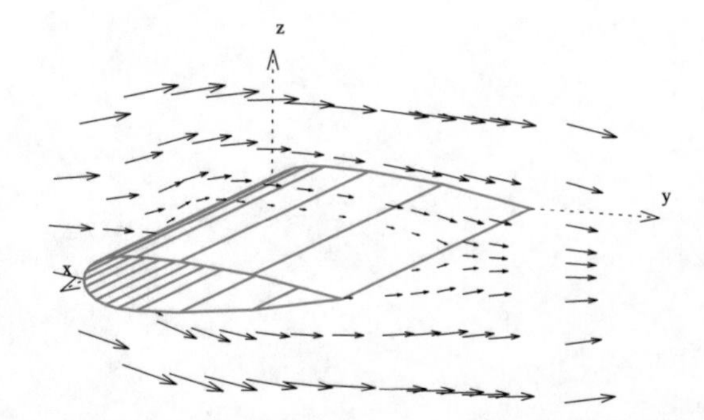

Fig. 15.27 Airfoil: flow vectors for Re = 10

Reynolds numbers that are higher than 500. It is surprising that for 3-D problems truncation, where no closed boundary is created, actually works reasonably well.

The advantage of the isogeometric BEM in being able to describe complex and smooth geometries with few parameters has been demonstrated on the example of an airfoil. The capability of our approach to capture non-linear behaviour near the boundary and the fact that the BEM can deal easily with infinite domains should be exploited further and its application to problems in fluid dynamics is still largely unexplored.

Chapter 16
Time Dependent Problems

So far we have only considered steady state problems, i.e. problems where time is not a factor and all results occur instantaneously. In other words we have neglected the influence of dynamic viscosity or mass. Here we re-introduce these effects. Time effects can be either transient or harmonic. We discuss transient problems for potential flow in 2-D first and then show an example of a harmonic problem, namely acoustics.

16.1 Transient Potential Problems

Transient problems include the solution of integrals over time. There are basically two main methods for dealing with the time integrals. Time marching methods and Laplace transform. In the former we discretise in time as well as in space in the latter time discretisation is avoided by transforming from the time domain into the Laplace domain and back.

16.1.1 Time Marching Method

Recalling the derivation in Sect. 2.1.2 the collocation integral equation can be written for time τ_M as:

$$\begin{aligned} c(\tilde{\boldsymbol{x}}_n)\, u(\tilde{\boldsymbol{x}}_n, \tau_M) &= \int_\Gamma [\mathsf{U} * t](\tilde{\boldsymbol{x}}_n, \tau_M, \hat{\boldsymbol{x}}) d\Gamma(\hat{\boldsymbol{x}}) \\ &- \int_\Gamma [\mathsf{T} * u](\tilde{\boldsymbol{x}}_n, \tau_M, \hat{\boldsymbol{x}}) d\Gamma(\hat{\boldsymbol{x}}) \end{aligned} \tag{16.1}$$

where $*$ denotes convolution.

G. Beer et al., *The Isogeometric Boundary Element Method*, Lecture Notes in Applied and Computational Mechanics 90,
https://doi.org/10.1007/978-3-030-23339-6_16

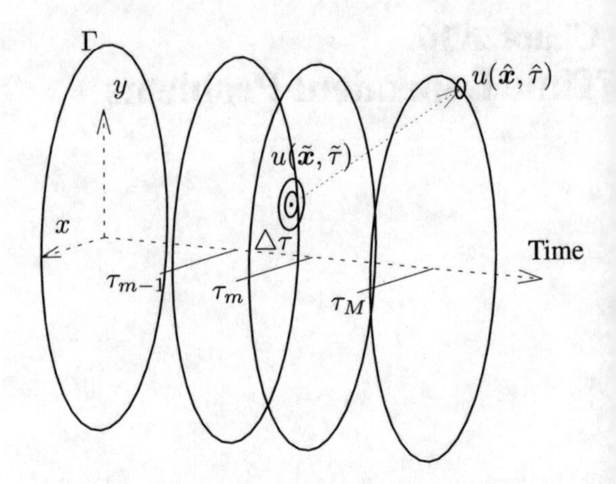

Fig. 16.1 Visualisation of time step procedure showing space-time domain and the source and field points (which are now location and time dependent)

To solve Eq. (16.1), we approximate the solution in the time domain. In the simplest case, we can divide the time into equal time steps $\triangle\tau$:

$$\tau_m = m\,\triangle\tau \tag{16.2}$$

where m is the time step number and assume the values u and t to be constant within one time step. The idea is to replace the integral over time with a sum of integrals over each time step $\triangle\tau$ as illustrated in Fig. 16.1.

Since u and t are assumed to be constant within the time step, they can be taken out of the time integral. The convolution integrals for the Mth time step can then be written as a sum over all time steps until time step M:

$$\mathsf{U} * t = \sum_{m=1}^{M} t(\hat{\boldsymbol{x}}, \tau_m)\mathsf{G}^m(\tilde{\boldsymbol{x}}, \hat{\boldsymbol{x}}) \tag{16.3}$$

$$\mathsf{T} * u = \sum_{m=1}^{M} u(\hat{\boldsymbol{x}}, \tau_m)\mathsf{F}^m(\tilde{\boldsymbol{x}}, \hat{\boldsymbol{x}}) \tag{16.4}$$

with

$$\mathsf{G}^m(\tilde{\boldsymbol{x}}, \hat{\boldsymbol{x}}) = \int_{\hat{\tau}=\tau_{m-1}}^{\tau_m} \mathsf{U}(\tilde{\boldsymbol{x}}_n, \tilde{\tau}, \hat{\boldsymbol{x}}, \hat{\tau})d\hat{\tau} \tag{16.5}$$

$$\mathsf{F}^m(\tilde{\boldsymbol{x}}, \hat{\boldsymbol{x}}) = \int_{\hat{\tau}=\tau_{m-1}}^{\tau_m} \mathsf{T}(\tilde{\boldsymbol{x}}_n, \tilde{\tau}, \hat{\boldsymbol{x}}, \hat{\tau})d\hat{\tau} \tag{16.6}$$

The boundary integral equation for time $\tau_M = M\triangle\tau$ can be written as

$$c(\tilde{\boldsymbol{x}}_n)u(\tilde{\boldsymbol{x}}_n,\tau_M) = \sum_{m=1}^{M} \int_{\Gamma} [\mathsf{G}^m(\tilde{\boldsymbol{x}}_n,\hat{\boldsymbol{x}})t(\hat{\boldsymbol{x}},\tau_m) - \mathsf{F}^m(\tilde{\boldsymbol{x}}_n,\hat{\boldsymbol{x}})u(\hat{\boldsymbol{x}},\tau_m)]\, d\Gamma(\hat{\boldsymbol{x}}). \tag{16.7}$$

After applying spatial discretisation (see Chap. 6), we obtain a system of equations:

$$\sum_{m=1}^{M} [\mathbf{G}]_{Mm} \{u\}_m = \sum_{m=1}^{M} [\mathbf{F}]_{Mm} \{t\}_m \tag{16.8}$$

where $\{u\}_m$ and $\{t\}_m$ are vectors containing values of parameters for the primary and secondary variable at the collocation points and at time m. $[\mathbf{G}]_{Mm}$ and $[\mathbf{F}]_{Mm}$ are assembled matrices that link time step M to time step m.

We can perform the integration of the convolution integrals in Eq. (16.5) analytically for each time step. For the Kernel F^m, we need the fundamental solution for the flux T, which is for plane problems

$$\mathsf{T}(\tilde{\boldsymbol{x}}_n,\tilde{\tau},\hat{\boldsymbol{x}},\hat{\tau}) = \frac{k\, r\, \frac{\partial r}{\partial n}}{8\pi\kappa^2(\hat{\tau}-\tilde{\tau})^2} \exp\left(-\frac{r^2}{4\kappa(\hat{\tau}-\tilde{\tau})}\right). \tag{16.9}$$

The analytical integration over a time step is now

$$\begin{aligned}
\mathsf{F}^m(\tilde{\boldsymbol{x}},\hat{\boldsymbol{x}}) &= \int_{\hat{\tau}=\tau_{m-1}}^{\tau_m} \mathsf{T}(\tilde{\boldsymbol{x}}_n,\tilde{\tau},\hat{\boldsymbol{x}},\hat{\tau})d\hat{\tau} \\
&= \frac{k\, r\, \frac{\partial r}{\partial n}}{8\pi\kappa^2} \int_{\tau_{m-1}}^{\tau_m} \frac{1}{\bar{\tau}^2} \exp\left(-\frac{r^2}{4\kappa\bar{\tau}}\right) d\hat{\tau} \\
&= \frac{k\, \frac{\partial r}{\partial n}}{2\pi\kappa r} [\exp(-a_m) - \exp(-a_{m-1})]
\end{aligned} \tag{16.10}$$

with $\kappa = k/(\rho c_h)$, $\bar{\tau} = \hat{\tau} - \tilde{\tau}$ and $a_m = r^2/(4\kappa(\tau_M - \tau_m))$.

The fundamental solution for the temperature of a two-dimensional problem is

$$\mathsf{U}(\tilde{\boldsymbol{x}}_n,\tilde{\tau},\hat{\boldsymbol{x}},\hat{\tau}) = \frac{1}{4\pi\kappa(\hat{\tau}-\tilde{\tau})} \exp\left(-\frac{r^2}{4\kappa(\hat{\tau}-\tilde{\tau})}\right) \tag{16.11}$$

and the analytical integration over a time step is

$$\begin{aligned}
\mathsf{G}^m(\tilde{\boldsymbol{x}},\hat{\boldsymbol{x}}) &= \int_{\hat{\tau}=\tau_{m-1}}^{\tau_m} \mathsf{U}(\tilde{\boldsymbol{x}}_n,\tilde{\tau},\hat{\boldsymbol{x}},\hat{\tau})d\hat{\tau} \\
&= \frac{1}{\pi r^2} \int_{\hat{\tau}=\tau_{m-1}}^{\tau_m} \frac{r^2}{4\kappa\bar{\tau}} \exp(-\frac{r^2}{4\kappa\bar{\tau}})d\hat{\tau}.
\end{aligned} \tag{16.12}$$

Substituting $z = \frac{r^2}{4\kappa\tau}$ we obtain

$$\mathsf{G}^m(\tilde{\boldsymbol{x}}, \hat{\boldsymbol{x}}) = \frac{1}{4\pi\kappa} \int_{z_{m-1}}^{z_m} \frac{e^{-z}}{z} dz = \frac{1}{4\pi\kappa} \int_{z_{m-1}}^{\infty} \frac{e^{-z}}{z} dz - \frac{1}{4\pi\kappa} \int_{z_m}^{\infty} \frac{e^{-z}}{z} dz$$
$$= \frac{1}{4\pi\kappa} \left[E_1(a_m) - E_1(a_{m-1}) \right]. \tag{16.13}$$

E_1 is the exponential integral function which is available in MATLAB.

We note that for $M = m$ both G and F are singular. G has a weak singularity of logarithmic type and can be numerically integrated using Gauss–Laguerre. F is strongly singular and in this case we can apply the regularisation method that has been used for steady state problems.

To solve for time step M, we re-arrange Eq. (16.8) so that known values of previous time steps are moved to the right-hand side

$$[\mathbf{G}]_{MM} \{u\}_M = [\mathbf{F}]_{MM} \{t\}_M + \{\mathbf{F}\}_M \tag{16.14}$$

where

$$\{\mathbf{F}\}_M = - \sum_{m=1}^{M-1} [\mathbf{G}]_{Mm} \{u\}_m + \sum_{m=1}^{M-1} [\mathbf{F}]_{Mm} \{t\}_m. \tag{16.15}$$

Separating known and unknown boundary values, we obtain for the unknown values at time $\tau_M = M \triangle \tau$

$$[\mathbf{L}]_{MM} \{x\}_M = \{\mathbf{R}\}_M + \{\mathbf{F}\}_M \tag{16.16}$$

where for example for a pure Dirichlet problem ($\{x\} = \{u\}$) we have $\{\mathbf{R}\}_M = [\mathbf{F}]_{MM} \{t\}_M$.

The algorithm is shown in Algorithm 10.

Algorithm 10 Algorithm for transient solution

1: Require time step $\triangle\tau$ and number of time steps Mt
2: Compute $[\mathbf{L}]_{11}$ and $\{\mathbf{R}\}_1$
3: Solve $\{x\}_1 = [\mathbf{L}]_{11}^{-1} \{\mathbf{R}\}_1$
4: Extract $\{u\}_1, \{t\}_1$
5: **for** $M = 2$ to number of time steps Mt **do**
6: Compute $[\mathbf{L}_{MM}]$ and $\{\mathbf{R}\}_M$
7: $\{\mathbf{F}\}_M = 0$
8: **for** $m = 1$ to $M - 1$ **do**
9: $\{\mathbf{F}\}_M = \{\mathbf{F}\}_M + \{\mathbf{F}\}_{Mm}\{t\}_m - [\mathbf{G}]_{Mm}\{u\}_m$
10: **end for**
11: Solve $\{x\}_M = [\mathbf{L}]_{MM}^{-1}(\{\mathbf{R}_M\} + \{\mathbf{F}\}_M)$
12: Extract $\{u\}_M, \{t\}_M$
13: **end for**

After the solution we can obtain the values of potential u at a point $\boldsymbol{x}$ inside the domain

$$u(\boldsymbol{x},\tau_M) = \sum_{m=1}^{M} \int_{\Gamma} [\mathsf{G}^m(\boldsymbol{x},\hat{\boldsymbol{x}})\, t(\hat{\boldsymbol{x}},\tau_m) - \mathsf{F}^m(\boldsymbol{x},\hat{\boldsymbol{x}})\, u(\hat{\boldsymbol{x}},\tau_m)]\, d\Gamma(\hat{\boldsymbol{x}}). \quad (16.17)$$

The flow vector $\mathbf{q} = \begin{pmatrix} q_x \\ q_y \end{pmatrix}$ can be computed by

$$\mathbf{q}(\boldsymbol{x},\tau_M) = \sum_{m=1}^{M} \int_{\Gamma} [\mathsf{S}^m(\boldsymbol{x},\hat{\boldsymbol{x}})\, t(\hat{\boldsymbol{x}},\tau_m) - \mathsf{R}^m(\boldsymbol{x},\hat{\boldsymbol{x}})\, u(\hat{\boldsymbol{x}},\tau_m)]\, d\Gamma(\hat{\boldsymbol{x}}) \quad (16.18)$$

where S^m and R^m are given by

$$\mathsf{S}_i^m(\tilde{\boldsymbol{x}},\hat{\boldsymbol{x}}) = \frac{\kappa r_{,i}}{2\pi r}\left[\exp(-a_m) - \exp(-a_{m-1})\right] \quad (16.19)$$

$$\begin{aligned} \mathsf{R}_i^m(\tilde{\boldsymbol{x}},\hat{\boldsymbol{x}}) = & \frac{r_{,i}\frac{\partial r}{\partial n}}{4\pi}\left[\frac{\exp(-a_m)}{(\tau_M-\tau_m)} - \frac{\exp(-a_{m-1})}{(\tau_M-\tau_{m-1})}\right] \\ & - \frac{\kappa n_i}{2r^2\pi}\left[\exp(-a_m) - \exp(-a_{m-1})\right] \\ & + \frac{\kappa r_{,i}\frac{\partial r}{\partial n}}{r^2\pi}\left[\exp(-a_m) - \exp(-a_{m-1})\right]. \end{aligned} \quad (16.20)$$

16.1.1.1 Test Example: Heat Conduction Across a Plate

The example relates to the transient heat conduction across a slab, initially at zero temperature. The bottom of the slab is perfectly insulated. At time zero the temperature at the top of the slab is increased to T_0 and maintained at that level. There is an exact solution for a slab of infinite extension and thickness L [41].

The temperature history at the bottom of the plate is given by:

$$T(\check{\tau}) = 1 - \frac{4}{\pi}\sum_{n=0}^{\infty}\left[\frac{(-1)^n}{2n+1}\exp\left(-(2n+1)^2\,\pi^2\,\check{\tau}/4\right)\right] \quad (16.21)$$

where the non-dimensional time is defined as $\check{\tau} = \kappa\tau/L^2$. Except for a very small time step at the beginning the sum converges fast with $n < 10$.

The test geometry including boundary conditions is shown in Fig. 16.2. The description of the geometry with 4 linear patches is shown in Fig. 16.3. For the simulation the basis functions for the two vertical patches were order elevated from linear to quadratic, resulting in the location of the collocation points depicted in the figure. The simulation has only 6 degrees of freedom. The test was run with $\kappa = 1$

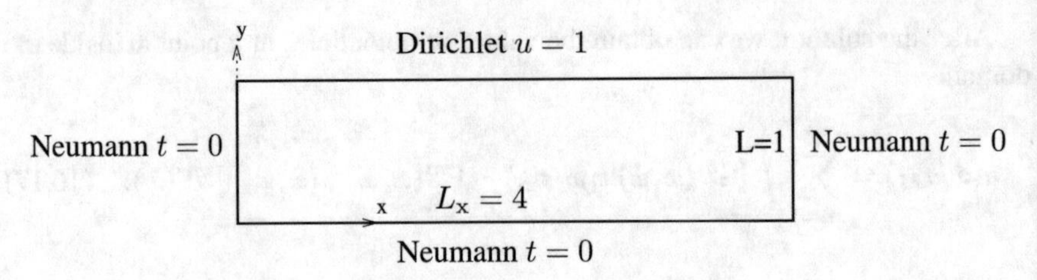

Fig. 16.2 Test example: sketch of the geometry and boundary conditions

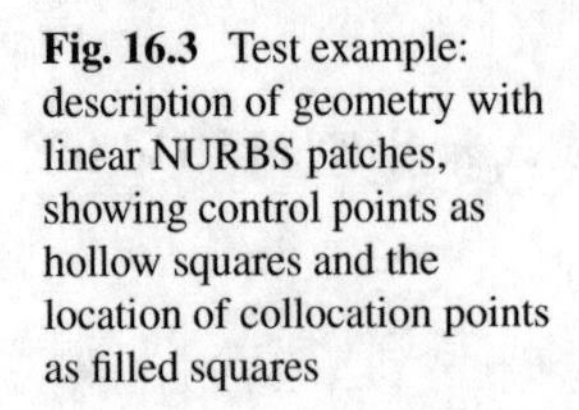

Fig. 16.3 Test example: description of geometry with linear NURBS patches, showing control points as hollow squares and the location of collocation points as filled squares

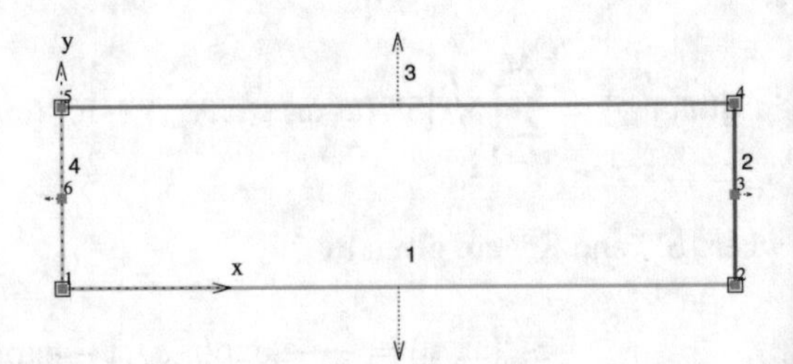

and $L = 1$ so $\tau = \check{\tau}$ in this case. The time history of the temperature at the bottom of the plate is shown in Fig. 16.4 for different time steps. It can be seen that the simulation agrees well with the analytical solution for small time steps $\triangle\tau$ but drifts for larger ones. The problem has been solved with conventional boundary elements in [51]. The mesh they used had 17 degrees of freedom and the reported results are very close to the ones obtained here. Here we only introduced the simplest time discretisation. Of course much more sophisticated time discretisation schemes are available and this is an active research topic (see for example [55]).

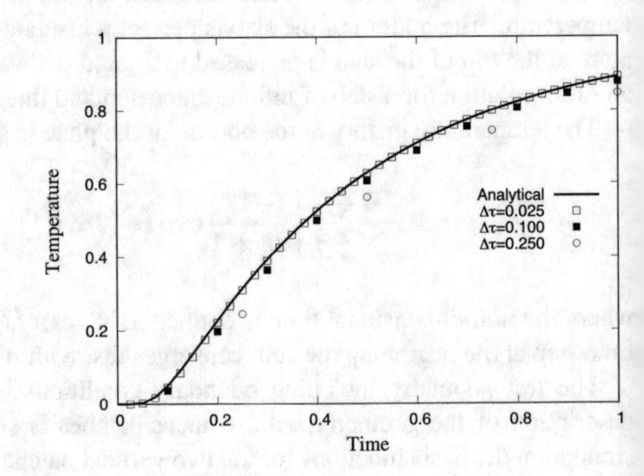

Fig. 16.4 Test example: comparison of computed temperature history at the bottom of the plate with the analytical result

16.1.2 Laplace Domain

Here we transform the variable $u(\hat{\boldsymbol{x}},\hat{\tau})$ to $u^*(\hat{\boldsymbol{x}},s)$ by

$$u^*(\hat{\boldsymbol{x}},s) = \mathcal{L}[u(\hat{\boldsymbol{x}},s)] = \int_0^\infty u(\hat{\boldsymbol{x}},\hat{\tau})\exp^{-s\hat{\tau}}\,d\hat{\tau} \tag{16.22}$$

where s is a Laplace transform parameter.

The diffusion equation now becomes

$$ku^*_{,ii}(\hat{\boldsymbol{x}},s) + \rho c_h\, s\, u^*(\hat{\boldsymbol{x}},s) = 0. \tag{16.23}$$

The integral equations become

$$\begin{aligned} c(\tilde{\boldsymbol{x}})\, u^*(\tilde{\boldsymbol{x}}) = &\int_\Gamma \mathsf{U}^*(\tilde{\boldsymbol{x}},\hat{\boldsymbol{x}},s)\, t^*(\hat{\boldsymbol{x}},s)\, d\Gamma(\hat{\boldsymbol{x}}) \\ &- \int_\Gamma \mathsf{T}^*(\tilde{\boldsymbol{x}},\hat{\boldsymbol{x}},s)\, u^*(\hat{\boldsymbol{x}},s)\, d\Gamma(\hat{\boldsymbol{x}}) \end{aligned} \tag{16.24}$$

where

$$t^*(\hat{\boldsymbol{x}},s) = -k\, u^*_{,i}(\hat{\boldsymbol{x}},s) n_i(\hat{\boldsymbol{x}}). \tag{16.25}$$

The fundamental solutions are:

$$\mathsf{U}^*(\tilde{\boldsymbol{x}},\hat{\boldsymbol{x}},s) = \frac{1}{4\pi k r}\exp\left(-r\sqrt{\frac{s}{\kappa}}\right) \quad \text{for 3-D} \tag{16.26}$$

$$\mathsf{U}^*(\tilde{\boldsymbol{x}},\hat{\boldsymbol{x}},s) = \frac{1}{2\pi k}K_0\left(r\sqrt{\frac{s}{\kappa}}\right) \quad \text{for 2-D} \tag{16.27}$$

$$\mathsf{T}^*(\tilde{\boldsymbol{x}},\hat{\boldsymbol{x}},s) = \frac{1}{4\pi r^2}\frac{\partial r}{\partial n}\left(1 + r\sqrt{\frac{s}{k}}\right)\exp\left(-r\sqrt{\frac{s}{k}}\right) \quad \text{for 3-D} \tag{16.28}$$

$$\mathsf{T}^*(\tilde{\boldsymbol{x}},\hat{\boldsymbol{x}},s) = \frac{1}{2\pi r}\frac{\partial r}{\partial n}\sqrt{\frac{s}{\kappa}}K_1\left(r\sqrt{\frac{s}{\kappa}}\right) \quad \text{for 2-D} \tag{16.29}$$

where K_0 and K_1 are modified Bessel functions of the second kind that are available as MATLAB functions ($K_n(Z) = besselk(n,Z)$).

The solution proceeds exactly as with steady state potential problems with $u(\hat{\boldsymbol{x}})$ replaced with $u^*(\hat{\boldsymbol{x}},s)$. We solve for several values of the Laplace transform parameter s. After the solution for $u^*(\hat{\boldsymbol{x}},s)$ we perform an inverse Laplace transform

$$u(\hat{\boldsymbol{x}},\hat{\tau}) = \mathcal{L}^{-1}[u^*(\hat{\boldsymbol{x}},s)] \tag{16.30}$$

to get the value of u at time $\hat{\tau}$. This step has to be performed numerically and is not trivial. Several methods have been proposed (see for example [52]), but all require some knowledge of the underlying physics and cannot be used as a black box. Therefore, even if the solution process is very attractive since a steady state program can be used for solving a transient problem by just changing the fundamental solutions, it is not really suitable for solving transient potential problems. The lack of numerical examples in the literature attest this.

16.2 Acoustics

The integral equations for acoustics and the fundamental solutions have been derived in Sect. 2.1.3 as

$$c(\tilde{\boldsymbol{x}})u(\tilde{\boldsymbol{x}}) = \int\limits_{\Gamma} \mathsf{U}(\tilde{\boldsymbol{x}},\hat{\boldsymbol{x}})t(\hat{\boldsymbol{x}})d\Gamma(\hat{\boldsymbol{x}}) - \int\limits_{\Gamma} \mathsf{T}(\tilde{\boldsymbol{x}},\hat{\boldsymbol{x}})u(\hat{\boldsymbol{x}})d\Gamma(\hat{\boldsymbol{x}}). \tag{16.31}$$

We can restate the equation in a form suitable for acoustic scattering

$$c(\tilde{\boldsymbol{x}})u(\tilde{\boldsymbol{x}}) + \int\limits_{\Gamma} \mathsf{T}(\tilde{\boldsymbol{x}},\hat{\boldsymbol{x}})u(\hat{\boldsymbol{x}})d\Gamma(\hat{\boldsymbol{x}}) = \int\limits_{\Gamma} \mathsf{U}(\tilde{\boldsymbol{x}},\hat{\boldsymbol{x}})t(\hat{\boldsymbol{x}})d\Gamma(\hat{\boldsymbol{x}}) + u_{\text{inc}}(\tilde{\boldsymbol{x}}) \tag{16.32}$$

where

$$u = u_s + u_{\text{inc}} \tag{16.33}$$

is the total acoustic potential, u_s is the scattered acoustic potential and u_{inc} is the incident wave acoustic potential. Incident plane waves are expressed as

$$u_{\text{inc}}(\boldsymbol{x}) = Ae^{i\boldsymbol{\alpha}\boldsymbol{x}}. \tag{16.34}$$

where A is the amplitude of the wave and $\boldsymbol{\alpha} = \alpha\mathbf{d}$ is the wave vector in the direction specified by unit vector $\mathbf{d}$.

Because the integral involving T is strongly singular and only exists as a Cauchy principal value we need to regularise Eq. (16.32). To do this we first subtract the static fundamental solution $\bar{\mathsf{T}}$ (calculated with $\alpha = 0$) and then add it

$$\begin{aligned} c(\tilde{\boldsymbol{x}})u(\tilde{\boldsymbol{x}}) &+ \int\limits_{\Gamma} \left[\mathsf{T}(\tilde{\boldsymbol{x}},\hat{\boldsymbol{x}})u(\hat{\boldsymbol{x}}) - \bar{\mathsf{T}}(\tilde{\boldsymbol{x}},\hat{\boldsymbol{x}})u(\tilde{\boldsymbol{x}})\right] d\Gamma(\hat{\boldsymbol{x}}) \\ &+ \bar{\mathsf{T}}(\tilde{\boldsymbol{x}},\hat{\boldsymbol{x}})u(\tilde{\boldsymbol{x}}) + u_{\text{inc}}(\tilde{\boldsymbol{x}}) = \int\limits_{\Gamma} \mathsf{U}(\tilde{\boldsymbol{x}},\hat{\boldsymbol{x}})t(\hat{\boldsymbol{x}})d\Gamma(\hat{\boldsymbol{x}}). \end{aligned} \tag{16.35}$$

The strong singularity of the first integral on the left-hand side has now been removed because as $\hat{\boldsymbol{x}}$ approaches $\tilde{\boldsymbol{x}}$ the integrand tends to zero.

The integral equation for the static problem can be written as

$$c(\tilde{\boldsymbol{x}})u(\tilde{\boldsymbol{x}}) = \int_\Gamma \bar{\mathsf{U}}(\tilde{\boldsymbol{x}},\hat{\boldsymbol{x}})t(\hat{\boldsymbol{x}})d\Gamma(\hat{\boldsymbol{x}}) - \int_\Gamma \bar{\mathsf{T}}(\tilde{\boldsymbol{x}},\hat{\boldsymbol{x}})u(\hat{\boldsymbol{x}})d\Gamma(\hat{\boldsymbol{x}}) \tag{16.36}$$

where $\bar{\mathsf{U}}$ and $\bar{\mathsf{T}}$ are the static fundamental solutions.

We can solve this for a constant value of $u = u(\tilde{\boldsymbol{x}})$ noting that for this case t is zero everywhere

$$c(\tilde{\boldsymbol{x}})u(\tilde{\boldsymbol{x}}) = \int_\Gamma \bar{\mathsf{T}}(\tilde{\boldsymbol{x}},\hat{\boldsymbol{x}})u(\hat{\boldsymbol{x}})d\Gamma(\hat{\boldsymbol{x}}). \tag{16.37}$$

Substituting this result into Eq. (16.35) we obtain

$$\int_\Gamma \left[\mathsf{T}(\tilde{\boldsymbol{x}},\hat{\boldsymbol{x}})u(\hat{\boldsymbol{x}}) - \bar{\mathsf{T}}(\tilde{\boldsymbol{x}},\hat{\boldsymbol{x}})u(\tilde{\boldsymbol{x}})\right] d\Gamma(\hat{\boldsymbol{x}}) + u_{\text{inc}}(\tilde{\boldsymbol{x}}) = \int_\Gamma \mathsf{U}(\tilde{\boldsymbol{x}},\hat{\boldsymbol{x}})t(\hat{\boldsymbol{x}})d\Gamma(\hat{\boldsymbol{x}}). \tag{16.38}$$

The discretised equation is

$$\sum_{e=1}^{E}\int_{\Gamma_e} \mathsf{T}(\tilde{\boldsymbol{x}}_n,\hat{\boldsymbol{x}})u^e(\hat{\boldsymbol{x}})d\Gamma_e(\hat{\boldsymbol{x}}) - \left[\sum_{e=1}^{E}\int_{\Gamma_e} \bar{\mathsf{T}}(\tilde{\boldsymbol{x}}_n,\hat{\boldsymbol{x}})d\Gamma_e(\hat{\boldsymbol{x}})\right] u(\tilde{\boldsymbol{x}}_n) = \tag{16.39}$$

$$\sum_{e=1}^{E}\int_{\Gamma_e} \mathsf{U}(\tilde{\boldsymbol{x}}_n,\hat{\boldsymbol{x}})\, t^e(\hat{\boldsymbol{x}})\, d\Gamma_e(\hat{\boldsymbol{x}}) + u_{inc}(\tilde{\boldsymbol{x}}).$$

16.2.1 Pulsating Sphere Example

The following example has been solved in [168]. In the pulsating sphere problem a constant acoustic radial velocity $v_0 = \frac{\partial u}{\partial n}$ is prescribed over the entire surface of a sphere. An exact solution is available for this problem. For a sphere of radius a, the acoustic potential at any point with a radial distance r is given by [134]:

$$u(r,\alpha) = \frac{v_0 a^2}{1 - i\alpha a}\frac{e^{i\alpha(r-a)}}{r}. \tag{16.40}$$

For the example the following input values are chosen: $\rho = 1.0$, $c = 1.0$, $a = 0.5$ $v_0 = -1650i$ and $\alpha = 1$.

The definition of the geometry with 8 patches of order 2 is shown in Fig. 16.5. This describes the geometry of the sphere exactly. In this definition, the control points at the poles are coincident. This means that the Jacobian tends to zero as the poles are approached but this is of no consequence as Gauss points will not be located there.

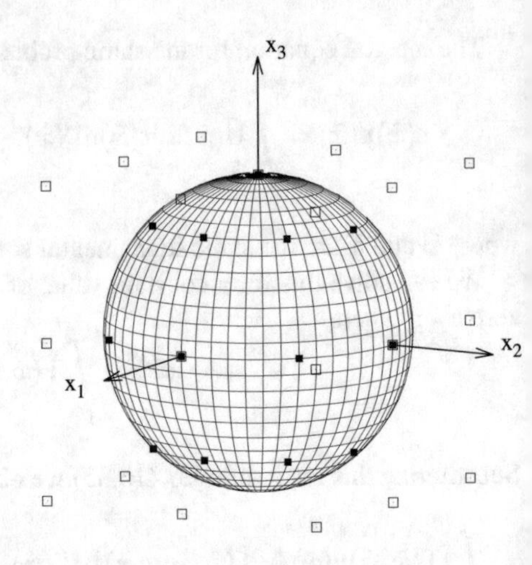

Fig. 16.5 Geometry definition of sphere showing control points as hollow squares and collocation points as filled squares

In fact, this geometry detail is beneficial for the computation of the nearly singular integral.

Since the exact solution of u is constant over the boundary and the geometry is exact, the numerical solution is obtained using the same basis functions as are used for describing the geometry. The associated collocation points are shown in Fig. 16.5. This discretisation has only 26 degrees of freedom. The accuracy of the solution now only depends on the specified precision of integration and the precision of the input data.

For the conventional BEM, a mesh has to be generated that approximates the geometry. The finer the mesh, the better the approximation of the geometry and the unknown. Figure 16.6 shows the discretisation into quadratic boundary elements and the error in the solution of u. The mesh has 450 degrees of freedom, but it can be

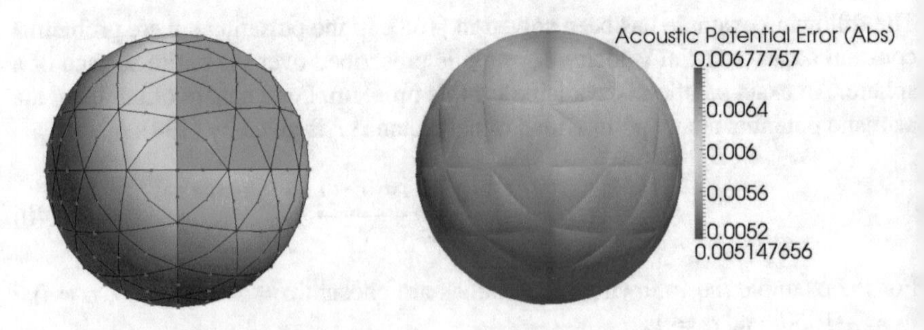

Fig. 16.6 Pulsating sphere, left: mesh of quadratic Lagrange elements. Right: contours of error, courtesy of Simpson et al. [168]

seen that even with this fairly fine mesh, the error is still significant. This is another example of the power of the isogeometric BEM.

It should be noted that, when the integral equation is applied to exterior problems, it suffers from instabilities when the wave number corresponds to an eigenfrequency. This problem can be alleviated by stabilisation according to Burton-Miller [38]. Further examples of the power of IGABEM in solving harmonic potential problems can be found in [167–169].

16.3 Summary

In this chapter, we dealt with the consideration of time in the simulation and with transient or harmonic problems. In transient problems, an additional dimension (time) is added, which means that fundamental solutions are not only dependent on space but also on time. The solution now proceeds in time steps and in the simplest case we can assume that the unknown is constant within a time step. An alternative to solving the transient problem is to use a Laplace transform, which means that time is eliminated in the simulation and the procedures for steady state problems can be used. However, the recovery of results using an inverse Laplace transform is a challenging problem. Regarding harmonic problems, we have shown a simulation in acoustics where an exact solution can be achieved with the IGABEM since the geometry is exactly represented.

Chapter 17
Summary and Outlook

The aim of the book was to introduce isogeometric concepts to the Boundary Element Method (BEM). In particular, B-splines or Non-Uniform Rational B-splines (NURBS) were used for the description of the boundary and for the approximation of the unknowns. At first glance this seems to be a trivial improvement, not warranting the publication of a book. However, as we have shown, this is not the case. The introduction of NURBS has significant implications in the implementation and an impact on the analysis. Firstly, due to the fact that NURBS can describe complex geometrical shapes with few parameters meant, that instead of approximating the boundary with a mesh of Lagrange boundary elements, we describe it exactly with NURBS patches. We have shown on many examples that this reduces the number of geometrical parameters, used to describe a boundary, drastically. In addition, we fully exploited the unique properties of NURBS for the approximation of the unknowns, introducing refinement options not available with Lagrange polynomials.

The use of a geometry independent field approximation, introduced by the authors in 2014 [112], makes the simulation more efficient. In this approach, we start the simulation with the basis functions for describing the geometry and use them also for the approximation of the unknown. We then refine the functions for the approximation only, keeping the geometry description – already very accurate or exact – unchanged. Therefore in the book, we have not only introduced the new basis functions to BEM technology, but also investigated how they can be used efficiently for the approximation of the unknowns. A comparison with Lagrange polynomials revealed that B-splines and NURBS perform significantly better.

It should be noted here that no attempt is made in this book to fit isogeometric analysis into the conventional analysis framework, as has been proposed for example through methods such as Bézier extraction [30]. Instead, we use NURBS patches directly and develop new analysis methods. This philosophy will eventually lead to a future tight connection between CAD and simulation, which is still subject to intensive research at the time of publishing.

G. Beer et al., *The Isogeometric Boundary Element Method*, Lecture Notes in Applied and Computational Mechanics 90,
https://doi.org/10.1007/978-3-030-23339-6_17

In general, the BEM is ideally suited to link analysis to CAD models since both rely on the same representation, namely volumes are defined by surfaces bounding the domain. Hence, challenging tasks like deriving volumetric discretisation are not required at all. Also, the fact that the BEM allows for discontinuous field approximation facilitates the interaction of CAD and simulation. This tight relation between the BEM and CAD models can be particularly beneficial in the early stage of virtual prototyping.

Two chapters have been devoted to the description of the geometry using CAD technology. An important aspect, that is considered, is that CAD programs use trimming, which means that the geometric models are not watertight and cannot be immediately used for simulation. This also means that the basis functions used for approximating the unknowns are trimmed as well.

The fact that patches cover a much larger surface area than conventional boundary elements and that NURBS are rational functions, meant that integration procedures have to be revised. A detailed error analysis of Gauss quadrature resulted in new integration tables for finding the proper number of Gauss points, that were used in the examples presented throughout the book. A comparison of our simulation results with analytical solutions and results from other publications reveal that the integration procedures work well.

Another innovative aspect of the book is that subdomains, where body forces may be present, are described using isogeometric methods. This is an attractive alternative to the generation of cell meshes, used in the conventional BEM. The consideration of body forces allows the simulation of problems with heterogeneous domains and non-linear material behaviour. Several examples of this have been presented, ranging from the treatment of elastic and in-elastic inclusions to the highly non-linear problem of viscous flow. In Chap. 16, we introduce time effects and show a problem in acoustics, where a comparison with an analytical solution showed that the error only depended on the integration error, due to the fact that the geometry of the boundary was exactly represented.

Throughout the book examples are presented that attest to the superiority of the isogeometric BEM (IGABEM) over the conventional BEM, being the number of parameters used to describe the boundary, the number of degrees of freedom used and the accuracy of the results. The range of applications discussed in this book is limited, mainly because, except for the acoustics example, the authors have only presented applications, which they have implemented and tested themselves. With the basis provided in this book, however, it should not be difficult to expand the IGABEM concepts to a range of other applications.

An application that, in the view of the authors, has great potential for the IGABEM is computational fluid dynamics, which should not only benefit from the ease with which the BEM can deal with infinite domains but also from the innovative methods presented here in dealing with the description of smooth boundaries and with highly non-linear effects near the boundary.

We are confident that the IGABEM will have a bright future.

Appendix A
Fundamental Solutions

Definitions:

- d … Cartesian dimension
- $\tilde{\boldsymbol{x}}$ … source point coordinates
- $\hat{\boldsymbol{x}}$ … field point coordinates
- $\mathbf{r} = \hat{\boldsymbol{x}} - \tilde{\boldsymbol{x}} = \{r_i\}_{i=1}^{d}$
- $r = |\hat{\boldsymbol{x}} - \tilde{\boldsymbol{x}}|$
- $r_{,i} = \frac{1}{r} r_i$
- $\mathbf{n} = \{n_i\}_{i=1}^{d}$ … outward normal
- $\frac{\partial r}{\partial n} = \frac{1}{r}\mathbf{r} \cdot \mathbf{n}$
- δ_{ij} … Kronecker Delta
- k … conductivity (isotropic)
- $\mathbf{K} = K_{ij}$ …conductivity tensor (anisotropic)
- $\bar{r} = \mathbf{r}^{\mathrm{T}}\mathbf{K}\,\mathbf{r}$
- ρ … density
- c_h … dynamic viscosity or specific heat
- $\kappa = k/(\rho c_h)$
- α … wave number
- E … modulus of elasticity
- ν … Poisson's ratio
- G … shear modulus
- μ … viscosity
- $\mathbf{b} = \{b_i\}_{i=1}^{d}$ … body force
- $\frac{\partial r}{\partial b} = \frac{1}{r}\mathbf{b} \cdot \mathbf{r}$
- $\frac{\partial b}{\partial n} = \mathbf{b} \cdot \mathbf{n}$

G. Beer et al., *The Isogeometric Boundary Element Method*, Lecture Notes in Applied and Computational Mechanics 90,
https://doi.org/10.1007/978-3-030-23339-6

A.1 2-D Potential Flow

A.1.1 Isotropic Fundamental Solution

$$\mathsf{U}(\tilde{\boldsymbol{x}}, \hat{\boldsymbol{x}}) = \frac{1}{2\pi k} \ln\left(\frac{1}{r}\right) \tag{A.1}$$

$$\mathsf{T}(\tilde{\boldsymbol{x}}, \hat{\boldsymbol{x}}) = -\frac{1}{2\pi} \frac{\partial r}{\partial n} \left(\frac{1}{r}\right) \tag{A.2}$$

$$\mathsf{S}_i(\boldsymbol{x}, \hat{\boldsymbol{x}}) = -\frac{r_{,i}}{2\pi r} \tag{A.3}$$

$$\mathsf{R}_i(\boldsymbol{x}, \hat{\boldsymbol{x}}) = -\frac{k}{2\pi r^2}(2r_{,i}(r_{,j}n_j) - n_i) \tag{A.4}$$

A.1.2 Anisotropic Fundamental Solutions

$$\mathsf{U} = \frac{1}{2\pi \ \det(\mathbf{K})^{1/2}} \ln\left(\frac{1}{\bar{r}}\right) \tag{A.5}$$

$$\mathsf{T} = -\frac{1}{2\pi \ \det(\mathbf{K})^{1/2}} \frac{n_1 r_1 + n_2 r_2}{\bar{r}^2} \tag{A.6}$$

$$\frac{\partial \mathsf{U}}{\partial x_1} = \frac{1}{2\pi \ \det(\mathbf{K})^{3/2}} \frac{K_{12} r_2 - K_{22} r_1}{\bar{r}^2} \tag{A.7}$$

$$\frac{\partial \mathsf{U}}{\partial x_2} = \frac{1}{2\pi \ \det(\mathbf{K})^{3/2}} \frac{K_{12} r_1 - K_{11} r_2}{\bar{r}^2} \tag{A.8}$$

$$\mathsf{S}_i = -(K_{i1} \frac{\partial \mathsf{U}}{\partial x_1} + K_{i2} \frac{\partial \mathsf{U}}{\partial x_2}) \tag{A.9}$$

$$\frac{\partial \mathsf{T}}{\partial x_1} = \frac{K_{22} n_1 {r_1}^2 - K_{11} n_1 {r_2}^2 - 2K_{12} n_2 {r_2}^2 + 2K_{22} n_2 r_1 r_2}{2\pi \ \det(\mathbf{K})^{3/2} \ \bar{r}^4} \tag{A.10}$$

$$\frac{\partial \mathsf{T}}{\partial x_2} = \frac{K_{11} n_2 {r_2}^2 - K_{22} n_2 {r_1}^2 - 2K_{12} n_1 {r_1}^2 + 2K_{11} n_1 r_1 r_2}{2\pi \ \det(\mathbf{K})^{3/2} \ \bar{r}^4} \tag{A.11}$$

$$\mathsf{R}_i = K_{i1} \frac{\partial \mathsf{T}}{\partial x_1} + K_{i2} \frac{\partial \mathsf{T}}{\partial x_2} \tag{A.12}$$

A.2 3-D Potential Flow

A.2.1 *Isotropic Fundamental Solution*

$$\mathsf{U}(\tilde{\boldsymbol{x}}, \hat{\boldsymbol{x}}) = \frac{1}{4\pi k}\left(\frac{1}{r}\right) \tag{A.13}$$

$$\mathsf{T}(\tilde{\boldsymbol{x}}, \hat{\boldsymbol{x}}) = -\frac{1}{4\pi}\frac{\partial r}{\partial n}\left(\frac{1}{r^2}\right) \tag{A.14}$$

$$\mathsf{S}_i(\boldsymbol{x}, \hat{\boldsymbol{x}}) = -\frac{r_{,i}}{4\pi r^2} \tag{A.15}$$

$$\mathsf{R}_i(\boldsymbol{x}, \hat{\boldsymbol{x}}) = -\frac{k}{4\pi r^3}(3r_{,i}(r_{,j}n_j) - n_i) \tag{A.16}$$

A.2.2 *Anisotropic Fundamental Solutions*

$$\mathsf{U} = \frac{1}{4\pi\ \det(\mathbf{K})^{1/2}}\frac{1}{\bar{r}} \tag{A.17}$$

$$\mathsf{T} = -\frac{1}{4\pi\ \det(\mathbf{K})^{1/2}}\ \frac{n_1 r_1 + n_2 r_2 + n_3 r_3}{\bar{r}^3} \tag{A.18}$$

A.3 Transient Potential Flow

$$\mathsf{U}(\tilde{\boldsymbol{x}}, \tilde{\tau}, \hat{\boldsymbol{x}}, \hat{\tau}) = \frac{1}{[4\pi\kappa(\hat{\tau} - \tilde{\tau})]^{d/2}}\exp\left(\frac{-r^2}{4\kappa(\hat{\tau} - \tilde{\tau})}\right) \tag{A.19}$$

$$\mathsf{T}(\tilde{\boldsymbol{x}}, \tilde{\tau}, \hat{\boldsymbol{x}}, \hat{\tau}) = \frac{k\, r}{2^{(d+1)}\ \pi^{d/2}\ [\kappa\ (\hat{\tau} - \tilde{\tau})]^{(d+2)/2}}\frac{\partial r}{\partial n}\exp\left(\frac{-r^2}{4\kappa(\hat{\tau} - \tilde{\tau})}\right) \tag{A.20}$$

A.4 3-D Acoustics

$$\mathsf{U}(\tilde{\boldsymbol{x}}, \hat{\boldsymbol{x}}) = \frac{e^{i\alpha r}}{4\pi r} \tag{A.21}$$

$$\mathsf{T}(\tilde{\boldsymbol{x}}, \hat{\boldsymbol{x}}) = \frac{e^{i\alpha r}}{4\pi r^2}\left(i\alpha r - \frac{\partial r}{\partial n}\right) \tag{A.22}$$

Table A.1 Constants for fundamental solutions

Constant	Plane strain	Plane stress	3-D
d	1	1	2
C	$\frac{1}{8\pi G(1-\nu)}$	$\frac{(1+\nu)}{8\pi G}$	$\frac{1}{16\pi G(1-\nu)}$
C_1	$3-4\nu$	$\frac{3-\nu}{1+\nu}$	$3-4\nu$
C_2	$\frac{1}{4\pi(1-\nu)}$	$\frac{1+\nu}{4\pi}$	$\frac{1}{8\pi(1-\nu)}$
C_3	$1-2\nu$	$\frac{1-\nu}{1+\nu}$	$1-2\nu$
C_4	2	2	3
C_5	4	4	5

A.5 Elasticity

$$\mathsf{U}(\tilde{\boldsymbol{x}}, \hat{\boldsymbol{x}}) = C(C_1 \ln \frac{1}{r}\delta_{ij} + r_{,i}r_{,j}) \quad \text{for 2-D} \tag{A.23}$$

$$\mathsf{U}(\tilde{\boldsymbol{x}}, \hat{\boldsymbol{x}}) = C\frac{1}{r}(C_1\delta_{ij} + r_{,i}r_{,j}) \quad \text{for 3-D} \tag{A.24}$$

$$\mathsf{T}(\tilde{\boldsymbol{x}}, \hat{\boldsymbol{x}}) = -\frac{C_2}{r^d}\left[(C_3\delta_{ij} + (d+1)r_{,i}r_{,j})\frac{\partial r}{\partial n} - C_3(n_j r_{,i} - n_i r_{,j})\right] \tag{A.25}$$

The derived solutions for the strain evaluation are:

$$\hat{\mathsf{S}}(\boldsymbol{x}, \hat{\boldsymbol{x}}) = \frac{C}{r^d}\left[C_3(r_{,j}\delta_{ik} + r_{,i}\delta_{jk}) - r_{,k}\delta_{ij} + C_4 r_{,i}r_{,j}r_{,k}\right] \tag{A.26}$$

$$\begin{aligned}\hat{\mathsf{R}}(\boldsymbol{x}_i, \hat{\boldsymbol{x}}) = \frac{C_2}{r^{d+1}}\Big[&C_4\frac{\partial r}{\partial n}(\nu(r_{,j}\delta_{ik} + r_{,i}\delta_{jk}) + r_{,k}\delta_{ij} - C_5 r_{,i}r_{,j}r_{,k}) \\ &+ C_3(n_j\delta_{ik} - n_k\delta_{ij} + n_i\delta_{jk} + C_4 r_{,i}r_{,j}n_k) \\ &+ C_4\nu(n_j r_{,i}r_{,k} + n_i r_{,j}r_{,k})\Big]\end{aligned} \tag{A.27}$$

For the stress evaluation the solutions are:

$$\mathsf{S}(\boldsymbol{x}, \hat{\boldsymbol{x}}) = \frac{C_2}{r^d}\left[C_3(r_{,j}\delta_{ik} + r_{,i}\delta_{jk} - r_{,k}\delta_{ij}) + C_4 r_{,i}r_{,j}r_{,k}\right] \tag{A.28}$$

$$\begin{aligned}\mathsf{R}(\boldsymbol{x}_i, \hat{\boldsymbol{x}}) = \frac{2GC_2}{r^{d+1}}\Big[&C_4\frac{\partial r}{\partial n}\Big(C_3 r_{,k}\delta_{ij} + \nu(r_{,j}\delta_{ik} + r_{,i}\delta_{jk}) - C_5 r_{,i}r_{,j}r_{,k}\Big) \\ &+ C_4\nu(n_i r_{,j}r_{,k} + n_j r_{,i}r_{,k}) - (1-4\nu)n_k\delta_{ij} \\ &+ C_3(C_4 n_k r_{,i}r_{,j} + n_j\delta_{ik} + n_i\delta_{jk})\Big]\end{aligned} \tag{A.29}$$

A.5.1 Coordinate Transformation

The transformation of a vector $\mathbf{v}$ in the Cartesian coordinate system x, y, z (with the basis vectors $\mathbf{e}_1, \mathbf{e}_2, \mathbf{e}_3$) to $\mathbf{v}'$ in an orthogonal coordinate system x', y', z' (with the basis vectors $\mathbf{e}_1', \mathbf{e}_2', \mathbf{e}_3'$) is given by

$$\mathbf{v}' = \mathbf{T}\,\mathbf{v} = \begin{bmatrix} \cos(\alpha_{11}) & \cos(\alpha_{12}) & \cos(\alpha_{13}) \\ \cos(\alpha_{21}) & \cos(\alpha_{22}) & \cos(\alpha_{23}) \\ \cos(\alpha_{31}) & \cos(\alpha_{32}) & \cos(\alpha_{33}) \end{bmatrix} \mathbf{v} \tag{A.30}$$

where α_{ij} is the angle between $\mathbf{e}_j$ and $\mathbf{e}_i'$. Moreover, the transformation matrix $\mathbf{T}$ has the property $\mathbf{T}\,\mathbf{T}^{\mathrm{T}} = \mathbf{I}$ and hence,

$$\mathbf{v} = \mathbf{T}^{\mathrm{T}}\,\mathbf{v}'. \tag{A.31}$$

A.5.2 Transversal Isotropy

The transformation of the locally defined constitutive matrix $\mathbf{D}'$ to the globally defined one $\mathbf{D}$ is given by:

$$\mathbf{D} = \mathbf{T}^{\mathrm{T}}\,\mathbf{D}'\,\mathbf{T} \tag{A.32}$$

To derive the transformation matrix $\mathbf{T}$, unit vectors $\mathbf{v}_i$ pointing to the direction of the local coordinate axes x_i' are defined by α and β as:

$$\mathbf{v}_1 = \begin{bmatrix} \sin(\alpha) \\ \cos(\alpha) \\ 0 \end{bmatrix} \quad \mathbf{v}_2 = \begin{bmatrix} \cos(\beta)\cos(\alpha) \\ -\cos(\beta)\sin(\alpha) \\ -\sin(\beta) \end{bmatrix} \quad \mathbf{v}_3 = \begin{bmatrix} -\sin(\beta)\cos(\alpha) \\ \sin(\beta)\sin(\alpha) \\ -\cos(\beta) \end{bmatrix} \tag{A.33}$$

Using these vectors $\mathbf{T}$ is given by:

$$\mathbf{T} = \begin{bmatrix} \mathrm{v}_{1x}^2 & \mathrm{v}_{1y}^2 & \mathrm{v}_{1z}^2 & \mathrm{v}_{1x}\mathrm{v}_{1y} & \mathrm{v}_{1y}\mathrm{v}_{1z} & \mathrm{v}_{1x}\mathrm{v}_{1z} \\ \mathrm{v}_{2x}^2 & \mathrm{v}_{2y}^2 & \mathrm{v}_{2z}^2 & \mathrm{v}_{2x}\mathrm{v}_{2y} & \mathrm{v}_{2y}\mathrm{v}_{2z} & \mathrm{v}_{2x}\mathrm{v}_{2z} \\ \mathrm{v}_{3x}^2 & \mathrm{v}_{3y}^2 & \mathrm{v}_{3z}^2 & \mathrm{v}_{3x}\mathrm{v}_{3y} & \mathrm{v}_{3y}\mathrm{v}_{3z} & \mathrm{v}_{3x}\mathrm{v}_{3z} \\ 2\mathrm{v}_{1x}\mathrm{v}_{2x} & 2\mathrm{v}_{1y}\mathrm{v}_{2y} & 2\mathrm{v}_{1z}\mathrm{v}_{2z} & \mathrm{v}_{1x}\mathrm{v}_{2y}+\mathrm{v}_{1y}\mathrm{v}_{2x} & \mathrm{v}_{1y}\mathrm{v}_{2z}+\mathrm{v}_{1z}\mathrm{v}_{2y} & \mathrm{v}_{1x}\mathrm{v}_{2z}+\mathrm{v}_{1z}\mathrm{v}_{2x} \\ 2\mathrm{v}_{2x}\mathrm{v}_{3x} & 2\mathrm{v}_{2y}\mathrm{v}_{3y} & 2\mathrm{v}_{2z}\mathrm{v}_{3z} & \mathrm{v}_{2x}\mathrm{v}_{3y}+\mathrm{v}_{2y}\mathrm{v}_{3x} & \mathrm{v}_{2y}\mathrm{v}_{3z}+\mathrm{v}_{2z}\mathrm{v}_{3y} & \mathrm{v}_{2x}\mathrm{v}_{3z}+\mathrm{v}_{2z}\mathrm{v}_{3x} \\ 2\mathrm{v}_{1x}\mathrm{v}_{3x} & 2\mathrm{v}_{1y}\mathrm{v}_{3y} & 2\mathrm{v}_{1z}\mathrm{v}_{3z} & \mathrm{v}_{1x}\mathrm{v}_{3y}+\mathrm{v}_{1y}\mathrm{v}_{3x} & \mathrm{v}_{1y}\mathrm{v}_{3z}+\mathrm{v}_{1z}\mathrm{v}_{3y} & \mathrm{v}_{1x}\mathrm{v}_{3z}+\mathrm{v}_{1z}\mathrm{v}_{3x} \end{bmatrix} \tag{A.34}$$

A.5.3 *Constant Body Force*

$$\mathsf{G}(\tilde{\boldsymbol{x}}_n, \boldsymbol{x}) = \frac{1}{8\pi G}\left(2\ln\frac{1}{r} - 1\right)\left(b_i\frac{\partial r}{\partial n} - \frac{1}{2(1-\nu)}n_i\frac{\partial r}{\partial b}\right) \quad \text{for 2-D} \tag{A.35}$$

$$\mathsf{G}(\tilde{\boldsymbol{x}}_n, \boldsymbol{x}) = \frac{1}{8\pi G r}\left(b_i\frac{\partial r}{\partial n} - \frac{1}{2(1-\nu)}n_i\frac{\partial r}{\partial b}\right) \quad \text{for 3-D} \tag{A.36}$$

The derived solution for 3-D is:

$$\hat{\mathsf{G}}(\tilde{\boldsymbol{x}}_n, \boldsymbol{x}) = \frac{1}{8\pi r}\left\{\begin{array}{c} \frac{\partial r}{\partial n}(b_i r_{,j} + b_j r_{,i}) + \frac{1}{1-\nu}\nu\delta_{ij}\left(\frac{\partial r}{\partial n}\frac{\partial r}{\partial b} - \frac{\partial b}{\partial n}\right) \\ -0.5\left[\frac{\partial r}{\partial b}(n_i r_{,j} + n_j r_{,i}) + (1-\nu)(b_i n_j + b_j n_i)\right] \end{array}\right\} \tag{A.37}$$

For plane strain problems we have

$$\hat{\mathsf{G}}(\tilde{\boldsymbol{x}}_n, \boldsymbol{x}) = \frac{1}{8\pi}\left(\begin{array}{c} 2\frac{\partial r}{\partial n}(b_i r_{,j} + b_j r_{,i}) \\ \frac{\nu\delta_{ij}\left[2\frac{\partial r}{\partial n}\frac{\partial r}{\partial b} - \frac{\partial b}{\partial n} + (1 - 2\ln\frac{1}{r})\frac{\partial r}{\partial n}\right] - \frac{\partial r}{\partial b}(n_i r_{,j} + n_j r_{,i})}{1-\nu} \\ -0.5\left[\frac{\partial r}{\partial b}(n_i r_{,j} + n_j r_{,i}) + (1-2\nu)(b_i n_j + b_j n_i)\right] \end{array}\right). \tag{A.38}$$

A.6 2-D Stokes Flow

$$\mathsf{U}(\tilde{\boldsymbol{x}}, \hat{\boldsymbol{x}}) = \frac{1}{4\pi\mu}\left(r_{,i}\, r_{,j} + \delta_{ij}\, \ln\left(\frac{1}{r}\right)\right) \tag{A.39}$$

$$\mathsf{T}(\tilde{\boldsymbol{x}}, \hat{\boldsymbol{x}}) = \frac{1}{\pi r}\, r_{,i}\, r_{,j}\, r_{,k}\, n_k \tag{A.40}$$

$$\mathsf{U}_{,k}(\tilde{\boldsymbol{x}}, \hat{\boldsymbol{x}}) = \frac{1}{4\pi\mu r}\left(\delta_{jk}\, r_i + \delta_{ik}\, r_j - \delta_{ij}\, r_k - 2\, r_i\, r_j\, r_k\right) \tag{A.41}$$

A.7 3-D Stokes Flow

$$\mathsf{U}(\tilde{\boldsymbol{x}}, \hat{\boldsymbol{x}}) = \frac{1}{8\pi\mu r}(r_{,i}\, r_{,j} + \delta_{ij}) \tag{A.42}$$

$$\mathsf{T}(\tilde{\boldsymbol{x}}, \hat{\boldsymbol{x}}) = \frac{1}{4\pi r^2} r_{,i}\, r_{,j}\, r_{,k}\, n_k \tag{A.43}$$

$$\mathsf{U}_{,k}(\tilde{\boldsymbol{x}}, \hat{\boldsymbol{x}}) = \frac{1}{8\pi\mu r^2}\left(-\delta_{ij}\, r_k + \delta_{jk}\, r_i + \delta_{ik}\, r_j - 3\, r_i\, r_j\, r_k\right) \tag{A.44}$$

References

1. Bacon, D.J., Barnett, D.M., Scattergood, R.O.: Anisotropic continuum theory of lattice defects. Prog. Mater. Sci. (Pergamon Press, Oxford) **23**, 51–262 (1980)
2. Banerjee, P.K., Butterfield, R. (eds.): Advanced implementation of the boundary element method in two- and three-dimensional elasto-statics. Developments in Boundary Element Methods. Elsevier Applied Science, London (1979)
3. Banerjee, P.K., Butterfield, R.: Boundary Element Methods in Engineering Science. McGraw-Hill, London (1981)
4. Banerjee, P.: The Boundary Element Method in Engineering. McGraw-Hill, London (1994)
5. Barnett, D.M.: The precise evaluation of derivatives of the anisotropic elastic Green's functions. Phys. Status Solidi (b) **49**(2), 741–748 (1972)
6. Barrett, R., Berry, M., Chan, T.F., Demmel, J., Donato, J.M., Dongarra, J., Eijkhout, V., Pozo, R., Romine, C., Van der Vorst, H.: Templates for the Solution of Linear Systems: Building Blocks for Iterative Methods. Society for Industrial and Applied Mathematics, Philadelphia (1993)
7. Baumgart, B.G.: Geometric modeling for computer vision. Ph.D. thesis, Stanford University, Stanford, CA (1974)
8. Bazilevs, Y., Calo, V.M., Cottrell, J.A., Evans, J.A., Hughes, T.J.R., Lipton, S., Scott, M.A., Sederberg, T.W.: Isogeometric analysis using T-splines. Comput. Methods Appl. Mech. Eng. **199**(5–8), 229–263 (2010)
9. Beer, G.: Advanced Numerical Simulation Methods - From CAD Data Directly to Simulation Results. CRC Press/Balkema, Boca Raton (2015)
10. Beer, G.: Mapped infinite patches for the NURBS based boundary element analysis in geomechanics. Comput. Geotech. **66**, 66–74 (2015)
11. Beer, G., Duenser, C.: New algorithms for the simulation of the sequential tunnel excavation with the boundary element method. In: Eberhardsteiner, J., et al. (eds.) Computational Methods in Tunnelling (EURO:TUN 2007) ECCOMAS (2007)
12. Beer, G., Duenser, C.: Technology Innovation in Underground Construction. CRC Press, Boca Raton (2010)
13. Beer, G., Duenser, C.: Advanced 3-D boundary element analysis of underground excavations. Comput. Geotech. **101**, 196–207 (2018)
14. Beer, G., Smith, I., Duenser, C.: The Boundary Element Method with Programming. Springer, Wien (2008)

G. Beer et al., *The Isogeometric Boundary Element Method*, Lecture Notes in Applied and Computational Mechanics 90,
https://doi.org/10.1007/978-3-030-23339-6

15. Beer, G., Marussig, B., Duenser, C.: Isogeometric boundary element method for the simulation of underground excavations. Géotech. Lett. **3**, 108–111 (2013)
16. Beer, G., Marussig, B., Zechner, J.: A simple approach to the numerical simulation with trimmed CAD surfaces. Comput. Methods Appl. Mech. Eng. **285**, 776–790 (2015)
17. Beer, G., Marussig, B., Zechner, J., Duenser, C., Fries, T.-P.: Isogeometric boundary element analysis with elasto-plastic inclusions. Part 1: plane problems. Comput. Methods Appl. Mech. Eng. **308**, 552–570 (2016)
18. Beer, G., Mallardo, V., Ruocco, E., Duenser, C.: Isogeometric boundary element analysis of steady incompressible viscous flow. Part 1: plane problems. Comput. Methods Appl. Mech. Eng. **326**, 51–69 (2017)
19. Beer, G., Mallardo, V., Ruocco, E., Marussig, B., Zechner, J., Duenser, C., Fries, T.-P.: Isogeometric boundary element analysis with elasto-plastic inclusions. Part 2: 3-D problems. Comput. Methods Appl. Mech. Eng. **315**, 418–433 (2017)
20. Beer, G.: Finite element, boundary element and coupled analysis of unbounded problems in elastostatics. Int. J. Numer. Methods Eng. **11**, 355–376 (1977)
21. Beer, G., Watson, J.: Introduction to Finite and Boundary Element Methods for Engineers. Wiley, New York (1992)
22. Beer, G., Mallardo, V., Ruocco, E., Duenser, C.: Isogeometric boundary element analysis of steady incompressible viscous flow. Part 2: 3-D problems. Comput. Methods Appl. Mech. Eng. **332**, 440–461 (2018)
23. Beskos, D.E. (ed.): The boundary element method. Some early history – a personal view. Boundary Element Methods in Structural Analysis, pp. 1–16. ASCE, New York (1989)
24. Biswas, A., Fenves, S.J., Shapiro, V., Sriram, R.: Representation of heterogeneous material properties in the Core Product Model. Eng. Comput. **24**(1), 43–58 (2008)
25. Bommes, D., Zimmer, H., Kobbelt, L.: Mixed-integer quadrangulation. ACM Trans. Graph. **28**(3), 77:1–77:10 (2009)
26. Bommes, D., Campen, M., Ebke, H.-C., Alliez, P., Kobbelt, L.: Integer-grid maps for reliable quad meshing. ACM Trans. Graph. **32**(4), 98:1–98:12 (2013)
27. Bonnet, M.: Boundary Integral Equation Methods for Solids and Fluids. Wiley, New York (1995)
28. de Boor, C.: A Practical Guide to Splines, vol. 27. Springer, New York (2001)
29. de Boor, C., Fix, G.J.: Spline approximation by quasiinterpolants. J. Approx. Theory **8**(1), 19–45 (1973)
30. Borden, M.J., Scott, M.A., Evans, J.A., Hughes, T.J.R.: Isogeometric finite element data structures based on Bézier extraction of NURBS. Int. J. Numer. Methods Eng. **87**(1–5), 15–47 (2011)
31. Bornemann, P.B., Cirak, F.: A subdivision-based implementation of the hierarchical b-spline finite element method. Comput. Methods Appl. Mech. Eng. **253**, 584–598 (2013)
32. de Borst, R., Chen, L.: The role of Bézier extraction in adaptive isogeometric analysis: local refinement and hierarchical refinement. Int. J. Numer, Methods Eng (2017)
33. Brebbia, C.A. (ed.): Applications in mining. Topics in Boundary Element Research, pp. 170–203. Springer, Berlin (1984)
34. Brebbia, C.A., Telles, J.C.: Boundary Element Techniques. Springer, Berlin (1983)
35. Brebbia, C.A., Walker, S.: Boundary Element Techniques in Engineering. Newnes-Butterworths, London (1980)
36. Breitenberger, M., Apostolatos, A., Philipp, B., Wüchner, R., Bletzinger, K.-U.: Analysis in computer aided design: nonlinear isogeometric B-Rep analysis of shell structures. Comput Methods Appl. Mech. Eng. **284**, 401–457 (2015)
37. Breitenberger, M.: CAD-integrated design and analysis of shell structures. Ph.D. thesis, Technische Universität München (2016)
38. Burton, J., Miller, G.F.: The application of integral equation methods to the numerical solution of some exterior boundary-value problems. Proc. R. Soc. Lond. **323**, 201–210 (1971)
39. Butterfield, R., Tomlin, G.: Integral techniques for solving zoned anisotropic continuum problems. Var. Methods Eng. **9**, 31–53 (1972)

40. C3D Kernel Documentation. C3D Labs
41. Carslaw, H.S., Jaeger, J.C.: Conduction of Heat in Solids. Oxford University Press, Oxford (1959)
42. Cirak, F., Long, Q.: Subdivision shells with exact boundary control and non-manifold geometry. Int. J. Numer. Methods Eng. **88**(9), 897–923 (2011)
43. Cividini, A., Gioda, G.: On the variable mesh finite element analysis of unconfined seepage problems. Geotechnique **2**, 251–267 (1989)
44. Cohen, E., Riesenfeld, R.F., Elber, G.: Geometric Modeling with Splines: An Introduction. A K Peters, Natick (2001)
45. Cormeau, I.: Numerical stability in quasi-static elasto-viscoplasticity. Int. J. Numer. Methods Eng. **9**(1) (1975)
46. Corney, J., Lim, T.: 3D Modeling with ACIS. Saxe-Coburg, Stirling (2001)
47. Cottrell, J.A., Hughes, T.J.R., Bazilevs, Y.: Isogeometric Analysis: Toward Integration of CAD and FEA. Wiley, New York (2009)
48. Crisfield, M.: Plasticity computations using the Mohr-Coulomb yield criterion. Eng. Comput. **4**(4), 300–308 (1987)
49. Crouch, S.L.: Solution of plane elasticity problems by the displacement discontinuity method. Int. J. Numer. Methods Eng. **10**, 301–343 (1976)
50. Crouch, S.L., Starfield, A.M.: Boundary Element Methods in Solid Mechanics. George Allen and Unwin, London (1983)
51. Dargush, G., Banerjee, P.: Application of the boundary element method to transient heat conduction. Int. J. Numer. Methods Eng. **28**, 2123–2142 (1989)
52. Davies, B., Martin, B.: Numerical inversion of the Laplace transform: a survey and comparison of methods. J. Comput. Phys. (1–32) (1979)
53. Deist, F.H., Georgiadis, E., Morris, J.P.E.: Computer applications in rock mechanics. J. S. Afr. Inst. Min. Metall. **73**, 265–272 (1972)
54. Ding, H., Liangjian, C.: The united point force solution for both isotropic and transversely isotropic media. Commun. Numer. Methods Eng. **13**(2), 95–102 (1997)
55. Dohr, S., Zapletal, J., Of, G., Merta, M., Kravčenko, M.: A parallel space-time boundary element method for the heat equation. Comput. Math, Appl (2019)
56. Dokken, T., Lyche, T., Pettersen, K.F.: Polynomial splines over locally refined box-partitions. Comput. Aided Geom. Des. **30**(3), 331–356 (2013)
57. Duenser, C., Beer, G.: Simulation of tunnelling with the boundary element method. In: IX International Congress on Numerical Methods in Engineering and Applied Sciences, Desarrollo y avances en methodos numericos para ingenieria y ciencias aplicadas CIMENICS 2008 (2008)
58. Duenser, C., Lindner, B., Beer, G.: Simulation of sequential tunnel excavation/construction with the boundary element method. In: Meschke, G., et al. (eds.) Computational Methods in Tunneling - EURO:TUN 2009, pp. 147–154. Aedificatio Publishers, Zürich (2009)
59. Farin, G.: Curves and Surfaces for CAGD: A Practical Guide, 5th edn. Morgan Kaufmann, San Francisco (2002)
60. Farouki, R.T.: The characterization of parametric surface sections. Comput. Vis. Graph. Image Process. **33**(2), 209–236 (1986)
61. Farouki, R.T.: Closing the gap between CAD model and downstream application. SIAM News **32**(5), 303–319 (1999)
62. Florez, W.F., Power, H.: Comparison between continuous and discontinuous boundary elements in the multidomain dual reciprocity method for the solution of the two-dimensional Navier-Stokes equations. Eng. Anal. Bound. Elem. **25**(1), 57–69 (2001)
63. Forsey, D.R., Bartels, R.H.: Hierarchical B-spline refinement. ACM SIGGRAPH Comput. Graph. **22**(4), 205–212 (1988)
64. Fries, T.-P., Omerović, S.: Higher-order accurate integration of implicit geometries. Int. J. Numer. Methods Eng. **106**(5), 323–371 (2016)
65. Gao, X., Davies, T.: Boundary Element Programming in Mechanics. Cambridge University Press, Cambridge (2011)

66. Gaul, L., Kögl, M., Wagner, M.: Boundary Element Methods for Engineers and Scientists. Springer, Berlin (2003)
67. Gel'fand, I.M., Graev, M.I., Vilenkin, N.Y.: Generalized functions. Integral Geom. Represent. Theory **5**, (1983)
68. Ghia, U., Ghia, K.N., Shin, C.T.: High-Re solutions for incompressible flow using the Navier-Stokes equations and multigrid method. J. Comput. Phys. **48**, 387–411 (1982)
69. Giannelli, C., Jüttler, B., Speleers, H.: THB-splines: the truncated basis for hierarchical splines. Comput. Aided Geom. Des. **29**(7), 485–498 (2012)
70. Giannelli, C., Jüttler, B., Speleers, H.: Strongly stable bases for adaptively refined multilevel spline spaces. Adv. Comput. Math. **40**, 459–490 (2014)
71. Glowinsky, E.Y. (ed.): Marriage a la mode - the best of both worlds (finite elements and boundary integrals). Energy Methods in Finite Element Analysis. Wiley, New York (1979)
72. Goldberg, M.: Recent developments in the numerical evaluation of particular solutions in the boundary element method. Appl. Math. Comput. **75**(1), 91–101 (1996)
73. Goldstein, B.L.M., Kemmerer, S.J., Parks, C.H.: A brief history of early product data exchange standards. NISTIR 6221, U.S. DEPARTMENT OF COMMERCE Technology Administration, Electronics and Electrical Engineering Laboratory, National Institute of Standards and Technology (1998)
74. Golovanov, N.: Geometric Modeling. Academia Publishing House, Praha (2014)
75. Greville, T.: Numerical procedures for interpolation by spline functions. J. Soc. Ind. Appl. Math. Ser. B. Numer, Anal (1964)
76. Guo, Y., Ruess, M., Schillinger, D.: A parameter-free variational coupling approach for trimmed isogeometric thin shells. Comput. Mech. **59**(4), 693–715 (2016)
77. Hackbusch, W.: A sparse matrix arithmetic based on $\mathcal{H}$-matrices. Computing **62**, 89–108 (1999)
78. Hackbusch, W.: Hierarchical Matrices: Algorithms and Analysis. Springer, Berlin (2015)
79. Hamann, B., Tsai, P.-Y.: A tessellation algorithm for the representation of trimmed NURBS surfaces with arbitrary trimming curves. Comput.-Aided Des. **28**(6–7), 461–472 (1996)
80. Harbrecht, H., Randrianarivony, M.: From computer aided design to wavelet BEM. Comput. Vis. Sci. **13**(2), 69–82 (2010)
81. Hardwick, M.F., Clay, R.L., Boggs, P.T., Walsh, E.J., Larzelere, A.R., Altshuler, A.: DART system analysis. Technical report SAND2005-4647, Sandia National Laboratories (2005)
82. Hass, J., Farouki, R.T., Han, C.Y., Song, X., Sederberg, T.W.: Guaranteed consistency of surface intersections and trimmed surfaces using a coupled topology resolution and domain decomposition scheme. Adv. Comput. Math. **27**(1), 1–26 (2007)
83. Hennig, P., Müller, S., Kästner, M.: Bézier extraction and adaptive refinement of truncated hierarchical NURBS. Comput. Methods Appl. Mech. Eng. **305**, 316–339 (2016)
84. Hiemstra, R.R., Calabrò, F., Schillinger, D., Hughes, T.J.R.: Optimal and reduced quadrature rules for tensor product and hierarchically refined splines in isogeometric analysis. Comput. Methods Appl. Mech, Eng (2016)
85. Hoffmann, C.M.: Geometric and Solid Modeling: An Introduction. Morgan Kaufmann, San Mateo (1989)
86. Höllig, K.: Finite Element Methods with B-Splines. Frontiers in Applied Mathematics, vol. 26. SIAM, Philadelphia (2003)
87. Hughes, T.J.R., Cottrell, J.A., Bazilevs, Y.: Isogeometric analysis: CAD, finite elements, NURBS, exact geometry and mesh refinement. Comput. Methods Appl. Mech. Eng. **194**(39–41), 4135–4195 (2005)
88. Hui, K.C., Wu, Y.-B.: Feature-based decomposition of trimmed surface. Comput.-Aided Des. **37**(8), 859–867 (2005)
89. Idelsohn, S. (ed.): Marriage a la mode (finite and boundary elements) revisited. Computational Mechanics - New Trends and Applications, CIMNE (1998)
90. Ingber, M.S., Mammoli, A.A., Brown, M.J.: A comparison of domain integral evaluation techniques for boundary element methods. Int. J. Numer. Methods Eng. **52**, 417–432 (2001)

91. Jaswon, M.A.: Integral equation methods in potential theory. Proc. R. Soc. A **275**, 23–32 (1963)
92. Johannessen, K.A., Kvamsdal, T., Dokken, T.: Isogeometric analysis using LR B-splines. Comput. Methods Appl. Mech. Eng. **269**, 471–514 (2014)
93. Kennicott, P.R., Morea, G., Reid, E., Parks, C., Rinaudot, G., Harrod, Jr, D.A., Gruttke, W.B.: Initial graphics exchange specification IGES 5.3. ANS US PRO/IPO-100-1996, U.S. Product Data Association (1996)
94. Kim, H.-J., Seo, Y.-D., Youn, S.-K.: Isogeometric analysis for trimmed CAD surfaces. Comput. Methods Appl. Mech. Eng. **198**(37–40), 2982–2995 (2009)
95. Kim, H.-J., Seo, Y.-D., Youn, S.-K.: Isogeometric analysis with trimming technique for problems of arbitrary complex topology. Comput. Methods Appl. Mech. Eng. **199**(45–48), 2796–2812 (2010)
96. Kim, J., Pratt, M.J., Iyer, R.G., Sriram, R.D.: Standardized data exchange of CAD models with design intent. Comput.-Aided Des. **40**(7), 760–777 (2008)
97. Kollmannsberger, S., Özcan, A., Baiges, J., Ruess, M., Rank, E., Reali, A.: Parameter-free, weak imposition of Dirichlet boundary conditions and coupling of trimmed and non-conforming patches. Int. J. Numer. Methods Eng. **101**(9), 670–699 (2015)
98. Kraft, R.: Adaptive and linearly independent multilevel B-splines. SFB 404, Geschäftsstelle (1997)
99. Kudela, L.: Highly accurate subcell integration in the context of the finite cell method. Master's thesis, Technical University Munich (2013)
100. Kudela, L., Zander, N., Bog, T., Kollmannsberger, S., Rank, E.: Efficient and accurate numerical quadrature for immersed boundary methods. Adv. Model. Simul. Eng. Sci. **2**(1), 1–22 (2015)
101. Kudela, L., Zander, N., Kollmannsberger, S., Rank, E.: Smart octrees: accurately integrating discontinuous functions in 3D. Comput. Methods Appl. Mech. Eng. **306**, 406–426 (2016)
102. LaCourse, D.E.: Handbook of Solid Modeling. McGraw-Hill Inc, London (1995)
103. Lachat, J.C., Watson, J.: Effective numerical treatment of boundary integral equations: a formulation for three dimensional elastostatics. Int. J. Numer. Methods Eng. **10**, 991–1005 (1976)
104. Lasserre, J.: Integration on a convex polytope. Proc. Am. Math. Soc. **126**(8), 2433–2441 (1998)
105. Li, K., Qian, X.: Isogeometric analysis and shape optimization via boundary integral. Comput.-Aided Des. **43**(11), 1427–1437 (2011)
106. Liggett, J.A., Liu, P.L.F.: The Boundary Integral Equation Method for Porous Media Flow. George Allen and Unwin, London (1983)
107. Liu, W., Mann, S.: An optimal algorithm for expanding the composition of polynomials. ACM Trans. Graph. **16**(2), 155–178 (1997)
108. Liu, Y.: Fast Multipole Boundary Element Method. Cambridge University Press, Cambridge (2006)
109. Lorenzo, G., Scott, M.A., Tew, K., Hughes, T.J.R., Gomez, H.: Hierarchically refined and coarsened splines for moving interface problems, with particular application to phase-field models of prostate tumor growth. Comput. Methods Appl. Mech. Eng. **319**, 515–548 (2017)
110. Martin, R.C.: Clean Code: A Handbook of Agile Software Craftsmanship, 1st edn. Pearson Education Inc, Paris (2009)
111. Martin, T., Cohen, E., Kirby, M.: Volumetric parameterization and trivariate B-spline fitting using harmonic functions. In: Proceedings of the 2008 ACM Symposium on Solid and Physical Modeling, SPM'08, pp. 269–280. ACM, New York (2008)
112. Marussig, B., Zechner, J., Beer, G., Fries, T.-P.: Fast isogeometric boundary element method based on independent field approximation. Comput. Methods Appl. Mech. Eng. **284**, 458–488 (2015)
113. Marussig, B.: Seamless Integration of Design and Analysis Through Boundary Integral Equations. Structural Analysis Verlag der Technischen Universität Graz, Monographic Series TU Graz (2016)

114. Marussig, B., Hughes, T.J.R.: A review of trimming in isogeometric analysis: challenges, data exchange and simulation aspects. Arch. Comput. Methods Eng. **25**(4), 1059–1127 (2018)
115. Marussig, B., Duenser, C., Beer, G.: Isogeometric boundary element methods for tunneling. Symposium of the International Association for Boundary Element Methods, IABEM **2013**, 100–106 (2013)
116. Marussig, B., Zechner, J., Beer, G., Fries, T.-P.: Stable isogeometric analysis of trimmed geometries. Comput. Methods Appl. Mech. Eng. **316**, 497–521 (2016)
117. Marussig, B., Hiemstra, R., Hughes, T.J.R.: Improved conditioning of isogeometric analysis matrices for trimmed geometries. Comput. Methods Appl. Mech. Eng. **334**, 79–110 (2018)
118. Massarwi, F., Elber, G.: A B-spline based framework for volumetric object modeling. Comput.-Aided Des. **78**, 36–47 (2016)
119. Michelitsch, T., Levin, V.M.: Green's function for the infinite two-dimensional orthotropic medium. Int. J. Fract. **107**(2), 33–38 (2001)
120. Mortenson, M.E.: Geometric Modeling, 2nd edn. Wiley, New York (1997)
121. Mukherjee, S.: Boundary Element Methods for Creep and Fracture. Applied Science, London (1982)
122. Nagel, R.N., Braithwaite, W.W., Kennicott, P.R.: Initial graphics exchange specification (IGES) version 1.0. NBSIR 80-1978 (R), National Bureau of Standards (1980)
123. Nagy, A.P., Benson, D.J.: On the numerical integration of trimmed isogeometric elements. Comput. Methods Appl. Mech. Eng. **284**, 165–185 (2015)
124. Omerović, S., Fries, T.-P.: Conformal higher-order remeshing schemes for implicitly defined interface problems. Int. J. Numer, Methods Eng (2016)
125. Pan, Y.-C., Chou, T.-W.: Point force solution for an infinite transversely isotropic solid. J. Appl. Mech. **43**, 608–612 (1976)
126. Parreira, P.: On the accuracy of continuous and discontinuous boundary elements. Eng. Anal. **5**(4), 205–211 (1988)
127. Partridge, P., Brebbia, C.A., Wrobel, L.C.: The Dual Reciprocity Boundary Element Method. Computational Mechanics Publications, Southampton (1992)
128. Patrikalakis, N.M., Maekawa, T.: Shape Interrogation for Computer Aided Design and Manufacturing. Springer Science & Business Media, New York (2009)
129. Peng, S., Zhang, J.: Engineering Geology for Underground Rocks. Springer, Berlin (2007)
130. Perzyna, P.: Fundamental Problems in Viscoplasticity. Advances in Applied Mechanics, vol. 9, pp. 243–377. Elsevier, Amsterdam (1966)
131. Philipp, B., Breitenberger, M., D'Auria, I., Wüchner, R., Bletzinger, K.-U.: Integrated design and analysis of structural membranes using the isogeometric B-Rep analysis. Comput. Methods Appl. Mech. Eng. **303**, 312–340 (2016)
132. Piegl, L., Tiller, W.: The NURBS Book, 2nd edn. Springer, New York (1997)
133. Piegl, L.A.: Ten challenges in computer-aided design. Comput.-Aided Des. **37**(4), 461–470 (2005)
134. Pierce, A.D.: Acoustics: An Introduction to Its Physical Principles and Applications. Acoustical Society of America, Melville (1989)
135. Pratt, M.J.: Introduction to ISO 10303-the STEP standard for product data exchange. Technical report, National Institute of Standards and Technology, Manufacturing Systems Integration Division, Gaithersburg (2001)
136. Pratt, M.J., Anderson, B.D., Ranger, T.: Towards the standardized exchange of parameterized feature-based CAD models. Comput.-Aided Des. **37**(12), 1251–1265 (2005)
137. Prazeres, P.G.C.: Simulation of NATM tunnel construction with the BEM. Ph.D. thesis, Graz University of Technology, Austria (2010)
138. Randrianarivony, M.: Geometric processing of CAD data and meshes as input of integral equation solvers. Ph.D. thesis, Computer Science Faculty Technische Universität Chemnitz (2006)
139. Randrianarivony, M.: On global continuity of Coons mappings in patching CAD surfaces. Comput.-Aided Des. **41**(11), 782–791 (2009)

140. Rank, E., Ruess, M., Kollmannsberger, S., Schillinger, D., Düster, A.: Geometric modeling, isogeometric analysis and the finite cell method. Comput. Methods Appl. Mech. Eng. **249–252**, 104–115 (2012)
141. Renner, G., Weiß, V.: Exact and approximate computation of B-spline curves on surfaces. Comput.-Aided Des. **36**(4), 351–362 (2004)
142. Requicha, A.A.G.: Representations for rigid solids: theory, methods, and systems. ACM Comput. Surv. **12**(4), 437–464 (1980)
143. Requicha, A.A.G., Voelcker, H.B.: Solid modeling: a historical summary and contemporary assessment. IEEE Comput. Graph. Appl. **2**(2), 9–24 (1982)
144. Riederer, K.: Modelling of ground support in tunnelling using the BEM. Ph.D. thesis, Graz University of Technology (2010)
145. Riederer, K., Duenser, C., Beer, G.: Analysis of rock bolts and inhomogeneities in tunneling with the BEM. In: WCCM8, 8th World Congress on Computational Mechanics; ECCOMAS 2008, 5th European Congress on Computational Methods in Applied Sciences and Engineering (2008)
146. Riederer, K., Duenser, C., Beer, G.: Simulation of linear inclusions with the BEM. Eng. Anal. Bound. Elem. **33**(7), 959–965 (2009)
147. Riffnaller-Schiefer, A., Augsdörfer, U.H., Fellner, D.W.: Isogeometric shell analysis with NURBS compatible subdivision surfaces. Appl. Math. Comput. Part 1 **139–147**, (2016)
148. Rizzo, F.J.: An integral equation approach to boundary value problems of classical elastostatics. Q. Appl. Math. **25**, 83–95 (1967)
149. Rogers, D.F., Adams, J.A.: Mathematical Elements for Computer Graphics, 2nd edn. McGraw-Hill Higher Education, New York (1990)
150. Rüberg, T., Cirak, F.: Subdivision-stabilised immersed b-spline finite elements for moving boundary flows. Comput. Methods Appl. Mech. Eng. **209–212**, 266–283 (2012)
151. Ruess, M., Schillinger, D., Bazilevs, Y., Varduhn, V., Rank, E.: Weakly enforced essential boundary conditions for NURBS-embedded and trimmed NURBS geometries on the basis of the finite cell method. Int. J. Numer. Methods Eng. **95**(10), 811–846 (2013)
152. Ruess, M., Schillinger, D., Özcan, A.I., Rank, E.: Weak coupling for isogeometric analysis of non-matching and trimmed multi-patch geometries. Comput. Methods Appl. Mech. Eng. **269**, 46–71 (2014)
153. Runge, C.: Über empirische Funktionen und die Interpolation zwischen äquidistanten Ordinaten. Zeitschrift für Mathematik und Physik **46**(224–243), 20 (1901)
154. STEP Application Handbook ISO 10303 Version 3. SCRA (2006)
155. Salamon, M.D.G.: Elastic analysis of displacements and stresses induced by the mining of seam or reef deposits. J. S. Afr. Inst. Min. Metall. **64**(4), 129–149 (1963)
156. Schillinger, D., Ruess, M.: The finite cell method: a review in the context of higher-order structural analysis of CAD and image-based geometric models. Arch. Comput. Methods Eng. **22**(3), 391–455 (2015)
157. Schillinger, D., Evans, J.A., Reali, A., Scott, M.A., Hughes, T.J.R.: Isogeometric collocation: cost comparison with Galerkin methods and extension to adaptive hierarchical NURBS discretizations. Comput. Methods Appl. Mech. Eng. **267**, 170–232 (2013)
158. Schmidt, R., Wüchner, R., Bletzinger, K.-U.: Isogeometric analysis of trimmed NURBS geometries. Comput. Methods Appl. Mech. Eng. **241–244**, 93–111 (2012)
159. Scott, M.A., Thomas, D.C., Evans, E.J.: Isogeometric spline forests. Comput. Methods Appl. Mech. Eng. **269**, 222–264 (2014)
160. Scott, M.A., Simpson, R.N., Evans, J.A., Lipton, S., Bordas, S.P.A., Hughes, T.J.R., Sederberg, T.W.: Isogeometric boundary element analysis using unstructured T-splines. Comput. Methods Appl. Mech. Eng. **254**, 197–221 (2013)
161. Sederberg, T.W., Anderson, D.C., Goldman, R.N.: Implicit representation of parametric curves and surfaces. Comput. Vis. Graph. Image Process. **28**(1), 72–84 (1984)
162. Sederberg, T.W., Li, X., Lin, H.W., Ipson, H.: Watertight trimmed NURBS. ACM Trans. Graph. **27**(3), 79:1–79:8 (2008)

163. Sederberg, T.W.: Implicit and parametric curves and surfaces for computer aided geometric design. Ph.D. thesis, Purdue University (1983)
164. Sederberg, T.W., Zheng, J., Bakenov, A., Nasri, A.: T-splines and T-NURCCs. ACM Trans. Graph. **22**(3), 477–484 (2003)
165. Shah, J.J., Mäntylä, M.: Parametric and Feature based CAD/CAM. Wiley, New York (1995)
166. Parasolid XT Format Reference. Siemens Product Lifecycle Management Software Inc. (2008)
167. Simpson, R.N., Liu, Z.: Acceleration of isogeometric boundary element analysis through a black-box fast multipole method. Eng. Anal. Bound. Elem. **66**, 168–182 (2016)
168. Simpson, R.N., Scott, M.A., Taus, M., Thomas, D.C., Lian, H.: Acoustic isogeometric boundary element analysis. Comput. Methods Appl. Mech. Eng. **269**, 265–290 (2014)
169. Simpson, R.N., Liu, Z., Vazquez, R., Evans, J.A.: An isogeometric boundary element method for electromagnetic scattering with compatible B-spline discretizations. J. Comput. Phys. **362**, 264–289 (2018)
170. Skytt, V., Haenisch, J.: Extension of ISO 10303 with isogeometric model capabilities. ISO TC 184/SC 4/WG 12, ISO (2013)
171. Smith, I.M., Griffiths, D.V., Margetts, L.: Programming the Finite Element Method. Wiley, New York (2013)
172. Stanford, J.W., Fries, T.-P.: A higher-order conformal decomposition finite element method for plane B-rep geometries. Comput, Struct (2018)
173. Steidl, J.W.: Trimmed NURBS: implementation of an element type discrimination algorithm. Institute for Structural Analysis at Graz University of Technology, Master Project (2013)
174. Stroud, A., Secrest, D.: Gaussian Quadrature Formulas. Prentice-Hall, Englewood Cliffs (1966)
175. Stroud, I.: Boundary Representation Modelling Techniques. Springer, London (2006)
176. Tassey, G., Brunnermeier, S.B., Martin, S.A.: Interoperability cost analysis of the U.S. automotive supply chain. Technical report, Research Triangle Institute (1999)
177. Telles, J.C.F., Brebbia, C.A.: Elasto-plastic boundary element analysis. In: Wunderlich, W., Stein, E., Bathe, K.-J. (eds.) Nonlinear Finite Element Analysis in Structural Mechanics, pp. 403–434. Springer, Berlin (1981)
178. Telles, J.C.F., Carrer, J.A.M.: Implicit procedures for the solution of elastoplastic problems by the boundary element method. Math. Comput. Model. **15**, 303–311 (1991)
179. Timoshenko, S.P.: On the correction factor for shear of the differential equation for transverse vibrations of bars of uniform cross-section. Philos. Mag. **744**, (1921)
180. Trefethen, L.N.: Approximation Theory and Approximation Practice, vol. 128. SIAM, Philadelphia (2013)
181. Trefftz, E.: Ein Gegenstück zum Ritzschen Verfahren. In: Proceedings of the 2nd International Congress in Applied Mechanics (1926)
182. Urick, B.: Reconstruction of tensor product spline surfaces to integrate surface-surface intersection geometry and topology while maintaining inter-surface continuity. Ph.D. thesis, The University of Texas at Austin (2016)
183. Urick, B., Marussig, B.: Why CAD surface geometry is inexact (2017). https://blog.pointwise.com/2017/11/29/why-cad-surface-geometry-is-inexact/#more-11007
184. da Veiga, L.B., Buffa, A., Sangalli, G., Vázquez, R.: Mathematical analysis of variational isogeometric methods. Acta Numer. **23**, 157–287 (2014)
185. Venturini, W.S.: Boundary Element Method in Geomechanics. Springer, Berlin (1983)
186. Vuong, A.V., Giannelli, C., Jüttler, B., Simeon, B.: A hierarchical approach to adaptive local refinement in isogeometric analysis. Comput. Methods Appl. Mech. Eng. **200**(49), 3554–3567 (2011)
187. Wang, Y.-W., Huang, Z.-D., Zheng, Y., Zhang, S.-G.: Isogeometric analysis for compound B-spline surfaces. Comput. Methods Appl. Mech. Eng. **261–262**, 1–15 (2013)
188. Wang, Y., Benson, D.J.: Geometrically constrained isogeometric parameterized level-set based topology optimization via trimmed elements. Front. Mech. Eng. **1–16**, (2016)

189. Wang, Y., Benson, D.J., Nagy, A.P.: A multi-patch nonsingular isogeometric boundary element method using trimmed elements. Comput. Mech. **56**(1), 173–191 (2015)
190. Wang, Z., Gu, Y.: The method of fundamental solutions for general orthotropic materials. Int. J. Appl. Exp. Math. **1**(109), 1–4 (2016)
191. Wei, X., Zhang, Y., Hughes, T.J.R., Scott, M.A.: Truncated hierarchical Catmull-Clark subdivision with local refinement. Comput. Methods Appl. Mech. Eng. **291**, 1–20 (2015)
192. Wendland, W.L. (ed.): Boundary Element Topics. Springer, Berlin (1997)
193. Wilson, R.B., Cruse, T.A.: Efficient implementation of anisotropic three dimensional boundary-integral equation stress analysis. Int. J. Numer. Methods Eng. **12**(9), 1383–1397 (1978)
194. Xia, S., Qian, X.: Isogeometric analysis with Bézier tetrahedra. Comput. Methods Appl. Mech. Eng. **316**, 782–816 (2016)
195. Zechner, J.: Fast elasto-plastic BEM with hierarchical matrices. Ph.D. thesis, TU Graz (2012)
196. Zechner, J., Marussig, B., Beer, G., Fries, T.-P.: The isogeometric Nyström method. Comput. Methods Appl. Mech. Eng. **308**, 212–237 (2016)
197. Zhang, Y.J.: Geometric Modeling and Mesh Generation from Scanned Images. CRC Press, Boca Raton (2016)

Printed in the United States
By Bookmasters